大型互联网应用
轻量级架构实战

柳伟卫 ◎ 著

内 容 提 要

本书主要介绍基于 Maven、Jetty、Spring 框架、Spring MVC、Spring Security、MyBatis、MySQL、Angular、NGINX、Redis、Spring Boot 等开源技术栈搭建应用框架并快速实现互联网应用的过程。这些技术并未完全遵守传统的 Java EE 所提供的规范，却被广大互联网公司所采用，其原因正是一种反传统的"轻量级"开发模式已经深入人心。所谓轻量，指的是抛弃墨守成规、面向规范的臃肿开发方式，转而采用开源的、重视解决具体问题的技术框架。

本书将带领读者从零开始搭建一个可以实现 Web 互联网应用的开发框架，命名为"Lite"。通过学习 Lite 轻量级框架的搭建过程，便于读者深刻理解 Spring、MyBatis、MySQL 等技术原理及实现方式，使读者具备架构软件的能力。同时，在 Lite 框架的基础上，还会继续实现一个"新闻头条"大型互联网应用，真正实现技术落地。

本书主要面向对 Web 开发、分布式系统、微服务开发感兴趣的计算机专业的学生、软件开发人员和系统架构师。

图书在版编目(CIP)数据

大型互联网应用轻量级架构实战 / 柳伟卫著. — 北京：北京大学出版社，2019.12
ISBN 978-7-301-30944-5

Ⅰ.①大… Ⅱ.①柳… Ⅲ.①互联网络－架构 Ⅳ.①TP393.4

中国版本图书馆CIP数据核字(2019)第267604号

书　　　名	大型互联网应用轻量级架构实战 DAXING HULIANWANG YINGYONG QINGLIANGJI JIAGOU SHIZHAN
著作责任者	柳伟卫　著
责 任 编 辑	吴晓月
标 准 书 号	ISBN 978-7-301-30944-5
出 版 发 行	北京大学出版社
地　　　址	北京市海淀区成府路205 号　100871
网　　　址	http://www.pup.cn　　新浪微博：@北京大学出版社
电 子 信 箱	pup7@ pup.cn
电　　　话	邮购部 010-62752015　发行部 010-62750672　编辑部 010-62570390
印 刷 者	河北滦县鑫华书刊印刷厂
经 销 者	新华书店
	787毫米×1092毫米　16开本　28印张　888千字 2019年12月第1版　2019年12月第1次印刷
印　　　数	1-3000册
定　　　价	89.00 元

未经许可，不得以任何方式复制或抄袭本书之部分或全部内容。
版权所有，侵权必究
举报电话：010-62752024　电子信箱：fd@pup.pku.edu.cn
图书如有印装质量问题，请与出版部联系。电话：010-62756370

本书献给我的父母，愿他们健康长寿！

前言

写作背景

随着云计算的普及，Cloud Native 应用开发模式逐渐深入人心，这意味着未来的应用将会朝着快速迭代、分布部署、独立运行等方向发展，敏捷、轻量的框架也必将受到更多开发者的青睐。为此笔者开发了 Lite 框架，以用于简化 Web 开发。

Lite 框架抛弃了墨守成规的臃肿开发方式，转而采用开源的、重视解决具体问题的技术。这些技术包括 Maven、Jetty、Spring 框架、Spring MVC、Spring Security、MyBatis、MySQL、Angular、NGINX、Redis、Spring Boot 等，可以说都是当今互联网公司使用的主流应用技术，经受住了大规模商业实践的考验。

读者通过学习 Lite 的框架搭建及开发基于 Lite 的互联网应用的过程，能够深刻领会上述技术的底层原理，掌握架构软件的能力。

全书内容分为四部分。

1. 基础：介绍 Spring、MyBatis、MySQL 等开源技术栈。涉及技术包括 Maven、Jetty、Spring 框架、Spring MVC、Spring Security、MyBatis、MySQL、Angular、NGINX、Redis、Spring Boot 等。
2. 进阶：从零开始搭建一个可以实现 Web 互联网应用的开发框架"Lite"。
3. 实战：基于 Lite 实现一个"新闻头条"互联网应用。
4. 提升：介绍 Spring Boot 等技术，通过 Spring Boot 来继续简化 Lite 框架。

源代码

本书提供源代码下载，地址为 https://github.com/waylau/lite-book-demos。也可扫描下方二维码关注微信公众号，根据提示获取资源。

本书所采用的技术及相关版本

技术的版本是非常重要的，因为不同版本之间存在兼容性问题，而且不同版本的软件所对应的功能也是不同的。本书所列出的技术在版本上相对较新，都是经过笔者大量测试的。读者在自行编写代码时，可以参考本书所列出的版本，从而避免版本不兼容所产生的问题。建议读者将相关开发环境设置得跟本书一致，或者不低于本书所列的配置。详细的版本配置，可以参阅"附录"中的内容。

本书示例采用 Eclipse 编写，但示例源码与具体的 IDE 无关，读者可以选择适合自己的 IDE，如 IntelliJ IDEA、NetBeans 等。运行本书示例，请确保 JDK 版本不低于 JDK 9。

勘误和交流

本书如有勘误，会在 https://github.com/waylau/lite-book-demos/issues 上进行发布。笔者在编写本书的过程中，虽已竭尽所能地为读者呈现最好、最全的实用功能，但错漏之处在所难免，欢迎读者批评指正，也可以通过以下方式联系我们。

- 博　客：https://waylau.com
- 邮　箱：waylau521@gmail.com
- 微　博：http://weibo.com/waylau521
- 开　源：https://github.com/waylau

致谢

感谢北京大学出版社的各位工作人员为本书的出版所做的努力。

感谢我的父母、妻子和两个女儿。由于撰写本书，牺牲了很多陪伴家人的时间，在此感谢家人对我工作的理解和支持。

柳伟卫

目 录
Contents

第1章 轻量级架构概述 1
1.1 大型互联网应用的特征 2
1.2 传统企业级应用技术的不足 5
1.3 Lite 框架简介 7

第2章 Servlet 10
2.1 Servlet 概述 11
2.2 请求 16
2.3 Servlet 上下文 19
2.4 响应 24
2.5 实战：基于 Servlet 的 Web 程序 26
2.6 Jetty 31
2.7 实战：在应用中内嵌 Jetty 容器 36

第3章 Spring 基础 42
3.1 Spring 概述 43
3.2 IoC 47
3.3 AOP 67
3.4 资源处理 76
3.5 表达式语言 SpEL 80

第4章 Spring 单元测试 90
4.1 Mock 对象 91
4.2 测试工具类 91

第 5 章　Spring 集成测试 93

5.1　集成测试概述 94
5.2　测试相关的注解 95
5.3　Spring TestContext 框架 106
5.4　Spring MVC Test 框架 118

第 6 章　Spring 事务管理 128

6.1　事务管理概述 129
6.2　通过事务实现资源同步 132
6.3　声明式事务管理 133
6.4　编程式事务管理 146

第 7 章　Spring Web MVC 148

7.1　Spring Web MVC 概述 149
7.2　DispatcherServlet 149
7.3　过滤器 153
7.4　控制器 155
7.5　异常处理 161
7.6　CORS 处理 163
7.7　HTTP 缓存 167
7.8　MVC 配置 169
7.9　实战：基于 Spring Web MVC 的 REST 接口 177

第 8 章　Spring Security 183

8.1　基于角色的权限管理 184
8.2　Spring Security 基础 187
8.3　实战：基于 Spring Security 的安全认证功能 194

第 9 章　MyBatis 基础 202

9.1　MyBatis 概述 203
9.2　与 Hibernate 对比 203
9.3　四大核心概念 204
9.4　生命周期及作用域 207

第 10 章 MyBatis 高级应用 209
- 10.1 配置文件 210
- 10.2 Mapper 映射文件 227
- 10.3 动态 SQL 240
- 10.4 常用 API 244

第 11 章 Lite 技术集成 253
- 11.1 技术集成概述 254
- 11.2 MySQL 的安装及基本操作 254
- 11.3 Spring 与 MyBatis 集成 258
- 11.4 集成 Spring Web MVC 264
- 11.5 集成 Spring Security 272
- 11.6 集成日志框架 276

第 12 章 Lite 架构分层 280
- 12.1 分层架构概述 281
- 12.2 数据访问层 286
- 12.3 事务处理 287
- 12.4 权限验证 287
- 12.5 接口访问层 288
- 12.6 实战：Lite 框架的搭建 290
- 12.7 发布 Lite 框架到 Maven 中央仓库 306

第 13 章 实战：基于 Lite 框架的互联网应用 318
- 13.1 lite-news 概述 319
- 13.2 模型设计 322
- 13.3 接口设计与实现 324
- 13.4 实现权限管理 332
- 13.5 前端 lite-news-ui 设计 337
- 13.6 实现 lite-news-ui 原型 338
- 13.7 实现路由器 344
- 13.8 实现用户登录 349
- 13.9 实现新闻编辑器 353
- 13.10 实现新闻列表展示 358
- 13.11 实现新闻详情展示 361
- 13.12 总结 364

第 14 章　使用 NGINX 实现高可用 365
14.1　NGINX 概述 366
14.2　部署 Angular 应用 370
14.3　实现负载均衡及高可用 372

第 15 章　使用 Redis 实现高并发 377
15.1　为什么需要缓存 378
15.2　了解 Redis 服务器 378
15.3　使用 Redis 384
15.4　lite-news 实现缓存 389

第 16 章　Spring Boot 概述 395
16.1　构建 RESTful 服务 396
16.2　Spring Boot 的配置详解 405
16.3　内嵌 Servlet 容器 408
16.4　实现安全机制 409

第 17 章　基于 Spring Boot 的 Lite 框架 421
17.1　Lite Spring Boot Starter 项目搭建 422
17.2　集成 Jetty 427
17.3　集成 Spring Security 428
17.4　集成 MyBatis 431
17.5　总结 436

附录 437

参 考 文 献 438

第1章
轻量级架构概述

本章介绍大型互联网应用的特征,同时指出传统技术无法适应当前互联网应用开发的原因。最后以Lite框架为例,介绍什么样的技术架构能够适用当前大型互联网应用的开发。

1.1 大型互联网应用的特征

当今互联网应用呈现出高速发展的趋势。

（1）互联网理财用户规模持续扩大。越来越多的网民选择在网上购买理财产品。

（2）全国网络零售交易再创新高。2018 年"双 11"当天，全国网络零售交易额突破 3000 亿元，其中天猫全天成交额为 2135 亿元。

（3）移动支付使用率持续增长。无论是网上购物，还是实体店购物，大多数用户都选择微信或支付宝等移动支付软件。

（4）短视频应用"异军突起"。大多数网民都使用过短视频应用（如快手、抖音等），以满足碎片化的娱乐需求。

（5）在线政务应用发展迅速。支付宝和微信均提供了城市服务平台，以对接政务服务。政府也积极出台相关政策，推动政务线上化发展。

还有很多其他互联网应用，也都深刻影响了人们的出行、饮食、健康等方面。人们已经无法适应没有智能手机的时代了。这些大型互联网应用大多都具有以下特征。

1.1.1 快！快！快！

"快"是所有互联网公司产品的特征。互联网公司要生存，就必须要与时间赛跑，与同行竞速。可以说，哪家公司能够率先推出产品，就能在同类市场上掌握极大的主动权。

推出产品，非常重要的一环就是产品的开发。只有更快地开发出产品，才能更快地推出产品。

1.1.2 渐进式开发

正如上面所说，"快"是所有互联网公司的诉求。那么如何才能实现快速开发呢？业界比较推崇的开发模式是渐进式开发。

所谓"渐进式"，是相对于传统的"瀑布模型"而言的。瀑布模型是一种软件开发方式，其开发过程是通过设计一系列阶段顺序展开的。从系统需求分析到产品发布和维护，每个阶段都会循环反馈。因此，如果有信息未被覆盖或发现了问题，那么可以返回上一个阶段并进行适当的修改。

瀑布模型有一个非常致命的弱点，就是会拉长整个产品推出的周期。毕竟如果每个阶段都考虑得非常全面的话，那么整个项目在最后的时间节点才能推出产品，这个时间点离最初的立项时间有可能间隔了数年。且不论开发、测试过程中可能会发现问题，导致"回滚"到上一个阶段，谁又能保证一款开发时间达数年之久的产品，最终还能够被用户所接受呢？用户在这么长的时间内想法不会产生变化吗？市场不会产生变化吗？

因为无法预料变化，所以应对变化的最好方式就是缩短周期。不去想若干年后的需求，把一个长期的大需求分解为若干个短期的小需求，即只考虑近几个月或近几个星期的需求。针对短期需求进行设计，而后开发，这就是所谓的"渐进式开发"方式。

渐进式开发方式能够及时感知需求产生的变化，从而调整开发策略。

1.1.3 拥抱变化

互联网公司的产品是不断变化的，唯一一成不变的产品，就是那些已经被市场淘汰了的。所以，如果一款产品还在不断地推出更新包，起码从侧面反映出该产品仍然在积极地做调整，同时也证明了该产品的需求量及开发的活跃度都比较高。

因此，只有拥抱变化的产品，才能够及时满足用户的需求。

用户的需求有时是明确的，但有时也是不明确的。只有在用户真正使用了产品之后，才能进一步完善需求。所以很多时候，变更需求都是在用户使用产品的过程中提出的。

1.1.4 敏捷之道

互联网公司大多采用敏捷开发方式。敏捷开发与渐进式开发有异曲同工之妙。

敏捷开发是将产品的开发周期划分为若干个"迭代"。一般一个迭代为 4 周或 2 周。开发团队全力完成当前迭代所确定的目标。在迭代完成之后，会发布一个可用的版本，交付给用户使用。

敏捷开发方式保障了产品能够及时交付给用户，同时也保障了开发团队能够及时从用户那里获取产品的使用反馈。这些反馈有可能是正面的，也可能是负面的。开发团队对用户反馈的内容进行整理，并将其纳入到下一个迭代的开发中去。

这样，开发团队与用户之间就形成了正向的闭环，可以使产品越来越趋近于用户的真实需求。

1.1.5 开源技术

开源技术相对于闭源技术而言，有其优势：一方面，开源技术源码是公开的，互联网公司在考察某项技术是否符合自身开发需求时，可以对源码进行分析；另一方面，开源技术商所花费的成本相对较低，这对于很多初创的互联网公司而言，可以节省一大笔技术投入。

当然，开源技术是把双刃剑，能够看到源码，并不意味着可以解决所有问题。开源技术在技术支持上不能与闭源技术相提并论，毕竟闭源技术有成熟的商业模式，可以提供足够的商业支持，而开源技术更多的是依赖于开源社区的支持。如果在使用开源技术的过程中发现了问题，可以反馈给开源社区，但开源社区不能保证什么时候、什么版本能够修复发现的问题。所以，使用开源技术，需要开发团队对其要有足够的了解。最好能够"吃透"源码，这样在发现源码中的问题时，能够及时解决。

例如，在关系型数据库方面，同属于 Oracle 公司的 MySQL 数据库和 Oracle 数据库，就是开源与闭源技术的两大代表，两者占据了全球数据库使用率的前两名[1]。MySQL 数据库主要是被中小企业和云计算供应商广泛采用，而 Oracle 数据库则由于其稳定、高性能的特性，深受政府和银行等客户的信赖。图 1-1 所示为 2019 年 2 月的数据库排名。

Rank Feb 2019	Rank Jan 2019	Rank Feb 2018	DBMS	Database Model	Score Feb 2019	Score Jan 2019	Score Feb 2018
1	1	1	Oracle	Relational DBMS	1264.02	-4.82	-39.26
2	2	2	MySQL	Relational DBMS	1167.29	+13.02	-85.18
3	3	3	Microsoft SQL Server	Relational DBMS	1040.05	-0.21	-81.98
4	4	4	PostgreSQL	Relational DBMS	473.56	+7.45	+85.18
5	5	5	MongoDB	Document store	395.09	+7.91	+58.67
6	6	6	IBM Db2	Relational DBMS	179.42	-0.43	-10.55
7	7	↑8	Redis	Key-value store	149.45	+0.43	+22.43
8	8	↑9	Elasticsearch	Search engine	145.25	+1.81	+19.93
9	9	↓7	Microsoft Access	Relational DBMS	144.02	+2.41	+13.95
10	10	↑11	SQLite	Relational DBMS	126.17	-0.63	+8.89

图1-1　数据库排名

1.1.6　微服务

相比敏捷开发方式将开发周期划分为若干个开发阶段，微服务架构则侧重于把整个软件划分为若干个不可分割的"原子"服务，这类原子服务就是微服务。

微服务架构采用 DDD（领域驱动设计）方式来进行业务建模，将每个微服务都设计成一个 DDD 边界上下文（Bounded Context）。这为系统内的微服务提供了一个逻辑边界，每个独立的团队负责一个逻辑上定义好的系统切片，负责与一个领域或业务功能相关的全部功能的开发，这样，团队开发出的最终代码会更易于理解和维护。

微服务架构可以理解为是 SOA（面向服务的架构）的一种特殊形式。这些服务之间由定义良好的接口和契约进行连接。接口采用中立的、与平台无关的方式进行定义，所以其能够跨越不同的硬件平台、操作系统和编程语言。

如果读者想完整了解微服务架构的设计，可以参阅笔者所著的《Spring Cloud 微服务架构开发实战》。

1.1.7　高并发

大型互联网应用往往有着非常高的并发量，如何抵御突如其来的流量洪峰，是每个运维人员需要思考的问题。其中一种常见的解决方案是，将经常需要访问的数据缓存起来，这样在下次查询的

---

[1] 数据来源于 DB-Engines，可见 https://db-engines.com/en/ranking。

时候就能快速地找到这些数据。

缓存的使用与系统的时效性有着非常大的关系。当应用的时效性要求不高时，则选择使用缓存是极好的。当系统要求的时效性比较高时，则不适合使用缓存。

本书后续章节会以 Redis 缓存的应用为例做详细介绍。

1.1.8 高可用

大型互联网应用往往会设置服务集群，多实例部署的方式可以提高整个系统的可用性。本书后续章节会就高可用展开详细讨论，同时也会对 NGINX 负载均衡、反向代理技术进行介绍。

1.2 传统企业级应用技术的不足

作为一门受欢迎的编程语言，Java 在经历了 20 多年的发展后，已然成为企业级应用开发的首选"利器"。在 2018 年 11 月的 TIOBE 编程语言排行榜 [1] 中，Java 位居榜首，可见开发者对于 Java 的厚爱。

图 1-2 所示为 2019 年 2 月的编程语言排名。

Feb 2019	Feb 2018	Change	Programming Language	Ratings	Change
1	1		Java	15.876%	+0.89%
2	2		C	12.424%	+0.57%
3	4	^	Python	7.574%	+2.41%
4	3	v	C++	7.444%	+1.72%
5	6	^	Visual Basic .NET	7.095%	+3.02%
6	8	^	JavaScript	2.848%	-0.32%
7	5	v	C#	2.846%	-1.61%
8	7	v	PHP	2.271%	-1.15%
9	11	^	SQL	1.900%	-0.46%
10	20	^^	Objective-C	1.447%	+0.32%

图1-2 编程语言排名

大型互联网公司也大多选择 Java 作为主力开发语言，包括 Google、IBM、Oracle 等外企，以

---

[1] 数据来源于 TIOBE，可见 https://www.tiobe.com/tiobe-index/。

及京东、百度、阿里巴巴等国内名企。Java 以其稳定性而著称，特别是"Write Once, Run Anywhere"（一次编写，各处运行）的特性，非常符合互联网企业快速推出产品、部署产品的需求。

但是，传统 Java 企业级应用所使用的技术，并不能适应当前互联网公司的发展需求，其不足之处总结如下。

1.2.1 规范不实用

Java 针对企业级应用市场推出的规范称为 Java EE，目前最新的版本是 Java EE 8。但曾几何时，Java EE 是"复杂、难用"的代名词。

传统的 Java EE 系统框架是臃肿、低效和脱离现实的。当时，SUN 公司推崇以 EJB 为核心的 Java EE 开发方式。但 EJB 本身是一种复杂的技术，虽然很好地解决了一些问题（如分布式事务），但在许多情况下增加了比其商业价值更大的复杂性问题。

传统 Java EE 应用的开发效率是低下的，应用服务器厂商对各种技术的支持并没有真正统一，导致 Java EE 应用并没有真正实现"Write Once, Run Anywhere"的承诺。

而出现这些问题的原因，是 Java EE 和 EJB 的设计都是"以规范为驱动"的。但其所遵循的这些规范并没有针对性地解决问题，反而在实际开发中引入了很多复杂性。毕竟，成功的标准都是从实践中得来的，而不是由哪个委员会创造出来的。

1.2.2 学习成本太高

传统 Java EE 的很多规范都是违反"帕累托法则"的。"帕累托法则"也称"二八定律"，是指花较少的成本（10%~20%）解决大部分问题（80%~90%），而架构的价值在于为常见的问题找到好的解决方案，而不是一心想要解决更复杂、更罕见的问题。EJB 的问题就在于，它违背了这个法则——为了满足少数情况下的特殊要求，给大多数使用者强加了不必要的复杂性，使开发者难以上手。

早期的 EJB 2.1 规范中，EJB 的目标定位有 11 项之多，而这些目标没有一项是致力于简化 Java EE 开发的。同时，EJB 的编程模型是非常复杂的，要使用 EJB 需要继承非常多的接口，而这些接口在实际开发中并不是真正为了解决问题。

1.2.3 不够灵活

EJB 依赖于容器，所以 EJB 在编写业务逻辑时是与容器耦合的，编程模型不够灵活。与容器耦合的方式必然会导致开发、测试、部署的难度增大，同时也拉长了整个开发的周期。

因为编写程序需要依赖于具体的容器实现，所以"Write Once, Run Anywhere"变成了"一次编写，到处重写"。特别是实体 bean，基本上迁移了一个服务器，就需要重新编写，相应的测试工作量也增加了。

1.2.4 发展缓慢

EJB 规范中对实体映射的定义过于宽泛，导致每个厂商都有自己的 ORM 实现，需引入特定厂商的部署描述符。又因为 Java EE 中除 Web 外，类加载的定义没有明确，导致产生了特定厂商的类加载机制和打包方式。同时，特定厂商的服务查找方式也是有差异的，这在一定程度上加大了开发的难度，使得移植变得困难。

规范如果不能解决开发者的实际问题，开发者自然不会买账，这种规范迟早会被市场淘汰。事实上，尽管 JCP（Java 社区进程）在这方面做出了一些努力，但仍然无法赶上现代 IT 市场快速发展的步伐。从 2013 年 6 月发布 Java EE 7 以来，出现了很多新兴技术，如 NoSQL、容器、微服务和无服务器架构，但它们都未能被包含在 Java EE 中。

Oracle 公司也意识到了 Java EE 发展缓慢的问题，所以在 2017 年 9 月宣布将 Java EE 8 移交给开源组织 Eclipse 基金会管理，期望通过开源的方式来"活化"Java EE。

1.3 Lite框架简介

正是由于传统企业级应用技术的不足，迫使开发者将目光转向了开源社区。Rod Johnson 在其编著的 *Expert One-on-One J2EE Design and Development* 中，可以说一针见血地指出了当时 Java EE 架构在实际开发中的种种弊端，并推出了 Spring 框架来简化企业级应用的开发。

之后，开源社区日益繁荣，Hibernate、Structs 等轻量级框架相继推出，以替换 Java EE 中的"重量级"实现。

本书主要介绍如何从零开始学习市面上优秀的开源框架，来实现属于自己的轻量级框架。笔者称这种框架为"Lite"[1]，意味着开源、简单、轻量。同时，在本书的后半部分，笔者也会展示如何基于 Lite 框架，来开发实现一个真实的互联网应用。

那么，Lite 框架到底是什么样的呢？

---

[1] Lite 框架开源地址为 https://github.com/waylau/lite。

1.3.1 轻量级架构

Lite 框架是一套轻量级 Web 框架，基于 Lite 可以轻松实现企业级应用。Lite 具有非侵入性，依赖的东西非常少，占用资源也非常少，部署简单，启动快速，比较容易使用。

Lite 底层基于 Spring 框架来实现 bean 的管理，因此，只要有 Spring 的开发经验，上手 Lite 就非常简单。即便是 Spring 的新手，通过对本书第 3~7 章相关实例的学习，也能轻松入门 Spring。

1.3.2 符合二八定律

Lite 旨在通过较少的成本（10%~20%）解决大部分问题（80%~90%）。

Lite 专注于解决企业级应用中的场景问题，如对象管理、事务管理、认证与授权、数据存储、负载均衡、缓存等，这些场景基本上涵盖了企业级应用。

通过学习 Lite 框架，读者能够掌握互联网公司常用的技术，能够解决企业关注的大部分问题，有利于提升技术人员的核心竞争力。

1.3.3 基于开源技术

Lite 框架吸收了优秀开源框架的技术，取其精华，使自身功能变得更强大，但又简单、易于理解。

Lite 框架所使用的开源技术都是目前大型互联网公司所采用的成熟技术，主要包括以下内容。

（1）基于 Maven 实现模块化开发及项目管理。

（2）基于 Jetty 提供开箱即用的 Servlet 容器。

（3）使用 Spring 实现 IoC 和 AOP 机制。

（4）基于 Spring TestContext 实现开发过程中的单元测试。

（5）使用 Spring Web MVC 实现 RESTful 风格的架构。

（6）基于 Spring Security 实现认证与授权。

（7）使用 MySQL 实现数据的高效存储。

（8）使用 MyBatis 实现数据库的操作与对象关系映射。

（9）使用 NGINX 实现应用的负载均衡与高可用。

（10）使用 Redis 实现应用的高并发。

（11）使用 Spring Boot 简化应用的配置。

本书将会在后续章节中继续深入探讨上述开源技术。

1.3.4 支持微服务

在复杂的大型互联网应用架构中，倾向于使用微服务架构来划分不同的微服务。这些微服务面向特定的领域，所实现的功能也更有针对性。

Lite 框架支持微服务架构。Lite 非常轻量，启动速度很快。同时，Lite 倾向于将应用打包成"fat jar"[1]的形式，因而能够轻易在微服务架构的常用容器等环境中运行。

1.3.5 可用性和扩展性

由于 Lite 框架支持微服务架构，因此 Lite 很容易实现自身的横向扩展。

理论上，每个微服务都是独立部署的，且会部署多个实例，以保证可用性和扩展性。同时，独立部署微服务实例，有利于监控每个微服务实例运行的状态，方便在应用达到告警阈值时及时做出调整。

1.3.6 支撑大型互联网应用

正是由于 Lite 具有良好的可用性和扩展性，使其非常适用于大型互联网应用。因为大型互联网应用既要部署快、运行快，还要求在运维过程中能够及时处理突发事件。

图 1-3 所示为微服务实例自动扩展的场景。

图1-3 微服务实例自动扩展

从图中可以清楚地看到，监控程序会对应用进行持续监控，当现有的服务实例 CPU 超过了预设的阈值（60%）时，监控程序会做出自动扩展的决策，新启动一个"实例3"来加入到原有的系统中。

---

[1] fat jar 也称 uber jar，是一种可执行的 jar 包，它将自己的程序及其依赖的三方 jar 全部打到一个 jar 包中。

第2章 Servlet

Servlet可以说是Java EE中最为成功、应用最为广泛的规范了。开发Java Web应用离不开Servlet的支持。

本章详细介绍Servlet的概念及基本用法，以及如何在应用中使用高性能Servlet容器——Jetty。

2.1 Servlet概述

Servlet 是 Server Applet 的简称，指服务器端小程序或服务连接器，主要功能在于交互式地浏览和修改数据，生成动态 Web 内容。目前，最新的 Servlet 规范版本为 Servlet 4.0（JSR 369）。

2.1.1 Servlet架构

Java Servlet 是运行在 Web 服务器或应用服务器上的程序，是作为来自 Web 浏览器或其他 HTTP 客户端的请求和 HTTP 服务器的数据库或应用程序之间的中间层。

使用 Servlet 可以收集来自网页表单的用户输入，呈现来自数据库或其他源的记录，还可以动态创建网页。

图 2-1 所示为 Servlet 的整体架构。

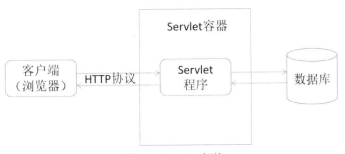

图2-1　Servlet架构

Servlet 主要执行以下任务。

（1）读取客户端（浏览器）发送的显式数据，包括网页上的 HTML 表单，或是自定义的 HTTP 客户端程序的表单。

（2）读取客户端（浏览器）发送的隐式 HTTP 请求数据，包括 Cookies、媒体类型和浏览器能理解的压缩格式等。

（3）处理数据并生成结果。这个过程可能需要访问数据库，执行 RMI 等远程过程调用，调用 Web 服务，或者直接计算得出对应的响应。

（4）发送显式数据（文档）到客户端（浏览器），数据格式可以是多种多样的，包括文本文件（HTML、XML 或 JSON）、二进制文件（GIF 图像）、Excel 表格等。

（5）发送隐式 HTTP 响应到客户端（浏览器），设置 Cookies 和缓存参数，以及其他类似的任务。

2.1.2 Servlet生命周期

Servlet 生命周期可被定义为从创建 Servlet 到其被销毁的整个过程，以下是 Servlet 的生命周期遵循的过程。

（1）Servlet 调用 init 方法进行初始化。

（2）Servlet 调用 service 方法来处理客户端的请求。

（3）Servlet 通过调用 destroy 方法终止。

（4）最后，Servlet 由 JVM 的垃圾回收器进行回收。

下面详细讨论生命周期中的方法。

1. init方法

init 方法被设计成只调用一次。其在第一次创建 Servlet 时被调用，在后续每次用户请求时不再调用。因此，它用于一次性初始化。

Servlet 创建于用户第一次调用对应该 Servlet 的 URL 时，也可以指定 Servlet 在服务器第一次启动时被加载。

当用户调用一个 Servlet 时，就会创建一个 Servlet 实例，每一个用户请求都会产生一个新的线程，在适当的时候移交给 doGet 方法或 doPost 方法。init 方法简单地创建或加载一些数据，这些数据将被用于 Servlet 的整个生命周期。

init 方法的定义如下。

```
public void init() throws ServletException {
  // 初始化代码…
}
```

2. service方法

service 方法是执行实际任务的主要方法。Servlet 容器调用 service 方法处理来自客户端（浏览器）的请求，并把格式化的响应写回给客户端。

每次服务器接收到一个 Servlet 请求时，都会产生一个新的线程并调用服务。service 方法用于检查 HTTP 请求类型（GET、POST、PUT、DELETE 等）。

下面是该方法的特征。

```
public void service(ServletRequest request,
                    ServletResponse response)
    throws ServletException, IOException{
}
```

service 方法由容器调用，它会在适当的时候调用 doGet、doPost、doPut、doDelete 等方法。所以不用对 service 方法做任何动作，只需要根据客户端的请求类型重写 doGet 方法、doPost 方法。

doGet 方法和 doPost 方法是每次服务请求中最常用的方法。

3. doGet方法

当 Servlet 容器接收到 GET 请求时，会将该请求交由 doGet 方法处理。处理逻辑写在重写的 doGet 方法中，代码如下。

```
public void doGet(HttpServletRequest request,
                  HttpServletResponse response)
        throws ServletException, IOException {
    // Servlet代码…
}
```

4. doPost方法

当 Servlet 容器接收到 POST 请求时，会将该请求交由 doPost 方法处理。处理逻辑写在重写的 doPost 方法中，代码如下。

```
public void doPost(HttpServletRequest request,
                   HttpServletResponse response)
        throws ServletException, IOException {
    // Servlet代码…
}
```

5. destroy方法

当 Servlet 容器确定 Servlet 应该从服务中移除时，将调用 Servlet 接口的 destroy 方法，以允许 Servlet 释放它自己使用的任何资源和保存任何持久化的状态。例如，当想要节省内存资源或 Servlet 被关闭时，Servlet 容器可以执行 destroy 方法。

在 Servlet 容器调用 destroy 方法之前，必须保证正在执行 service 方法的线程已经执行完毕，或者超过了服务器定义的时间限制。

一旦调用了 Servlet 实例的 destroy 方法，容器就无法再路由其他请求到该 Servlet 实例了。如果容器需要再次使用该 Servlet，则必须使用该 Servlet 类的一个新实例。在 destroy 方法执行完毕后，Servlet 容器必须释放 Servlet 实例以便被回收。

destroy 方法的定义如下。

```
public void destroy() {
    // 终止化代码…
}
```

2.1.3 常用方法

基本的 Servlet 接口定义了 service 方法用于处理客户端的请求。当有请求到达时，该方法由 Servlet 容器路由到一个 Servlet 实例来调用。

Web 应用的并发请求处理通常需要 Web 开发人员去设计适合多线程执行的 Servlet，从而保证

Service 方法能在一个特定时间点处理多线程并发执行。通常 Web 容器对于并发请求会使用同一个 Servlet 处理，并且在不同的线程中并发执行 Service 方法。

HttpServlet 抽象子类在基本的 Servlet 之上添加了一些协议相关的方法，并且这些方法能根据 HTTP 请求类型自动将 HttpServlet 中实现的 service 方法转发到相应协议的处理方法上。这些方法如下。

（1）doGet 处理 HTTP GET 请求。

（2）doPost 处理 HTTP POST 请求。

（3）doPut 处理 HTTP PUT 请求。

（4）doDelete 处理 HTTP DELETE 请求。

（5）doHead 处理 HTTP HEAD 请求。

（6）doOptions 处理 HTTP OPTIONS 请求。

（7）doTrace 处理 HTTP TRACE 请求。

一般情况下，开发基于 HTTP 的 Servlet 时，Servlet 开发人员只需实现 doGet 和 doPost 请求处理方法。如果开发人员想使用其他处理方法，那么使用方式跟之前类似，即 HTTP 编程都类似。

doPut 方法和 doDelete 方法允许 Servlet 开发人员让支持 HTTP/1.1 的客户端使用这些功能。HttpServlet 中的 doHead 方法可以认为是 doGet 方法的一种特殊形式，它仅返回由 doGet 方法产生的 header 信息。doOptions 方法返回当前 Servlet 支持的 HTTP 方法。doTrace 方法返回的响应包含 TRACE 请求的所有头信息。

2.1.4　Servlet 容器

Servlet 容器是 Web 服务器或应用服务器的一部分，用于提供基于请求 / 响应发送模式的网络服务，解码基于 MIME 的请求，以及格式化基于 MIME 的响应。Servlet 容器同时也包含并管理其生命周期中的 Servlet。

Servlet 容器可以嵌入宿主的 Web 服务器中，或者通过 Web 服务器的本地扩展 API 单独作为附加组件安装。Servlet 容器也可以内嵌或安装到启用 Web 功能的应用服务器中。

所有的 Servlet 容器都必须支持 HTTP 协议以处理请求和响应，但额外的基于请求 / 响应的协议，如 HTTPS（HTTP over SSL）的支持是可选的。对于 HTTP 规范要求的版本，容器必须支持 HTTP/1.1 和 HTTP/2。

在支持 HTTP/2 时，Servlet 容器必须支持"h2""h2c"协议标识符，这意味着所有 Servlet 容器都必须支持 ALPN。因为容器可能有缓存，可以在将协议标识符传递给 Servlet 之前修改来自客户机的请求，也可以在将 Servlet 发送到客户机之前修改响应，或者可以响应请求而不将其传递给 Servlet。

Java SE 8 是与 Servlet 4.0 一起使用的最低 Java 平台版本。

2.1.5 常用Servlet容器

常见的 Servlet 容器有闭源的也有开源的，包括 Tomcat、Jetty、Oracle Application Server、Oracle Weblogic Server、JBoss Application Server 等。其中，Tomcat、Jetty 在开源界比较有名，且在市场上占有率比较高。

下面就 Tomcat 和 Jetty 的异同点进行比较。

1. 相同点

Tomcat 和 Jetty 都是 Servlet 引擎，它们都支持标准的 Servlet 规范和 Java EE 规范；它们都是开源的，可以免费使用。

2. 不同点

（1）在架构上，Jetty 的架构比 Tomcat 的架构更为简单，具体如下。

①Jetty 的架构是基于 Handler 实现的，主要的扩展功能都可以用 Handler 来实现，扩展简单。

②Tomcat 的架构是基于容器设计的，进行扩展时需要了解 Tomcat 的整体设计结构，不易扩展。

（2）在性能上，Jetty 和 Tomcat 差异不大，具体如下。

①Jetty 可以同时处理大量连接且可以长时间保持连接，适合于 Web 聊天应用等。

②Jetty 的架构简单，因此作为服务器，Jetty 可以按需加载组件，减少不需要的组件，能减少服务器的内存开销，从而提高服务器性能。

③Jetty 默认采用 NIO 结束，在处理 I/O 请求上更占优势，在处理静态资源时性能较高。

④Tomcat 适合处理少数非常繁忙的连接，也就是说连接生命周期短的话，Tomcat 的总体性能更高。

⑤Tomcat 默认采用 BIO 处理 I/O 请求，在处理静态资源时性能较差。

（3）在其他方面，二者差异如下。

①Jetty 的应用更加快速，修改简单，对新的 Servlet 规范支持较好。

②Tomcat 目前应用比较广泛，对 Java EE 和 Servlet 的支持更加全面，很多特性可直接集成进来。

3. 总结

Jetty 的主要特性包括易用性、可扩展性及易嵌入性。可以把 Jetty 理解为一个嵌入式的 Web 服务器，Jetty 的运行速度较快，而且是轻量级的。Jetty 的轻量级也使其在处理高并发、细粒度请求的场景下显得更快速、高效。

Jetty 更灵活，体现在其可插拔性和可扩展性，更易于开发者对 Jetty 本身进行二次开发，定制一个符合自身需求的 Web 服务器。

Tomcat 支持的规范更全面，功能也更多，也显得更加"重量级"。所以，当面临大规模企业级应用时，Jetty 需要扩展，在这种场景下，使用 Tomcat 则会更加方便。

Jetty 更满足公有云的分布式环境的需求，而 Tomcat 更符合企业级环境。

综上所述，Lite 框架基于 Jetty 实现内嵌容器。在本书的后续章节中，还会对 Jetty 进行深入探讨。要了解更多有关 Servlet 的内容，可以参阅笔者所著的开源电子书《Java Servlet 3.1 规范》。

2.2 请求

请求对象封装了客户端请求的所有信息。在 HTTP 协议中，这些信息包含在从客户端发送到服务器请求的 HTTP 头部和消息体中。

2.2.1 HTTP协议参数

Servlet 请求参数以字符串的形式作为请求的一部分，从客户端发送到 Servlet 容器。当请求是一个 HttpServletRequest 对象，且符合"参数可用时"描述的条件时，容器从 URI 查询字符串和 POST 数据中填充参数。参数以一系列的名 - 值对（name-value）的形式保存。任何给定的参数名称可存在多个参数值。ServletRequest 接口的下列方法可访问这些参数。

- getParameter
- getParameterNames
- getParameterValues
- getParameterMap

getParameterValues 方法返回一个 String 对象的数组，包含了与参数名称相关的所有参数值。getParameter 方法的返回值必须是 getParameterValues 方法返回的 String 对象数组中的第一个值。getParameterMap 方法返回请求参数的一个 java.util.Map 对象，其中以参数名称作为 map 键，参数值作为 map 值。

查询字符串和 POST 请求的数据被汇总到请求参数集合中，查询字符串数据放在 POST 数据之前。例如，如果请求由查询字符串 a=hello 和 POST 数据 a=goodbye&a=world 组成，得到的参数集合顺序将是 a=(hello, goodbye, world)。

这些 API 不会暴露 GET 请求（HTTP 1.1 所定义的）的路径参数。路径参数必须从 getRequestURI 方法或 getPathInfo 方法返回的字符串值中解析。

将 POST 数据填充到参数集（Paramter Set）必须满足 4 个条件：第一，该请求是一个 HTTP 或 HTTPS 请求；第二，HTTP 方法是 POST；第三，内容类型是 application/x-www-form-urlencoded；第四，该 Servlet 已经对请求对象的任意 getParameter 方法进行了初始调用。如果不满足这些条件，且参数集中不包括 POST 数据，那么 Servlet 必须通过请求对象的输入流得到 POST 数据。如果满足这些条件，那么从请求对象的输入流中直接读取 POST 数据将不再有效。

2.2.2 属性

属性是与请求相关联的对象。属性可以由容器设置来表达信息，或者由 Servlet 设置将信息传达给另一个 Servlet（通过 RequestDispatcher）。属性通过 ServletRequest 接口的以下方法来访问。

- getAttribute
- getAttributeNames
- setAttribute

一个属性名称只能关联一个属性值。

前缀以"java.""javax."开头的属性名称是 Java 规范的保留定义；前缀以"sun.""com.sun.""oracle""com.oracle"开头的属性名称是 Oracle 公司的保留定义。

2.2.3 请求头

通过以下 HttpServletRequest 接口方法，Servlet 可以访问 HTTP 请求的头信息。

- getHeader
- getHeaders
- getHeaderNames

getHeader 方法返回给定头名称的头。多个头可以具有相同的名称，如 HTTP 请求中的 Cache-Control 头。如果多个头的名称相同，getHeader 方法返回请求中的第一个头。getHeaders 方法允许访问所有与特定头名称相关的头值，返回一个 String 对象的 Enumeration（枚举）。

头可包含 String 形式的 int 或 Date 数据。HttpServletRequest 接口提供以下方法访问这些类型的头数据。

- getIntHeader
- getDateHeader

如果 getIntHeader 方法不能转换为 int 的头值，则抛出 NumberFormatException 异常。如果 getDateHeader 方法不能把头转换成一个 Date 对象，则抛出 IllegalArgumentException 异常。

2.2.4 请求路径元素

引导 Servlet 服务请求的路径由许多重要部分组成：

```
URI = Context Path + Servlet Path + PathInfo
```

下面分别进行说明。

（1）Context Path：与 ServletContext 相关联的路径前缀是这个 Servlet 的一部分。如果这个上下文是基于 Web 服务器的 URL 命名空间的"默认"上下文，那么这个路径将是一个空字符串。如果上下文不是基于服务器的命名空间，那么这个路径就会以"/"字符开始，但不以"/"字符结束。

（2）Servlet Path：路径部分直接与激活请求的映射对应。这个路径以"/"字符开头，如果请求与"/*"或""（空字符串）模式匹配，它将是一个空字符串。

（3）PathInfo：请求路径的一部分，不属于 Context Path 或 Servlet Path。如果没有额外的路径，它要么是 null，要么是以"/"开头的字符串。

使用 HttpServletRequest 接口中的以下方法来访问这些信息。

- getContextPath
- getServletPath
- getPathInfo

表 2-1 所示为请求路径元素的使用例子。

表2-1　请求路径示例

请求路径	路径元素
/catalog/lawn/index.html	ContextPath: /catalog ServletPath: /lawn PathInfo: /index.html
/catalog/garden/implements/	ContextPath: /catalog ServletPath: /garden PathInfo: /implements/
/catalog/help/feedback.jsp	ContextPath: /catalog ServletPath: /help/feedback.jsp PathInfo: null

2.2.5　路径转换方法

在 Servlet API 中有两个简便的方法，允许开发者获得与某个特定路径等价的文件系统路径。具体如下。

- ServletContext.getRealPath
- HttpServletRequest.getPathTranslated

getRealPath 方法需要一个 String 参数，并返回一个 String 形式的路径，这个路径对应一个在本地文件系统上的文件。getPathTranslated 方法推断出请求的 PathInfo 的实际路径。

这些方法在 Servlet 容器无法确定一个有效文件路径的情况下必须返回 null，如 Web 应用程序不能访问远程文件系统。jar 文件中 META-INF/resources 目录下的资源，只有当调用 getRealPath 方法时，才认为容器已经从包含这些资源的 jar 文件中解压，在这种情况下，必须返回解压缩后的位置。

2.2.6　请求数据编码

目前，许多浏览器并不随着 Content-Type 头一起发送字符编码限定符，而是读取 HTTP 请求来确定字符编码。如果客户端没有指定请求默认的字符编码，容器用来创建请求读取器和解析 POST 数据的编码必须是"ISO-8859-1"。在客户端没有指定请求默认的字符编码的情况下，客户端发送字符编码失败，容器从 getCharacterEncoding 方法返回 null。

如果客户端没有设置字符编码，并使用不同的编码来请求数据，而不是使用上面描述的默认字符编码，那么可能会发生问题。为了避免这种情况，开发人员可以通过以下方法来覆盖由容器提供的字符编码。

（1）*ServletContext 提供的 setRequestCharacterEncoding(String enc)。

（2）web.xml 提供的元素。

（3）ServletRequest 接口提供的 setCharacterEncoding(String enc)。

必须在解析任何 POST 数据或从请求中读取任何输入之前调用上述方法。调用上述方法不会影响已经读取的数据编码。

2.3 Servlet上下文

本节将详细介绍 Servlet 上下文 ServletContext 接口的作用及配置方式。

2.3.1 ServletContext接口介绍

ServletContext 接口定义了 Servlet 运行在 Web 应用的视图。容器供应商负责提供 Servlet 容器的 ServletContext 接口实现。Servlet 可以使用 ServletContext 对象记录事件，获取 URL 引用的资源，存取当前上下文中其他 Servlet 可以访问的属性。

ServletContext 是 Web 服务器中已知路径的根。例如，Servlet 上下文可以从"http://www.waylau.com/catalog"找到，"/catalog"请求路径称为上下文路径，所有以其开头的请求都会被路由到与 ServletContext 相关联的 Web 应用中。

2.3.2 ServletContext接口作用域

每一个部署到容器的 Web 应用都有一个 ServletContext 接口的实例与之关联。当容器分布在多台虚拟机时，每个 JVM 的 Web 应用将会有一个 ServletContext 实例。

如果容器内的 Servlet 没有部署到 Web 应用中，则作为"默认"Web 应用的一部分，并有一个默认的 ServletContext。在分布式的容器中，默认的 ServletContext 是非分布式的，且仅存在于一个 JVM 中。

2.3.3 初始化参数

以下 ServletContext 接口方法允许 Servlet 访问由应用开发人员在 Web 应用的部署描述符中指定的上下文初始化参数。

- getInitParameter
- getInitParameterNames

应用开发人员使用初始化参数来表达配置信息。代表性的例子是一个网络管理员的 E-mail 地址，或保存关键数据的系统名称。

2.3.4 配置方法

下面的方法从 Servlet 3.0 开始添加到 ServletContext，可以启用编程方式定义 Servlet、Filter 和其映射到的 URL 模式。这些方法只能从 ServletContextListener 实现的 contexInitialized 方法，或者 ServletContainerInitializer 实现的 onStartup 方法进行应用初始化的过程中调用。除了添加 Servlet 和 Filter，也可以查找关联到 Servlet 或 Filter 的一个 Registration 对象实例，或者关联到 Servlet 或 Filter 的所有 Registration 对象的 Map。

如果 ServletContext 传到了 ServletContextListener 的 contextInitialized 方法，但该 ServletContext-Listener 既没有在 web.xml 或 web-fragment.xml 中声明，也没有使用 @WebListener 注解，则在 ServletContext 中定义的用于 Servlet、Filter 和 Listener 编程式配置的所有方法必须抛出 Unsupported-OperationException。

1. 编程式添加和配置Servlet

编程式添加 Servlet 到上下文对框架开发者是很有用的。例如，框架可以使用这个方法声明一个控制器 Servlet，该方法将返回一个 ServletRegistration 或 ServletRegistration.Dynamic 对象，并允许进一步配置如 init-params、url-mapping 等 Servlet 的其他参数。下面描述了该方法的 6 个重载版本。

（1）addServlet(String servletName, String className)：该方法允许应用以编程方式声明一个 Servlet。它添加给定的 Servlet 名称和 class 名称到 Servlet 上下文。

（2）addServlet(String servletName, Servlet servlet)：该方法允许应用以编程方式声明一个 Servlet。它添加给定的名称和 Servlet 实例的 Servlet 到 Servlet 上下文。

（3）addServlet(String servletName, Class <? extends Servlet> servletClass)：该方法允许应用以编程方式声明一个 Servlet。它添加给定的名称和 Servlet 类的一个实例的 Servlet 到 Servlet 上下文。

（4）T createServlet(Class clazz)：该方法实例化一个给定的 Servlet 类。该方法必须支持适用于 Servlet 的除 @WebServlet 外的所有注解。返回的 Servlet 实例通过调用上面定义的 addServlet (String, Servlet) 注册到 ServletContext 之前，可以进行进一步的定制。

（5）ServletRegistration getServletRegistration(String servletName)：该方法返回与给定名称的

Servlet 相关的 ServletRegistration，如果没有与其相关的 ServletRegistration 则返回 null。如果 ServletContext 传到了 ServletContextListener 的 contextInitialized 方法，但该 ServletContextListener 既没有在 web.xml 或 web-fragment.xml 中声明，也没有使用 javax.servlet.annotation.WebListener 注解，则必须抛出 UnsupportedOperationException。

（6）Map getServletRegistrations()：该方法返回 ServletRegistration 对象的 Map，由名称作为键并对应注册到 ServletContext 的所有 Servlet。如果没有 Servlet 注册到 ServletContext 则返回一个空的 Map。返回的 Map 包括所有声明和注解的 Servlet 对应的 ServletRegistration 对象，也包括那些使用 addServlet 方法添加的所有 Servlet 对应的 ServletRegistration 对象。返回的 Map 的任何改变都不影响 ServletContext。如果 ServletContext 传到了 ServletContextListener 的 contextInitialized 方法，但该 ServletContextListener 既没有在 web.xml 或 web-fragment.xml 中声明，也没有使用 javax.servlet.annotation.WebListener 注解，则必须抛出 UnsupportedOperationException。

2. 编程式添加和配置Filter

（1）addFilter(String filterName, String className)：该方法允许应用以编程方式声明一个 Filter。它添加给定的名称和类名称的 Filter 到 Web 应用。

（2）addFilter(String filterName, Filter filter)：该方法允许应用以编程方式声明一个 Filter。它添加给定的名称和 Filter 实例的 Filter 到 Web 应用。

（3）addFilter(String filterName, Class <? extends Filter> filterClass)：该方法允许应用以编程方式声明一个 Filter。它添加给定的名称和 Filter 类的一个实例的 Filter 到 Web 应用。

（4）T createFilter(Class clazz)：该方法实例化一个给定的 Filter。该方法必须支持适用于 Filter 的所有注解。返回的 Filter 实例通过调用上面定义的 addServlet(String, Filter) 注册到 ServletContext 之前，可以进行进一步的定制。给定的 Filter 类必须定义一个用于实例化的空参构造器。

（5）FilterRegistration getFilterRegistration(String filterName)：该方法返回与给定名称的 Filter 相关的 FilterRegistration，如果没有与其相关的 FilterRegistration 则返回 null。如果 ServletContext 传到了 ServletContextListener 的 contextInitialized 方法，但该 ServletContextListener 既没有在 web.xml 或 web-fragment.xml 中声明，也没有使用 javax.servlet.annotation.WebListener 注解，则必须抛出 UnsupportedOperationException。

（6）Map getFilterRegistrations()：该方法返回 FilterRegistration 对象的 Map，由名称作为键并对应注册到 ServletContext 的所有 Filter。如果没有 Filter 注册到 ServletContext，则返回一个空的 Map。返回的 Map 中包括所有声明和注解的 Filter 对应的 FilterRegistration 对象，也包括那些使用 addFilter 方法添加的所有 Servlet 对应的 ServletRegistration 对象。返回的 Map 的任何改变都不影响 ServletContext。如果 ServletContext 传到了 ServletContextListener 的 contextInitialized 方法，但该 ServletContextListener 既没有在 web.xml 或 web-fragment.xml 中声明，也没有使用 javax.servlet.annotation.WebListener 注解，则必须抛出 UnsupportedOperationException。

3. 编程式添加和配置Listener

（1）void addListener(String className)：往 ServletContext 添加给定类名的监听器。ServletContext 将使用由与应用关联的 classloader 加载的该给定名称的 class，且它们必须实现一个或多个以下接口。

- javax.servlet.ServletContextAttributeListener
- javax.servlet.ServletRequestListener
- javax.servlet.ServletRequestAttributeListener
- javax.servlet.http.HttpSessionListener
- javax.servlet.http.HttpSessionAttributeListener
- javax.servlet.http.HttpSessionIdListener

（2）void addListener(T t)：往 ServletContext 添加一个给定的监听器。给定的监听器实例必须实现一个或多个以下接口。

- javax.servlet.ServletContextAttributeListener
- javax.servlet.ServletRequestListener
- javax.servlet.ServletRequestAttributeListener
- javax.servlet.http.HttpSessionListener
- javax.servlet.http.HttpSessionAttributeListener
- javax.servlet.http.HttpSessionIdListener

（3）void addListener(Class <? extends EventListener> listenerClass)：往 ServletContext 添加给定类名的监听器。给定的监听器类必须实现一个或多个以下接口。

- javax.servlet.ServletContextAttributeListener
- javax.servlet.ServletRequestListener
- javax.servlet.ServletRequestAttributeListener
- javax.servlet.http.HttpSessionListener
- javax.servlet.http.HttpSessionAttributeListener
- javax.servlet.http.HttpSessionIdListener

（4）void createListener(Class clazz)：该方法实例化给定的 EventListener 类。给定的 EventListener 类必须实现至少一个以下接口。

- javax.servlet.ServletContextAttributeListener
- javax.servlet.ServletRequestListener
- javax.servlet.ServletRequestAttributeListener
- javax.servlet.http.HttpSessionListener
- javax.servlet.http.HttpSessionAttributeListener

- javax.servlet.http.HttpSessionIdListener

4. 用于编程式添加Servlet、Filter和Listener的注解处理需求

除了需要一个实例的 addServlet 外，当使用编程式 API 添加 Servlet 或创建 Servlet 时，以下类中的有关注解必须被内省，且其定义的元数据必须被使用，除非它被 ServletRegistration.Dynamic 或 ServletRegistration 中调用的 API 覆盖了。

2.3.5 上下文属性

Servlet 可以通过名称将对象属性绑定到上下文。同一个 Web 应用内的任何 Servlet 都可以使用绑定到上下文的属性，以下 ServletContext 接口中的方法允许访问此功能。

- setAttribute
- getAttribute
- getAttributeNames
- removeAttribute

在 JVM 中创建的上下文属性是本地的，这可以防止从一个分布式容器的共享内存储中获取 ServletContext 属性。当需要在运行在分布式环境的 Servlet 之间共享信息时，该信息应该被放到会话中或存储到数据库，或者设置到 Enterprise JavaBeans（企业级 JavaBean）组件中。

2.3.6 资源

ServletContext 接口提供了直接访问 Web 应用中仅是静态内容层次结构的文件的方法，包括 HTML、GIF 和 JPEG 文件等。

- getResource
- getResourceAsStream

getResource 方法和 getResourceAsStream 方法需要一个以 "/" 开头的 String 作为参数，给定的资源路径是相对于上下文的根，或者相对于 Web 应用的 WEB-INF/lib 目录下的 jar 文件中的 META-INF/resources 目录。这两个方法先根据请求的资源查找 Web 应用上下文的根，然后查找所有 WEB-INF/lib 目录下的 jar 文件。查找 WEB-INF/lib 目录中 jar 文件的顺序是不确定的。这种层次结构的文件可以存在于服务器的文件系统、Web 应用的归档文件、远程服务器或其他位置中。

需要注意的是，这两个方法不能用于获取动态内容。例如，在支持 JSP 的容器中，用 getResource("/index.jsp") 形式的方法调用这两个方法，将返回 JSP 源码而不是处理后的输出。

2.4 响应

响应对象封装了从服务器返回到客户端的所有信息。在 HTTP 协议中，这些信息包含在从服务器传输到客户端的 HTTP 头信息或响应的消息体中。

2.4.1 缓冲

出于性能的考虑，Servlet 容器允许（但不要求）从缓存中输出内容到客户端。一般情况下，服务器是默认执行缓存，但也允许 Servlet 指定缓存参数。

下面是 ServletResponse 接口允许 Servlet 访问和设置缓存信息的方法。

- getBufferSize
- setBufferSize
- isCommitted
- reset
- resetBuffer
- flushBuffer

不管 Servlet 使用的是一个 ServletOutputStream 还是一个 Writer，ServletResponse 接口提供的这些方法都允许执行缓冲操作。getBufferSize 方法返回使用的底层缓冲区大小，如果没有使用缓冲，该方法必须返回一个 int 值 0。Servlet 可以请求 setBufferSize 方法来设置一个最佳的缓冲大小。isCommitted 方法返回一个表示是否有任何响应字节返回客户端的 boolean 值。flushBuffer 方法强制刷出缓冲区的内容到客户端。当响应没有提交时，reset 方法用来清空缓冲区的数据。头信息、状态码，以及在调用 reset 之前，Servlet 调用 getWriter 或 getOutputStream 设置的状态也必须被清空。如果响应没有被提交，resetBuffer 方法将清空缓冲区中的内容，但不清空请求头和状态码。

如果响应已经提交并且 reset 或 resetBuffer 方法已被调用，则必须抛出 IllegalStateException，响应及它关联的缓冲区将保持不变。

当使用缓冲区时，容器必须立即刷出填满的缓冲区内容到客户端。

2.4.2 头

Servlet 可以通过下面 HttpServletResponse 接口的方法来设置 HTTP 响应头。

- setHeader
- addHeader

setHeader 方法通过给定的名称和值来设置头。前面的头会被后面的新头替换。如果已经存在同名头集合的值，则集合中的值会被清空并用新值替换。

addHeader 方法使用给定的名称添加一个头值到集合。如果没有头与给定的名称关联，则创建一个新集合。

头可能包含表示 int 或 Date 对象的数据。以下 HttpServletResponse 接口提供的便利方法允许 Servlet 对适当的数据类型用正确的格式设置一个头。

- setIntHeader
- setDateHeader
- addIntHeader
- addDateHeader

为了成功传回客户端，头必须在响应提交前设置。响应提交后的头设置将被 Servlet 容器忽略。

Servlet 程序员负责保证为 Servlet 生成的内容设置合适的响应对象的 Content-Type 头。HTTP 1.1 规范中没有要求在 HTTP 响应中设置此头。当 Servlet 程序员没有设置该类型时，Servlet 容器也不能设置默认的内容类型。

容器使用 X-Powered-By HTTP 头发布其实现信息。字段值应包含一个或多个实现类型，如"Servlet / 4.0"。也可以在括号内的实现类型之后添加容器和底层 Java 平台的补充信息，以下是设置该头的示例。

```
X-Powered-By: Servlet/4.0
```

```
X-Powered-By: Servlet/4.0 JSP/2.3 (GlassFish Server Open Source Edition 5.0 Java/Oracle Corporation/1.8)
```

2.4.3 方法

HttpServletResponse 提供了以下简便的方法。

- sendRedirect
- sendError

sendRedirect 方法用于设置适当的头和内容体，将客户端重定向到另一个地址。使用相对 URL 路径调用该方法是合法的，但是底层的容器必须将传回客户端的相对地址转换为全路径 URL。无论出于什么原因，如果给定的 URL 是不完整的，且不能转换为一个有效的 URL，那么该方法必须抛出 IllegalArgumentException。

sendError 方法用于设置适当的头和内容体，给客户端返回错误消息。可以使用 sendError 方法提供一个可选的 String 参数，用于指定错误的内容体。

如果响应已经提交并终止，这两个方法将对提交的响应产生副作用。调用这两个方法后，Servlet 将不会产生到客户端的后续输出，如果有数据继续写到响应，则这些数据将被忽略。如果数据已经写到响应的缓冲区，但没有返回客户端（如响应没有提交），则响应缓冲区中的数据必须被清空，并使用由这两个方法设置的数据替换。如果响应已提交，则必须抛出 IllegalStateException。

2.5 实战：基于Servlet的Web程序

本节将演示如何创建一个基于 Servlet 的 Web 程序。该程序源码可以在 hello-servlet 目录下找到。

2.5.1 创建动态Web项目

打开 Eclipse，选择"File"→"New"→"Dynamic Web Project"选项，创建一个动态 Web 项目，如图 2-2 所示。

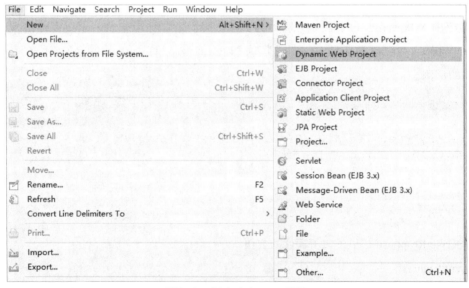

图2-2 创建动态Web项目

指定动态 Web 项目的名称为"hello-servlet"，如图 2-3 所示。

单击"Finish"按钮完成创建。

第 2 章　Servlet

图2-3　指定动态Web项目的名称

2.5.2 创建Servlet实现类

选择 hello-servlet 项目并右击，在弹出的快捷菜单中选择"New"→"Servlet"选项，创建一个 Servlet 实现类，如图 2-4 和图 2-5 所示。

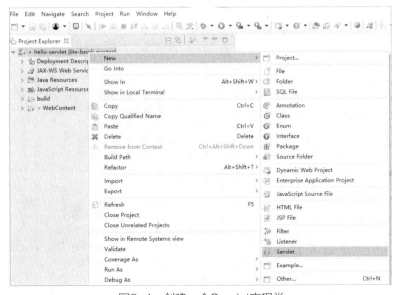

图2-4　创建一个Servlet实现类

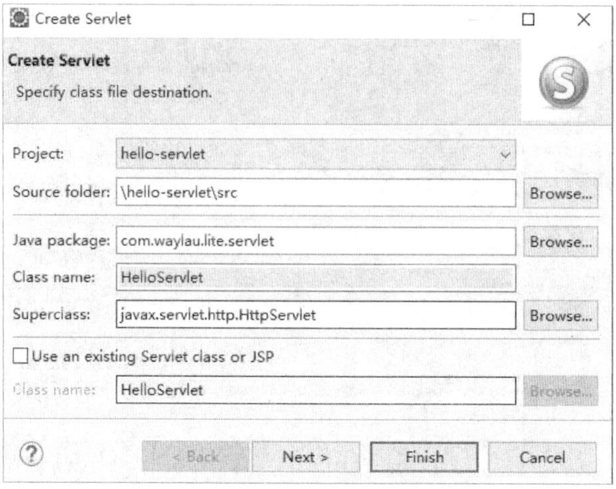

图2-5 给Servlet命名

此时，所创建的HelloServlet类会自动实现HttpServlet接口的doGet方法和doPost方法，其代码如下。

```
package com.waylau.lite.servlet;

import java.io.IOException;
import javax.servlet.ServletException;
import javax.servlet.annotation.WebServlet;
import javax.servlet.http.HttpServlet;
import javax.servlet.http.HttpServletRequest;
import javax.servlet.http.HttpServletResponse;
/**
 * Servlet implementation class HelloServlet
 */
@WebServlet("/HelloServlet")
public class HelloServlet extends HttpServlet {
    private static final long serialVersionUID = 1L;

    /**
     * @see HttpServlet#HttpServlet()
     */
    public HelloServlet() {
        super();
        // TODO Auto-generated constructor stub
    }

    /**
     * @see HttpServlet#doGet(HttpServletRequest request, HttpServletResponse response)
     */
    protected void doGet(HttpServletRequest request, HttpServletResponse
```

```
response) throws ServletException, IOException {
    // TODO Auto-generated method stub
    response.getWriter().append("Served at: ").append(request.getContextPath());
}

/**
 * @see HttpServlet#doPost(HttpServletRequest request, HttpServletResponse response)
 */
protected void doPost(HttpServletRequest request, HttpServletResponse response) throws ServletException, IOException {
    // TODO Auto-generated method stub
    doGet(request, response);
}

}
```

2.5.3 修改HelloServlet类

修改 HelloServlet 类，将应用的逻辑写在 doGet 中，重写的代码如下。

```
package com.waylau.lite.servlet;

import java.io.IOException;
import javax.servlet.ServletException;
import javax.servlet.annotation.WebServlet;
import javax.servlet.http.HttpServlet;
import javax.servlet.http.HttpServletRequest;
import javax.servlet.http.HttpServletResponse;

/**
 * Hello Servlet
 *
 * @since 1.0.0 2018年11月29日
 * @author <a href="https://waylau.com">Way Lau</a>
 */
@WebServlet("/HelloServlet")
public class HelloServlet extends HttpServlet {
    private static final long serialVersionUID = 1L;

    public HelloServlet() {
        super();
    }

    protected void doGet(HttpServletRequest request, HttpServletResponse response) throws ServletException, IOException {
        // 响应Hello World!
```

```
        response.getWriter().append("Hello World!");
    }
    protected void doPost(HttpServletRequest request, HttpServletResponse response) throws ServletException, IOException {
        doGet(request, response);
    }
}
```

HelloServlet 类的逻辑非常简单，当客户端访问"/HelloServlet"URL 时，就会响应"Hello World!"字样的文本内容给客户端。

2.5.4 运行应用

右击应用，在弹出的快捷菜单中选择"Run As"→"Rum on Server"选项，选择要部署应用的服务器，如图 2-6 和图 2-7 所示。

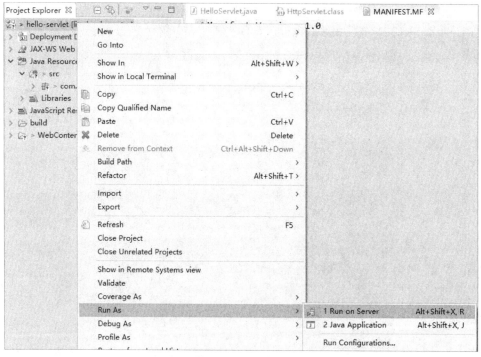

图2-6　运行应用

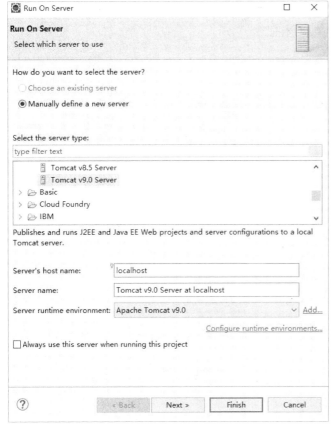

用浏览器访问 http://localhost:8080/hello-servlet/HelloServlet 时，应能看到图 2-8 所示的响应内容。

图2-7　选择服务器　　　　　　　　　图2-8　界面效果

2.6　Jetty

本节将介绍一款轻量级的 Servlet 容器——Jetty。Jetty 具有开源、轻量级、高性能、可拔插等特点，深受互联网公司的喜爱。

Jetty 实现了 Java EE 各个方面的规范，主要是 Servlet 规范。最近发布的 Java EE 平台引入了 Web Profile，虽然 Jetty 本身并不提供所有 Web Profile 技术，但 Jetty 架构可以插入第三方实现，以生成根据用户需求定制的容器。

2.6.1　高性能Servlet容器

Jetty 作为一款高性能的 Web 容器，非常适合大量连接和高并发的场景。其中，Jetty 使用 NIO（非阻塞 IO）模式，这在某种程度上是 Jetty 作为高性能服务器的有力支点。

Jetty NIO 组件由以下基本内容组成。

（1）EndPoint：网络上相互通信的对端实体抽象。

（2）Connection：网络实体通信连接的抽象。

（3）ByteBufferPool：缓冲区对象池。

（4）SelectChannelEndPoint：基于 NIO 模型描述的 EndPoint 封装。

（5）SelectorManager：选择器管理器。

整个 Jetty NIO 组件架构基于 Reactor 模型，并采用异步处理方式来模拟 Proactor 模式。

2.6.2 可拔插

Jetty 通过插件化的方式来增强或简化应用，官方支持以下插件。

- Proxy Servlet
- Balancer Servlet
- CGI Servlet
- Quality of Service Filter
- Denial of Service Filter
- Header Filter
- Gzip Handler
- Cross Origin Filter
- Resource Handler
- Debug Handler
- Statistics Handler
- IP Access Handler
- Moved Context Handler
- Shutdown Handler
- Default Handler
- Error Handler
- Rewrite Handler

使用这些插件也非常简单。例如，想使用 Proxy Servlet 插件，只需要在应用中添加以下依赖即可。

```
<dependency>
    <groupId>org.eclipse.jetty</groupId>
    <artifactId>jetty-proxy</artifactId>
    <version>${jetty.version}</version>
</dependency>
```

2.6.3 Jetty常用配置

将 Jetty 常用配置进行总结，具体内容如下。

1. httpConnector

httpConnector 是可选配置。如果没有设置，Jetty 将创建 ServerConnector 实例来监听 8080 端口。它可以在命令行上使用系统属性 "jetty.http.port" 来修改默认的端口配置，例如：

```
mvn -Djetty.http.port=9999 jetty:run
```

当然，也可以通过配置下面的属性来配置 ServerConnector。

- port：连接监听的端口，默认为 8080
- host：监听的主机，默认监听所有主机，即所有的主机都可以访问
- name：连接器的名称，在配置指定连接器来处理指定请求时有用
- idleTimeout：连接超时时间
- soLinger：socket 连接时间

同样可以在一个标准 Jetty 的 XML 配置文件中配置连接，并把配置文件的路径赋值给 jettyXml 参数。

2. jettyXml

jettyXml 是可选配置。通常可以把以逗号分隔的 jettyXml 配置文件的地址字符串增加到任何插件的配置参数中。如果有另一个 Web 应用或处理器，特别是连接器，就可以使用它。但是若有另一个 Jetty 对象，则不能通过插件得到配置信息。

3. scanIntervalSeconds

scanIntervalSeconds 为改变自动扫描文件并进行热部署的时间间隔，单位为秒。默认值为 0，代表禁用扫描及热部署，只有大于 0 的配置才可以使其生效。

4. reload

reload 为重新加载选项，默认值为 automatic（自动），一般用来和配置不为 0 的 scanIntervalSeconds 一同使用。默认配置下，当发现有文件改变会自动进行热部署。如果设置为 manual（手动），部署将会通过插件被手动触发。这对需要频繁改动文件的情况比较有用，它可忽略改动，直到做完所有改动。

5. dumpOnStart

dumpOnStart 为可选择的配置，默认为 false。如果设置为 true，那么 Jetty 会在启动时打印出 server 的结构。

6. loginServices

loginServices 为可选择的配置，是一系列 org.eclipse.jetty.security.LoginService 的实现类。注意，其没有指定默认的域，如果需要在 web.xml 中配置域，那么可以配置一个统一的域。当然，也可以

在 Jetty 的 xml 中进行配置，并把配置文件的地址增加到 jettyXml 中。

7. requestLog

requestLog 为可选择的配置，是一个实现了 org.eclipse.jetty.server.RequestLog 接口的请求日志记录。它有 3 种方式配置请求日志。

（1）在 jettyXml 中配置文件，并加到 jettyXml 参数中。

（2）在 context Xml 中配置文件，并加到 contextXml 参数中。

（3）在 webAPP 元素中配置文件。

8. server

server 为可选择的配置，通过配置 org.eclipse.jetty.server.Server 实例来支持插件的使用，然而通常是不需要配置的，因为插件会自动配置。特别是在使用 jettyXml 时，不会使用这个元素。

9. stopPort

stopPort 为可选择的配置，是一个用来监听停止命令的端口。

10. stopKey

stopKey 为可选择的配置，通常和 stopPort 结合使用。

11. systemProperties

systemProperties 为可选择的配置，它允许为执行插件而配置系统参数。

12. systemPropertiesFile

systemPropertiesFile 为可选择的配置，是一个包含执行插件系统参数的文件。默认情况下，在文件中设置的参数不会覆盖在命令行中写的参数，不管是通过 JVM，还是通过 POM 的 systemProperties。

13. skip

默认为 false，如果为 true，则插件会取消执行，它同样可以使用 -Djetty.skip 命令进行设置。

以下是配置示例，需要提醒的是，下面设置了 org.eclipse.jetty.server.HttpConfiguration 元素，也可以使用一个子 XML 文件来配置，这是一个相关的部分：

```
<New id="httpConfig" class="org.eclipse.jetty.server.HttpConfiguration">
    <Set name="secureScheme">https</Set>
    <Set name="securePort"><Property name="jetty.secure.port" default="8443" /></Set>
    <Set name="outputBufferSize">32768</Set>
    <Set name="requestHeaderSize">8192</Set>
    <Set name="responseHeaderSize">8192</Set>
    <Set name="sendServerVersion">true</Set>
    <Set name="sendDateHeader">false</Set>
    <Set name="headerCacheSize">512</Set>
</New>
```

14. jetty-ssl.xml

为 HTTPS 连接配置 SSL，下面的 jetty-ssl.xml 例子来自 jetty-distribution。

```xml
<?xml version="1.0"?>
<!DOCTYPE Configure PUBLIC "-//Jetty//Configure//EN" "http://www.eclipse.org/jetty/configure_9_4.dtd">

<!-- ============================================================= -->
<!-- SSL基础配置                                                    -->
<!-- 这个配置文件需要和至少一个或多个                                -->
<!-- etty-https.xml 或 jetty-http2.xml文件同时使用                   -->
<!-- ============================================================= -->
<Configure id="Server" class="org.eclipse.jetty.server.Server">

  <!-- ============================================================= -->
  <!-- 不使用协议工厂增加一个SSL连接                                   -->
  <!-- ============================================================= -->
  <Call name="addConnector">
    <Arg>
      <New id="sslConnector" class="org.eclipse.jetty.server.ServerConnector">
        <Arg name="server"><Ref refid="Server" /></Arg>
        <Arg name="acceptors" type="int"><Property name="jetty.ssl.acceptors" deprecated="ssl.acceptors" default="-1"/></Arg>
        <Arg name="selectors" type="int"><Property name="jetty.ssl.selectors" deprecated="ssl.selectors" default="-1"/></Arg>
        <Arg name="factories">
          <Array type="org.eclipse.jetty.server.ConnectionFactory">
            <!-- 注释掉的配置,是用于支持代理的方法
            <Item>
              <New class="org.eclipse.jetty.server.ProxyConnectionFactory"/>
            </Item>-->
          </Array>
        </Arg>

        <Set name="host"><Property name="jetty.ssl.host" deprecated="jetty.host" /></Set>
        <Set name="port"><Property name="jetty.ssl.port" deprecated="ssl.port" default="8443" /></Set>
        <Set name="idleTimeout"><Property name="jetty.ssl.idleTimeout" deprecated="ssl.timeout" default="30000"/></Set>
        <Set name="soLingerTime"><Property name="jetty.ssl.soLingerTime" deprecated="ssl.soLingerTime" default="-1"/></Set>
        <Set name="acceptorPriorityDelta"><Property name="jetty.ssl.acceptorPriorityDelta" deprecated="ssl.acceptorPriorityDelta" default="0"/></Set>
        <Set name="acceptQueueSize"><Property name="jetty.ssl.acceptQueueSize" deprecated="ssl.acceptQueueSize" default="0"/></Set>
      </New>
    </Arg>
  </Call>
```

```xml
<!-- ============================================================ -->
<!-- 基于定义在jetty.xml配置文件里的HttpConfiguration           -->
<!-- 创建一个基于TLS的HttpConfiguration                          -->
<!-- 增加一个SecureRequestCustomizer来管理证书和session信息     -->
<!-- ============================================================ -->
<New id="sslHttpConfig" class="org.eclipse.jetty.server.HttpConfiguration">
    <Arg><Ref refid="httpConfig"/></Arg>
    <Call name="addCustomizer">
      <Arg>
        <New class="org.eclipse.jetty.server.SecureRequestCustomizer">
          <Arg name="sniHostCheck" type="boolean"><Property name="jetty.ssl.sniHostCheck" default="true"/></Arg>
          <Arg name="stsMaxAgeSeconds" type="int"><Property name="jetty.ssl.stsMaxAgeSeconds" default="-1"/></Arg>
          <Arg name="stsIncludeSubdomains" type="boolean"><Property name="jetty.ssl.stsIncludeSubdomains" default="false"/></Arg>
        </New>
      </Arg>
    </Call>
</New>
</Configure>
```

2.7 实战：在应用中内嵌Jetty容器

Jetty 在互联网应用中能够广泛使用的一个非常重要的原因是，它能够通过内嵌的方式嵌入到应用中，使应用具备独立运行的能力。这种能力使 Jetty 非常适合在云环境中通过容器来部署应用。这也是 Jetty 被称为"Servlet Engine"的原因，用最少的代码，就可以使应用具备处理 HTTP 请求的能力。

一般来说，Jetty 可以通过以下两种方式嵌入到应用中。

2.7.1 Maven插件形式

下面创建一个名为"jetty-maven"的应用来演示如何将 Jetty 以 Maven 插件形式嵌入到应用中。

1. 配置POM文件

在应用中需要使用 Jetty 的 Maven 及 Servlet 的接口。编辑 pom.xml 文件，添加依赖内容如下。

```
<project xmlns="http://maven.apache.org/POM/4.0.0"
    xmlns:xsi="http://www.w3.org/2001/XMLSchema-instance"
```

```xml
    xsi:schemaLocation="http://maven.apache.org/POM/4.0.0
    http://maven.apache.org/xsd/maven-4.0.0.xsd">
  <modelVersion>4.0.0</modelVersion>
  <groupId>com.waylau.jetty</groupId>
  <artifactId>jetty-maven</artifactId>
  <version>1.0.0</version>
  <name>jetty-maven</name>
  <packaging>war</packaging>
  <organization>
    <name>waylau.com</name>
    <url>https://waylau.com</url>
  </organization>

  <properties>
      <jetty.version>9.4.14.v20181114</jetty.version>
  </properties>

  <dependencies>
    <dependency>
        <groupId>javax.servlet</groupId>
        <artifactId>javax.servlet-api</artifactId>
        <version>4.0.1</version>
        <scope>provided</scope>
    </dependency>
  </dependencies>

  <build>
    <plugins>
      <plugin>
         <groupId>org.eclipse.jetty</groupId>
         <artifactId>jetty-maven-plugin</artifactId>
         <version>${jetty.version}</version>
      </plugin>
    </plugins>
  </build>
</project>
```

2. 编写HelloServlet类

编写 HelloServlet 类的代码如下。

```
package com.waylau.jetty.servlet;

import java.io.IOException;
import javax.servlet.ServletException;
import javax.servlet.annotation.WebServlet;
import javax.servlet.http.HttpServlet;
import javax.servlet.http.HttpServletRequest;
import javax.servlet.http.HttpServletResponse;
```

```java
/**
 * Hello Servlet
 *
 * @since 1.0.0 2018年11月29日
 * @author <a href="https://waylau.com">Way Lau</a>
 */
@WebServlet("/HelloServlet")
public class HelloServlet extends HttpServlet {
    private static final long serialVersionUID = 1L;

    public HelloServlet() {
        super();
    }

    protected void doGet(HttpServletRequest request,
            HttpServletResponse response)
            throws ServletException, IOException {
        // 响应Hello World!
        response.getWriter().append("Hello World!");
    }

    protected void doPost(HttpServletRequest request,
            HttpServletResponse response)
            throws ServletException, IOException {
        doGet(request, response);
    }
}
```

HelloServlet 类的逻辑非常简单，当客户端访问"/HelloServlet"URL 时，会响应"Hello World!"字样的文本内容给客户端。

3. 运行应用

通过 Maven 命令行来运行应用，命令如下。

```
> mvn jetty:run
```

启动之后，在浏览器中访问 http://localhost:8080/HelloServlet，应能看到图 2-9 所示的响应内容。

图2-9 界面效果

2.7.2 编程方式

下面创建一个名为"jetty-server"的应用来演示如何将 Jetty 以编程方式嵌入到应用中。

1. 配置POM文件

在应用中需要使用 Jetty 的 Maven 及 Servlet 接口。编辑 pom.xml 文件，添加依赖内容如下。

```xml
<project xmlns="http://maven.apache.org/POM/4.0.0"
    xmlns:xsi="http://www.w3.org/2001/XMLSchema-instance"
    xsi:schemaLocation="http://maven.apache.org/POM/4.0.0
    http://maven.apache.org/xsd/maven-4.0.0.xsd">
    <modelVersion>4.0.0</modelVersion>
    <groupId>com.waylau.jetty</groupId>
    <artifactId>jetty-server</artifactId>
    <version>1.0.0</version>
    <name>jetty-server</name>
    <packaging>jar</packaging>
    <organization>
        <name>waylau.com</name>
        <url>https://waylau.com</url>
    </organization>

    <properties>
      <jetty.version>9.4.14.v20181114</jetty.version>
    </properties>
    <dependencies>
        <dependency>
            <groupId>org.eclipse.jetty</groupId>
            <artifactId>jetty-servlet</artifactId>
            <version>${jetty.version}</version>
            <scope>provided</scope>
        </dependency>
    </dependencies>
</project>
```

2. 编写HelloServlet类

编写 HelloServlet 类的代码如下。

```java
package com.waylau.servlet;

import java.io.IOException;
import javax.servlet.ServletException;
import javax.servlet.http.HttpServlet;
import javax.servlet.http.HttpServletRequest;
import javax.servlet.http.HttpServletResponse;
/**
 * Hello Servlet
 *
 * @since 1.0.0 2018年11月29日
 * @author <a href="https://waylau.com">Way Lau</a>
 */
```

```java
public class HelloServlet extends HttpServlet {
    private static final long serialVersionUID = 1L;

    public HelloServlet() {
        super();
    }

    protected void doGet(HttpServletRequest request,
            HttpServletResponse response)
            throws ServletException, IOException {
        // 响应Hello World!
        response.getWriter().append("Hello World!");
    }

    protected void doPost(HttpServletRequest request,
            HttpServletResponse response)
            throws ServletException, IOException {
        doGet(request, response);
    }
}
```

HelloServlet 类的逻辑非常简单，当客户端访问该 HelloServlet 时，会响应"Hello World!"字样的文本内容给客户端。

3. 编写Application类

Application 类用于启动 Jetty 服务器，其代码如下。

```java
/**
 * Welcome to https://waylau.com
 */
package com.waylau.servlet;

import org.eclipse.jetty.server.Server;
import org.eclipse.jetty.servlet.ServletHandler;

/**
 * Application Main.
 *
 * @since 1.0.0 2018年3月21日
 * @author <a href="https://waylau.com">Way Lau</a>
 */
public class Application {

    public static void main(String[] args) throws Exception {
        // 新建一个Jetty的服务器 并启动在8080端口
        Server server = new Server(8080);

        // 创建处理器
```

```
        ServletHandler handler = new ServletHandler();
        server.setHandler(handler);

        // 将URL映射到Servlet
        handler.addServletWithMapping(HelloServlet.class, "/HelloServ-
let");

        // 启动服务器
        server.start();
        server.join();
    }
}
```

其中，ServletHandler.addServletWithMapping 方法将 "/HelloServlet" URL 映射到了 HelloServlet 上。

4. 运行应用

在 IDE 中右击运行该应用。启动之后，在浏览器中访问 http://localhost:8080/HelloServlet，应能看到图 2-10 所示的响应内容。

图2-10　界面效果

第3章 Spring 基础

从本章开始，将踏上学习Spring框架之路。本章主要介绍Spring的基础知识，包括IoC、AOP、资源处理及SpEL表达式等内容。

3.1 Spring概述

自 Spring 诞生以来，在 Java 企业级应用中应用广泛，致力于简化传统 Java 企业级应用开发的烦琐过程。

其实 Spring 的意义远不止是一个应用框架，下面就来详细解读 Spring。

3.1.1 Spring的含义

Spring 有狭义与广义之说。

1. 狭义上的Spring——Spring Framework

狭义上的 Spring 特指 Spring 框架（Spring Framework）。Spring 框架是为了解决企业级应用开发的复杂性而创建的，它的主要优势之一就是分层架构。分层架构允许使用者选择使用哪一个组件，同时为 Java EE 应用程序开发提供集成的框架。Spring 框架使用基本的 POJO 来完成以前只可由 EJB 完成的事情。Spring 框架不局限于服务器端的开发，而且从简单性、可测试性和松耦合的角度来看，Java 应用开发也可以从 Spring 框架中获益。Spring 框架的核心是控制反转（IoC）和面向切面（AOP）。简单来说，Spring 框架是一个分层的、面向切面与 Java 应用的一站式轻量级开源框架。

Spring 框架的前身是 Rod Johnson 在 *Expert One-on-One J2EE Design and Development* 一书中发表的包含 3 万行代码的附件。在该书中，他展示了如何在不使用 EJB 的情况下构建高质量、可扩展的在线预留座位应用程序。为了构建该应用程序，他写了上万行的基础结构代码。这些代码包含许多可重用的 Java 接口和类，如 ApplicationContext 和 BeanFactory 等。由于 Java 接口是依赖注入的基本构件，因此他将类的根包命名为"com.interface21"，即这是一个提供给 21 世纪的参考。根据书中描述，这些代码已经在一些真实的金融系统中被使用。

由于该书影响甚广，当时有几名开发人员（如 Juergen Hoeller 及 Yann Caroff）联系 Rod Johnson，希望将 com.interface21 代码开源。Yann Caroff 将这个新框架命名为"Spring"，其含义为，Spring 就像一缕春风，扫平传统 J2EE 的寒冬。所以说，Rod Johnson、Juergen Hoeller 及 Yann Caroff 是 Spring 框架的共同创建者。

2003 年 2 月，Spring 0.9 发布，采用了 Apache 2.0 开源协议。2004 年 4 月，Spring 1.0 发布。截至目前，Spring 框架已经是第 5 个主要版本了。有关 Spring 5 方面的内容，可以参阅笔者所著的《Spring 5 开发大全》。

2. 广义上的 Spring——Spring 技术栈

广义上的 Spring 是指以 Spring 框架为核心的 Spring 技术栈。这些技术栈涵盖了从企业级应用到云计算等各个方面的内容，具体如下。

（1）Spring Data：Spring 框架中的数据访问模块对 JDBC 及 ORM 提供了很好的支持。随着

NoSQL 和大数据的兴起，出现了越来越多的新技术，如非关系型数据库、MapReduce 框架。Spring Data 正是为了让 Spring 开发者能更方便地使用这些新技术而诞生的，它由一系列小的项目组成，分别为不同的技术提供支持，如 Spring Data JPA、Spring Data Hadoop、Spring Data MongoDB、Spring Data Redis 等。通过 Spring Data，开发者可以用 Spring 提供的相对一致的方式，访问位于不同类型数据存储中的数据。在本书后续章节中，将会对 Spring Data Redis 的使用做详细介绍。

（2）Spring Batch：一款专门针对企业级系统中日常批处理任务的轻量级框架，能够帮助开发者方便地开发出高效的批处理应用程序。通过 SpringBatch 可以轻松构建轻量级的、高效的并口处理应用，并支持事务、并发、流程、监控、纵向和横向扩展，提供统一的接口管理和任务管理。Spring Batch 对批处理任务进行了一定的抽象，它的架构大致分为 3 层，自上而下分别是业务逻辑层、批处理执行环境层和基础设施层。Spring Batch 可以很好地利用 Spring 框架所带来的各种便利，同时也为开发者提供了相对熟悉的开发环境。有关 Spring Batch 方面的内容，可以参阅笔者所著的《Cloud Native 分布式架构原理与实践》。

（3）Spring Security：前身是 Acegi，是一款支持定制化的身份验证和访问控制框架。读者如果对该技术感兴趣，可以参阅笔者所著的开源书《Spring Security 教程》，以了解更多 Spring Security 方面的内容。在本书后续章节中，将会对 Spring Security 的使用做详细介绍。

（4）Spring Boot：指 Spring 团队提供的全新框架，其设计目的是简化新 Spring 应用的初始搭建及开发过程。该框架使用了特定的方式来进行配置，从而使开发人员不再需要定义样板化的配置。Spring Boot 为 Spring 平台及第三方库提供了"开箱即用"的设置，这样就可以有条不紊地进行应用的开发。多数 Spring Boot 应用只需要用到很少的 Spring 配置，其致力于在蓬勃发展的快速应用开发领域成为领导者。读者如果对该技术感兴趣，可以参阅笔者所著的《Spring Boot 企业级应用开发实战》及开源书《Spring Boot 教程》，以了解更多 Spring Boot 方面的内容。在本书后续章节中，将会对 Spring Boot 的使用做详细介绍。

（5）Spring Cloud：使用 Spring Cloud，开发人员可以"开箱即用"地实现分布式系统中常用的服务。这些服务可以在任何环境下运行，不仅包括分布式环境，还包括开发人员的笔记本电脑、裸机数据中心，以及 Cloud Foundry 等托管平台。Spring Cloud 基于 Spring Boot 来进行构建服务，并可以轻松地集成第三方类库来增强应用程序的行为。读者如果对该技术感兴趣，可以参阅笔者所著的《Spring Cloud 微服务架构开发实战》及开源书《Spring Cloud 教程》，以了解更多 Spring Cloud 方面的内容。

Spring 的技术栈还有很多，如果读者感兴趣，可以访问 Spring 项目页面（https://spring.io/projects）了解更多信息。

3. 约定

由于 Spring 是早期 Spring 框架的总称，因此，有时候"Spring"这个命名会给读者带来困扰。本书约定"Spring 框架"特指狭义上的 Spring，即 Spring Framework；而"Spring"特指广义上的

Spring，即 Spring 技术栈。

3.1.2　Spring框架总览

Spring 框架是整个 Spring 技术栈的核心。Spring 框架实现了对 bean 的依赖管理及 AOP 的编程方式，这些都极大地提升了 Java 企业级应用开发过程中的编程效率，降低了代码之间的耦合。

Spring 框架是很好的一站式构建企业级应用的轻量级解决方案。

1. 模块化的Spring框架

Spring 框架是模块化的，允许开发人员自由选择需要使用的部分。例如，可以在任何框架上使用 IoC 容器，也可以只使用 Hibernate 集成代码或 JDBC 抽象层。Spring 框架支持声明式事务管理，通过 RMI 或 Web 服务远程访问用户的逻辑，并支持通过多种选择来持久化用户的数据。它提供了一个全功能的 Spring Web MVC 及 Spring WebFlux 框架，同时也支持将 AOP 集成到软件中。

2. 使用 Spring 框架的好处

Spring 框架是一个轻量级的 Java 平台，能够提供完善的基础设施来支持 Java 应用程序的开发。Spring 负责实现基础设施功能，开发人员可以专注于应用逻辑的开发。Spring 可以使用 POJO 来构建应用程序，并将企业服务非侵入性地应用到 POJO。此功能适用于 Java SE 编程模型和完全或部分的 Java EE 模型。

作为一个 Java 应用程序的开发者，可以从 Spring 平台获得以下好处。

（1）使本地 Java 方法可以执行数据库事务，而无须自己处理事务 API。

（2）使本地 Java 方法可以执行远程过程，而无须自己处理远程 API。

（3）使本地 Java 方法成为 HTTP 端点，而无须自己处理 Servlet API。

（4）使本地 Java 方法拥有管理操作，而无须自己处理 JMX API。

（5）使本地 Java 方法可以执行消息处理，而无须自己处理 JMS API。

3.1.3　Spring框架常用模块

Spring 框架基本涵盖了企业级应用开发的各个方面，它由二十多个模块组成。

1. 核心容器

核心容器（Core Container）由 spring-core、spring-beans、spring-context、spring-context-support 和 spring-expression（Spring Expression Language）模块组成。

（1）spring-core 模块和 spring-beans 模块提供框架的基础部分，包括 IoC 和 Dependency Injection 功能。BeanFactory 是一个复杂工厂模式的实现，无须编程就能实现单例，并允许开发人员将配置和特定的依赖从实际程序逻辑中解耦。

（2）Context（spring-context）模块建立在 Core 和 Beans 模块提供的功能之上，它是一种在框架类型下实现对象存储操作的手段,有一点像 JNDI 注册。Context 继承了 Beans 模块的特性,并且增加了对国际化的支持(如用在资源包中)、事件广播、资源加载和创建上下文(如一个 Servlet 容器)。Context 模块也支持如 EJB、JMX 和基础远程访问的 Java EE 特性。ApplicationContext 接口是 Context 模块的主要表现形式。

(3) spring-context-support 模块提供了对常见第三方库的支持,以便集成到 Spring 应用上下文中,如缓存(EhCache、JCache)、调度(CommonJ、Quartz)等。

(4) spring-expression 模块提供了一种强大的表达式语言,用来在运行时查询和操作对象图。它是作为 JSP 2.1 规范所指定的统一表达式语言的一种扩展。这种语言支持对属性值、属性参数、方法调用、数组内容存储、收集器和索引、逻辑和算数的操作及命名变量,并且可以通过名称从 Spring 的控制反转容器中取回对象。表达式语言模块还支持列表投影、列表选择和通用列表聚合。

2. AOP及Instrumentation

spring-aop 模块提供 AOP(面向切面编程)的实现,从而能够实现方法拦截器和切入点完全分离。使用源码级别元数据的功能,也可以在代码中加入行为信息,在某种程度上类似于 .NET 属性。

单独的 spring-aspects 模块提供了集成使用 AspectJ 框架。spring-instrument 模块提供了类似于基础设施的支持。spring-instrument-tomcat 用于 Tomcat Instrumentation 代理。

3. 消息

自 Spring Framework 4 版本开始提供 spring-messaging 模块,主要包含从 Spring Integration 项目中抽象出来的,如 Message、MessageChannel、MessageHandler 模块,以及其他用来提供消息的基础服务。

spring-messaging 模块中还包括一组消息映射方法的注解,类似编程模型中 Spring MVC 的注解。

4. 数据访问/集成

数据访问/集成(Data Access/Integration)层由 JDBC、事务、ORM、OXM 和 JMS 模块组成。

(1) spring-jdbc 模块提供了一个 JDBC 抽象层,这样开发人员就能避免进行一些烦琐的 JDBC 编码和解析数据库供应商特定的错误代码。

(2) spring-tx 模块用于支持声明式事务管理。

(3) spring-orm 模块为流行的对象关系映射 API 提供集成层,包括 JPA 和 Hibernate。使用 spring-orm 模块可以将这些 O/R 映射框架与 Spring 提供的所有其他功能结合使用,如前面提到的声明式事务管理功能。

(4) spring-oxm 模块提供了一个支持 Object/XML 映射实现的抽象层,如 JAXB、Castor、JiBX 和 XStream。

(5) spring-jms 模块包含用于生成和使用消息的功能。从 Spring Framework 4.1 开始,它提供了与 spring-messaging 的集成。

5. Web

Web 层由 spring-web、spring-webmvc、spring-websocket 和 spring-webflux 模块组成。

（1）spring-web 模块提供了基本的面向 Web 开发的集成功能，如文件上传及用于初始化 IoC 容器的 Servlet 监听和 Web 开发应用程序上下文。它也包含对 HTTP 客户端及 Web 相关的 Spring 远程访问的支持。

（2）spring-webmvc 模块（Web Servlet 模块）包含 Spring 的 MVC 功能和 Rest 服务功能。

（3）spring-websocket 模块用于开发基于 WebSocket 协议通信的程序。

（4）spring-webflux 模块是 Spring 5 中新添加的支持响应式编程的 Web 开发框架。

6. 测试

spring-test 模块支持通过组合 JUnit 或 TestNG 来实现单元测试和集成测试等功能。它不仅提供 Spring ApplicationContexts 的持续加载，还能缓存这些上下文，而且提供了可用于孤立测试代码的模拟对象（mock objects）。

3.1.4　Spring设计模式

在 Spring 框架设计中，广泛使用了设计模式。Spring 使用以下设计模式使企业级应用开发变得简单和可测试。

（1）Spring 使用 POJO 模式的强大功能来实现企业应用程序的轻量级和最小侵入性的开发。

（2）Spring 使用依赖注入模式（DI 模式）实现松耦合，并使系统可以更加面向接口编程。

（3）Spring 使用 Decorator 和 Proxy 设计模式进行声明式编程。

（4）Spring 使用 Template 设计模式消除样板代码。

3.2　IoC

IoC 容器是 Spring 框架中非常重要的核心组件，它伴随了 Spring 的诞生和成长。Spring 通过 IoC 容器来管理所有 Java 对象（bean）及其依赖关系。

本节将全面讲解 IoC 容器的概念及用法。

3.2.1　依赖注入与控制反转

在 Java 应用程序中，不管是受限的嵌入式应用程序，还是多层架构的服务端企业级应用程序，对象在应用程序中均通过彼此依赖来实现功能。那么"依赖注入"与"控制反转"又是什么关系呢？

依赖注入（Dependency Injection）是 Martin Fowler 在 2004 年提出的关于控制反转的解释。[1] Martin Fowler 认为控制反转一词让人产生疑惑，无法直白地理解"到底哪方面的控制被反转了"。所以 Martin Fowler 建议采用依赖注入一词来代替控制反转。

依赖注入和控制反转其实就是一个事物的两种不同说法而已，本质上是一回事。依赖注入是一种程序设计模式和架构模型，有些时候也称为控制反转。尽管从技术上来讲，依赖注入是控制反转的特殊实现，但依赖注入还指一个对象应用另一个对象来实现一种特殊的能力。例如，把一个数据库连接以参数形式传到一个对象的结构方法中，而不是在那个对象内部自行创建一个连接。依赖注入和控制反转的基本思想就是把类的依赖从类内部转到外部，以减少依赖。利用控制反转，对象在被创建时，会由一个调控系统统一进行对象实例的管理，将该对象所依赖对象的引用通过调控系统进行传递。也可以说，依赖被注入对象中。所以控制反转是关于一个对象如何获取其所依赖对象的引用的过程，而这个过程是"谁来传递依赖的引用"这个职责的反转。

控制反转一般分为依赖注入（Dependency Injection）和依赖查找（Dependency Lookup）两种实现类型。其中依赖注入应用比较广泛，Spring 就是采用依赖注入这种方式来实现控制反转的。

3.2.2　IoC容器和bean

Spring 通过 IoC 容器来管理所有 Java 对象及其依赖关系。在软件开发过程中，系统的各个对象之间、各个模块之间、软件系统与硬件系统之间，或多或少都会存在耦合关系，如果一个系统的耦合度过高，就会造成难以维护的问题。但是完全没有耦合的代码是不能工作的，代码之间需要相互协作、相互依赖来完成功能。IoC 技术恰好解决了这类问题，各个对象之间不需要直接关联，而是在需要用到对方时，由 IoC 容器来管理对象之间的依赖关系。对于开发人员来说，只需维护相对独立的各个对象的代码即可。

IoC 是一个过程，即对象定义其依赖关系，而其他与之配合的对象只能通过构造函数参数、工厂方法的参数，或者在从工厂方法构造后或返回后在对象实例上设置的属性来定义依赖关系，随后，IoC 容器在创建 bean 时会注入这些依赖项。这个过程在职责上是反转的，就是把原先代码中需要实现的对象创建、依赖的代码反转给容器来帮忙实现和管理，所以称为控制反转。

IoC 的应用有以下两种模式。

（1）反射模式：在运行状态中，根据提供的类的路径或类名，通过反射来动态地获取该类的所有属性和方法。

（2）工厂模式：把 IoC 容器当作一个工厂，在配置文件或注解中给出定义，然后利用反射技术，根据给出的类名生成相应的对象。对象生成的代码及对象之间的依赖关系在配置文件中定义，这样

---

[1] 有关 Martin Fowler 的博客原文可见 https://martinfowler.com/articles/injection.html。

就实现了解耦。

org.springframework.beans 和 org.springframework.context 包是 Spring IoC 容器的基础。BeanFactory 接口提供了能够管理所有类型对象的高级配置机制。ApplicationContext 是 BeanFactory 的子接口，它更容易与 Spring 的 AOP 功能集成，进行消息资源处理、事件发布，以及作为应用层特定的上下文（如用于 Web 应用程序的 WebApplicationContext）。简言之，BeanFactory 提供了基本的配置功能，而 ApplicationContext 在此基础上增加了更多的企业特定功能。

在 Spring 应用中，bean 由 Spring IoC 容器进行实例化、组装，并受其管理。bean 和 Spring IoC 容器之间的依赖关系反映在容器使用的配置元数据中。

3.2.3 配置元数据

配置元数据（Configuration Metadata）描述了 Spring 容器在应用程序中是如何实例化、配置和组装对象的。

最初，Spring 用 XML 文件格式记录配置元数据，从而很好地实现了 IoC 容器本身与实际写入此配置元数据的格式完全分离。

当然，基于 XML 的元数据不是唯一允许的配置元数据形式。目前，比较流行的配置元数据的方式是采用注解或基于 Java 的配置。

以下示例展示了基于 XML 的配置元数据的基本结构。

```xml
<?xml version="1.0" encoding="UTF-8"?>
<beans xmlns="http://www.springframework.org/schema/beans"
xmlns:xsi="http://www.w3.org/2001/XMLSchema-instance"
xsi:schemaLocation="http://www.springframework.org/schema/beans
http://www.springframework.org/schema/beans/spring-beans.xsd">
    <bean id="…" class="…">
    <!-- 放置这个bean的协作者和配置 -->
    </bean>
    <bean id="…" class="…">
    <!-- 放置这个bean的协作者和配置 -->
    </bean>
    <!-- 省略更多的bean的配置-->
</beans>
```

在上面的 XML 文件中，id 属性是用于标识单个 bean 定义的字符串，它的值是协作对象；class 属性定义 bean 的类型，并使用完全限定的类名。

以下示例展示了基于注解的配置元数据的基本结构。

```java
@Configuration
public class AppConfig {
    @Bean
    public MyService myService() {
```

```
        return newMyServiceImpl();
    }
}
```

3.2.4 实例化容器

Spring IoC 容器需要在应用启动时进行实例化。在实例化过程中，IoC 容器会从各种外部资源（如本地文件系统、Java 类路径等）加载配置元数据，提供给 ApplicationContext 构造函数。

下面是从类路径中加载基于 XML 的配置元数据的示例。

```
ApplicationContext context =
    newClassPathXmlApplicationContext(newString[] {"services.xml","daos.xml"});
```

当系统规模比较大时，通常会将 bean 定义分到多个 XML 文件。这样，每个单独的 XML 配置文件通常能够表示系统结构中的逻辑层或模块，就像上面的示例所演示的那样。当某个构造函数需要多个资源位置时，可从外部文件加载 bean 的定义。例如：

```
<beans>
    <import resource="services.xml"/>
    <import resource="resources/messageSource.xml"/>
    <import resource="/resources/themeSource.xml"/>
    <bean id="bean1" class="…"/>
    <bean id="bean2" class="…"/>
</beans>
```

3.2.5 使用容器

ApplicationContext 是高级工厂的接口，能维护不同 bean 及其依赖项的注册表。其提供的方法 T getBean(String name, Class< T > requiredType)，可用于检索 bean 的实例。

ApplicationContext 读取 bean 定义并按以下方式访问它们。

```
// 创建并配置 bean
ApplicationContext context =
    newClassPathXmlApplicationContext("services.xml", "daos.xml");
// 检索配置了的 bean 实例
PetStoreService service = context.getBean("petStore", PetStoreService.class);
// 使用 bean 实例
List<String> userList = service.getUsernameList();
```

如果配置方式不是 XML 而是 Groovy，则将 ClassPathXmlApplicationContext 更改为 GenericGroovyApplicationContext 即可。GenericGroovyApplicationContext 是另一个 Spring 框架上下文的实现。

```
ApplicationContext context =
    newGenericGroovyApplicationContext("services.groovy", "daos.
groovy");
```

以上是使用 ApplicationContext 的 getBean 来检索 bean 实例的方式。ApplicationContext 接口还有其他检索 bean 的方法，但理想情况下，应用程序代码不应该使用它们。因为程序代码根本不需要调用 getBean 方法，就可以完全不依赖 Spring API。例如，Spring 与 Web 框架的集成为各种 Web 框架组件（如控制器和 JSF 托管的 bean）提供了依赖注入，允许开发人员通过元数据（如自动装配注入）声明对特定 bean 的依赖关系。

3.2.6　实例化bean的方式

每个 bean 都有标识符，这些标识符在托管 bean 的容器中必须是唯一的。一个 bean 通常只有一个标识符，如果它需要多个标识符，额外的标识符会被认为是别名。

在基于 XML 的配置元数据中，使用 id 或 name 属性来指定 bean 标识符。id 属性允许指定一个 id。通常，这些标识符的名称是字母，如 "myBean" "userService" 等，但也可能包含特殊字符。如果想为 bean 引入别名，也可以在 name 属性中指定，用 ","　";" 或空格分隔。由于历史原因，在 Spring 3.1 以前的版本中，id 属性被定义为一个 xsd:ID 类型，因此限制了可能的字符。从 Spring 3.1 开始，它被定义为一个 xsd:string 类型。值得注意的是，虽然类型做了更改，但 bean id 的唯一性仍由容器强制执行。

用户可以不为 bean 提供名称或标识符。如果没有显式名称或标识符，则容器会为该 bean 自动生成唯一的名称。但是如果要通过名称引用该 bean，则必须提供一个名称。

在命名 bean 时尽量使用标准 Java 约定。也就是说，bean 的名称要遵循以一个小写字母开头的驼峰法命名规则，如 "accountManager" "accountService" "userDao" "loginController" 等，这样命名的 bean 会让应用程序的配置更易于阅读和理解。

Spring 为未命名的组件生成 bean 名称，同样遵循以上规则。本质上，最简单的命名方式就是直接采用类名称并将其初始字符变为小写。但也有特例，当前两个字符或多个字符是大写时，可以不进行处理。例如，"URL" 类 bean 的名词仍然是 "URL"。这些命名规则定义在 java.beans.Introspector.decapitalize 方法中。

3.2.7　注入方式

所谓 bean 的实例化就是根据配置来创建对象的过程。如果是使用基于 XML 的配置方式，则在元素的 class 属性中指定需要实例化对象的类型（或类）。这个 class 属性在内部实现，通常是一个 BeanDefinition 实例的 class 属性。但也有例外情况，如使用工厂方法或 bean 定义继承进行实例化。

使用 class 属性有以下两种方式。

（1）通常容器本身是通过反射机制来调用指定类的构造函数，从而创建 bean。这与使用 Java 代码的 new 运算符相同。

（2）通过静态工厂方法创建的类中包含静态方法。通过调用静态方法返回对象的类型可能与 class 一样，也可能完全不一样。

如果想配置使用静态的内部类，必须用内部类的二进制名称。例如，在 com.waylau 包下有一个 User 类，而 User 类中有一个静态的内部类 Account，这种情况下 bean 定义的 class 属性应该是"com.waylau.User"字符，用来分隔外部类和内部类的名称。

概括起来，bean 的实例化有以下 3 种方式。

1. 通过构造函数实例化

Spring IoC 容器可以管理几乎所有用户想让它管理的类，而不仅仅是管理 POJO。大多数 Spring 用户更喜欢使用 POJO，但在容器中使用非 bean 形式的类也是可以的。例如，遗留系统中的连接池很显然与 JavaBean 规范不符，但 Spring 也能管理它。

当开发人员使用构造方法来创建 bean 时，Spring 对类来说并没有什么特殊性。也就是说，正在开发的类不需要实现任何特定的接口或以特定的方式进行编码。但是，根据所使用的 IoC 类型，很可能需要一个默认（无参）的构造方法。当使用基于 XML 的元数据配置文件时，可以按如下所示来指定 bean 类。

```xml
<bean id="exampleBean" class="waylau.ExampleBean"/>
<bean name="anotherExample" class="waylau.ExampleBeanTwo"/>
```

2. 使用静态工厂方法实例化

当采用静态工厂方法创建 bean 时，除了需要指定 class 属性，还需要通过 factory-method 属性来指定创建 bean 实例的工厂方法，Spring 将调用此方法返回实例对象。就此而言，它与通过普通构造器创建类实例没什么区别。下面的 bean 定义展示了如何通过工厂方法来创建 bean 实例。

以下是基于 XML 的元数据配置文件。

```xml
<bean id="clientService"
    class="waylau.ClientService"
    factory-method="createInstance"/>
```

以下是需要创建实例的类的定义。

```java
public class ClientService {
    private static ClientService clientService = newClientService();
    private ClientService() {}
    public static ClientService createInstance() {
        return clientService;
    }
}
```

注意，在此例中，createInstance() 必须是一个 static 方法。

3. 使用工厂实例方法实例化

通过调用工厂实例的非静态方法进行实例化，与通过静态工厂方法实例化类似。使用这种方式时，class 属性设置为空，而 factory-bean 属性必须指定为当前（或其祖先）容器中包含工厂方法的 bean 的名称，而该工厂 bean 的工厂方法本身必须通过 factory-method 属性来设定。

以下是基于 XML 的元数据配置文件。

```xml
<!-- 工厂bean  包含createInstance()方法 -->
<bean id="serviceLocator" class="waylau.DefaultServiceLocator">
<!-- 其他需要注入的依赖项 -->
</bean>
<!-- 通过工厂bean创建的bean -->
<bean id="clientService"
    factory-bean="serviceLocator"
    factory-method="createClientServiceInstance"/>
```

以下是需要创建实例的类的定义。

```java
public class DefaultServiceLocator {
    private static ClientService clientService = newClientServiceImpl();
    private DefaultServiceLocator() {}
    public ClientService createClientServiceInstance() {
        return clientService;
    }
}
```

当然，一个工厂类也可以有多个工厂方法。以下是基于 XML 的元数据配置文件。

```xml
<bean id="serviceLocator" class="waylau.DefaultServiceLocator">
<!-- 其他需要注入的依赖项 -->
</bean>
<bean id="clientService"
    factory-bean="serviceLocator"
    factory-method="createClientServiceInstance"/>
<bean id="accountService"
    factory-bean="serviceLocator"
    factory-method="createAccountServiceInstance"/>
```

以下是需要创建实例的类的定义。

```java
public class DefaultServiceLocator {
    private static ClientService clientService = newClientServiceImpl();
    private static AccountService accountService = newAccountServiceImpl();
    private DefaultServiceLocator() {}
    public ClientService createClientServiceInstance() {
        return clientService;
    }
```

```
    public AccountService createAccountServiceInstance() {
        return accountService;
    }
}
```

3.2.8　实战：依赖注入

本节将创建一个名为 dependency-injection 的应用，用于 Spring 框架的依赖注入。

该示例中采用了基于 XML 的配置方式，将会演示基于构造函数的依赖注入，同时也会演示如何解析构造函数的参数。

1. 添加依赖

要使用依赖注入功能，只需要添加 spring-context 模块即可。pom.xml 文件配置如下。

```
<project xmlns="http://maven.apache.org/POM/4.0.0"
xmlns:xsi="http://www.w3.org/2001/XMLSchema-instance"
xsi:schemaLocation="http://maven.apache.org/POM/4.0.0
http://maven.apache.org/xsd/maven-4.0.0.xsd">
    <modelVersion>4.0.0</modelVersion>
    <groupId>com.waylau.spring</groupId>
    <artifactId>dependency-injection</artifactId>
    <version>1.0.0</version>
    <name>dependency-injection</name>
    <packaging>jar</packaging>
    <organization>
        <name>waylau.com</name>
        <url>https://waylau.com</url>
    </organization>

    <dependencies>
        <dependency>
            <groupId>org.springframework</groupId>
            <artifactId>spring-context</artifactId>
            <version>5.1.3.RELEASE</version>
        </dependency>
    </dependencies>
</project>
```

2. 创建服务类

定义了消息服务接口 MessageService，该接口描述如下。

```
package com.waylau.spring.di.service;

public interface MessageService {
    String getMessage();    // 获取消息
}
```

MessageService 的主要职责是获取消息。

创建该消息服务类接口的实现 MessageServiceImpl，用来返回真实的、想要的业务消息。MessageServiceImpl 代码如下。

```
package com.waylau.spring.di.service;

public class MessageServiceImpl implements MessageService {

    private String username;
    private int age;

    public MessageServiceImpl(String username, int age) {
        this.username = username;
        this.age = age;
    }

    public String getMessage() {
        return "Hello World! " + username + ", age is " + age;
    }
}
```

其中，MessageServiceImpl 是带参构造函数，username、age 这两个参数的值将在 getMessage 方法中返回。

3. 创建打印器

创建打印器 MessagePrinter，用于打印消息。MessagePrinter 代码如下。

```
package com.waylau.spring.di;

import com.waylau.spring.di.service.MessageService;

public class MessagePrinter {

    final private MessageService service;

    public MessagePrinter(MessageService service) {
        this.service = service;
    }

    public void printMessage() {
        System.out.println(this.service.getMessage());
    }
}
```

执行 printMessage 方法后就能将消息内容打印出来，消息内容是依赖 MessageService 提供的。稍后会通过 XML 配置，将 MessageService 的实现注入到 MessagePrinter 中。

4. 应用主类

Application 是应用的入口类，参考如下代码对其进行修改。

```java
package com.waylau.spring.di;

import org.springframework.context.ApplicationContext;
import org.springframework.context.support.ClassPathXmlApplicationContext;

public class Application {
    public static void main(String[] args) {
        @SuppressWarnings("resource")
        ApplicationContext context =
            new ClassPathXmlApplicationContext("spring.xml");
        MessagePrinter printer =
            context.getBean(MessagePrinter.class);
        printer.printMessage();
    }
}
```

由于应用是基于 XML 的配置，因此这里需要 ClassPathXmlApplicationContext 类，该类是 Spring 上下文的一种实现，可以实现基于 XML 的配置加载。按照约定，Spring 应用的配置文件 spring.xml 放置在应用的 resources 目录下。

5. 创建配置文件

在应用的 resources 目录下创建一个 Spring 应用的配置文件 spring.xml，内容如下。

```xml
<?xml version="1.0" encoding="UTF-8"?>
<beans xmlns="http://www.springframework.org/schema/beans"
    xmlns:xsi="http://www.w3.org/2001/XMLSchema-instance"
    xmlns:context="http://www.springframework.org/schema/context"
    xsi:schemaLocation="
        http://www.springframework.org/schema/beans
        http://www.springframework.org/schema/beans/spring-beans.xsd
        http://www.springframework.org/schema/context
        http://www.springframework.org/schema/context/spring-context.xsd">

    <!-- 定义 bean -->
    <bean id="messageServiceImpl"
        class="com.waylau.spring.di.service.MessageServiceImpl">
        <constructor-arg name="username" value="Way Lau" />
        <constructor-arg name="age" value="30" />
    </bean>

    <bean id="messagePrinter"
```

```
            class="com.waylau.spring.di.MessagePrinter">
        <constructor-arg name="service"
            ref="messageServiceImpl" />
    </bean>
</beans>
```

在该 spring.xml 文件中,可以清楚地看到 bean 之间的依赖关系。messageServiceImpl 有两个构造函数的参数 username 和 age,其参数值在实例化时就解析了。messagePrinter 引用了 messageServiceImpl 作为其构造函数的参数。

6. 运行

运行 Application 类,就能在控制台中看到"Hello World! Way Lau, age is 30"字样的信息了。打印效果如图 3-1 所示。

图3-1 运行结果

3.2.9　bean scope

默认情况下,所有 Spring bean 都是单例的,也就是在整个 Spring 应用中,bean 的实例只有一个。开发人员可以在 bean 中添加 scope 属性来修改这个默认值。scope 属性可用值及描述如表 3-1 所示。

表3-1　scope属性可用值及描述

可用值	描述
singleton	每个Spring容器都有一个实例(默认值)
prototype	允许bean被多次实例化(使用一次就创建一个实例)
request	定义bean的scope是HTTP请求,每个HTTP请求都有自己的实例,只有在使用有Web功能的Spring上下文时才有效
session	定义bean的scope是HTTP会话,只有在使用有Web功能的Spring ApplicationContext时才有效
application	定义了每个ServletContext有一个实例
websocket	定义了每个WebSocket有一个实例,只有在使用有Web功能的Spring ApplicationContext时才有效

3.2.10 自定义scope

如果用户觉得 Spring 内置的几种 scope 不能满足需求，则可以定制自己的 scope，即实现 org.springframework.beans.factory.config.Scope 接口即可。Scope 接口定义了以下方法。

```
public interface Scope {
    Objectget(String name, ObjectFactory<?> objectFactory);
    @Nullable
    Objectremove(String name);
    void registerDestructionCallback(String name, Runnable callback);
    @Nullable
    ObjectresolveContextualObject(String key);
    @Nullable
    String getConversationId();
}
```

3.2.11 基于注解的配置

Spring 应用支持多种配置方式，除 XML 配置外，开发人员更加青睐基于注解的配置。

基于注解的配置方式允许开发人员将配置信息移入组件类本身，在相关的类、方法或字段上声明使用注解。

Spring 提供了非常多的注解。例如，Spring 2.0 引入的用 @Required 注解来强制所需属性不能为空，在 Spring 2.5 中可以使用相同的处理方法来驱动 Spring 的依赖注入。从本质上来说，@Autowired 注解提供了更细粒度的控制和更广泛的适用性。Spring 2.5 也添加了对 JSR-250 注解的支持，如 @Resource、@PostConstruct 和 @PreDestroy 等。Spring 3.0 添加了对 JSR-330 注解的支持，包含在 javax.inject 包下，如 @Inject、@Qualifier、@Named 和 @Provider 等。使用这些注解也需要在 Spring 容器中注册特定的 BeanPostProcessor。

下面将详细介绍这些注解的用法。

1. @Required

@Required 注解应用于 bean 属性的 setter 方法，如下面的示例。

```
public class SimpleMovieLister {
    private MovieFinder movieFinder;

    @Required
    public void setMovieFinder(MovieFinder movieFinder) {
        this.movieFinder = movieFinder;
    }
    // …
}
```

这个注解只是表明受影响的 bean 的属性必须在 bean 的定义中或自动装配中通过明确的属性值

在配置时来填充。如果受影响的 bean 的属性没有被填充，那么容器就会抛出异常。这就是通过快速失败的机制来避免 NullPointerException。

2. @Autowired

如下面的示例，可以将 @Autowired 注解应用于传统的 setter 方法。

```
public class SimpleMovieLister {
    private MovieFinder movieFinder;

    @Autowired
    public void setMovieFinder(MovieFinder movieFinder) {
        this.movieFinder = movieFinder;
    }
    // ...
}
```

JSR-330 的 @Inject 注解可以代替以上示例中 Spring 的 @Autowired 注解。

下面是可以将注解应用于任意名称和（或）多个参数的方法的示例。

```
public class MovieRecommender {
    private MovieCatalog movieCatalog;
    private CustomerPreferenceDao customerPreferenceDao;

    @Autowired
    public void prepare(MovieCatalog movieCatalog,
        CustomerPreferenceDao customerPreferenceDao) {
        this.movieCatalog = movieCatalog;
        this.customerPreferenceDao = customerPreferenceDao;
    }
    // ...
}
```

也可以将它用于构造方法和字段：

```
public class MovieRecommender {
    @Autowired
    private MovieCatalog movieCatalog;
    private CustomerPreferenceDao customerPreferenceDao;

    @Autowired
    public MovieRecommender(CustomerPreferenceDao customerPreference-
Dao) {
        this.customerPreferenceDao = customerPreferenceDao;
    }
    // ...
}
```

还可以通过添加注解到期望类型的数组的字段或方法上，从而提供 ApplicationContext 中特定类型的所有 bean。

```java
public class MovieRecommender {
    @Autowired
    private MovieCatalog[] movieCatalogs;
    // …
}
```

同样，也可以用于特定类型的集合：

```java
public class MovieRecommender {
    private Set<MovieCatalog> movieCatalogs;

    @Autowired
    public void setMovieCatalogs(Set<MovieCatalog> movieCatalogs) {
        this.movieCatalogs = movieCatalogs;
    }
    // …
}
```

默认情况下，当出现 0 个候选 bean 时，自动装配就会失败。默认的行为是将被注解的方法、构造方法和字段作为需要的依赖关系。这种行为也可以通过以下做法来改变。

```java
public class SimpleMovieLister {
    private MovieFinder movieFinder;
    @Autowired(required=false)
    public void setMovieFinder(MovieFinder movieFinder) {
        this.movieFinder = movieFinder;
    }
    // …
}
```

推荐使用 @Autowired 的 required 属性（不是 @Required 注解）。一方面，required 属性表示了属性对于自动装配目的不是必需的，如果它不能被自动装配，那么属性就会被忽略。另一方面，@Required 更健壮一些，它强制由容器支持的各种方式的属性设置，如果没有注入任何值，就会抛出对应的异常。

3. @Primary

因为通过类型的自动装配可能有多个候选者，所以在选择过程中通常需要更多控制。达成这个目的的一种做法是使用 Spring 的 @Primary 注解。当一个依赖有多个候选者 bean 时，@Primary 会指定一个优先提供的特殊 bean。当多个候选者 bean 中存在一个明确指定的 @Primary 的 bean 时，就会自动装载这个 bean。例如：

```java
@Configuration
public class MovieConfiguration {
    @Bean
    @Primary
    public MovieCatalog firstMovieCatalog() { … }
    @Bean
```

```
    public MovieCatalog secondMovieCatalog() { … }
    // …
}
```

对于上面的配置,下面的 MovieRecommender 将会使用 firstMovieCatalog 自动注解。

```
public class MovieRecommender {
    @Autowired
    private MovieCatalog movieCatalog;
    // …
}
```

4. @Qualifier

达成更多控制目的的另一种做法是使用 Spring 的 @Qualifier 注解。开发者可以用特定的参数来关联限定符的值,缩小类型的匹配范围,这样就能选择指定的 bean 了。其用法如下。

```
public class MovieRecommender {
    @Autowired
    @Qualifier("main")
    private MovieCatalog movieCatalog;
    // …
}
```

@Qualifier 注解也可以在独立的构造方法的参数中指定。

```
public class MovieRecommender {
    private MovieCatalog movieCatalog;
    private CustomerPreferenceDao customerPreferenceDao;

    @Autowired
    public void prepare(@Qualifier("main")MovieCatalog movieCatalog,
    CustomerPreferenceDao customerPreferenceDao) {
        this.movieCatalog = movieCatalog;
        this.customerPreferenceDao = customerPreferenceDao;
    }
    // …
}
```

5. @Resource

Spring 也支持使用 JSR-250 的 @Resource 注解在字段或 bean 属性的 setter 方法上注入。这在 Java EE 5 和 Java EE 6 中是一个通用的模式,如在 JSF 1.2 中管理的 bean 或 JAX-WS 2.0 端点。Spring 也为其所管理的对象支持这种模式。

@Resource 使用 name 属性,默认情况下 Spring 解析这个值作为要注入的 bean 的名称。也就是说,如果遵循 by-name 语义,就如同以下示例。

```
public class SimpleMovieLister {
    private MovieFinder movieFinder;
```

```
@Resource(name="myMovieFinder")
public void setMovieFinder(MovieFinder movieFinder) {
    this.movieFinder = movieFinder;
}
}
```

如果没有明确地指定 name 值，那么默认的名称就从字段名称或 setter 方法中派生出来。如果是字段，就会选用字段名称；如果是 setter 方法，就会选用 bean 的属性名称。所以在下面的示例中，名称为"movieFinder"的 bean 通过 setter 方法来注入。

```
public class SimpleMovieLister {
    private MovieFinder movieFinder;

    @Resource
    public void setMovieFinder(MovieFinder movieFinder) {
        this.movieFinder = movieFinder;
    }
}
```

6. @PostConstruct和@PreDestroy

CommonAnnotationBeanPostProcessor 不但能识别 @Resource 注解，还能识别 JSR-250 生命周期注解。在以下示例中，初始化后缓存会预先填充，并在销毁后清理。

```
public class CachingMovieLister {
    @PostConstruct
    public void populateMovieCache() {
        //在初始化时缓存电影信息…
    }

    @PreDestroy
    public void clearMovieCache() {
        //在销毁时清空电影信息…
    }
}
```

3.2.12 类路径扫描及组件管理

本小节将介绍另一种通过扫描类路径来隐式检测候选组件的方式。候选组件是匹配过滤条件的类库，并在容器中注册对应 bean 的定义。这就可以不用 XML 来执行 bean 的注册，那么开发人员可以使用注解（如 @Component）、AspectJ 风格的表达式，或者自定义过滤条件来选择哪些类在容器中注册 bean。

自 Spring 3.0 开始，很多由 Spring JavaConfig 项目提供的特性成为 Spring 核心框架的一部分。这就允许开发人员使用 Java（而不是传统的 XML 文件）来定义 bean。可以参考 @Configuration、

@Bean、@Import 和 @DependsOn 注解的例子来了解如何使用它们的新特性。

1. @Component及其同义的注解

在 Spring 2.0 版本之后，@Repository 注解是用于数据仓库（如数据访问对象 DAO）的类标记。这个标记有多种用途，其中之一就是异常自动转换。

Spring 2.5 引入了更多的典型注解，如 @Repository、@Component、@Service 和 @Controller。@Component 注解是对受 Spring 管理的组件的通用注解。@Repository、@Service 和 @Controller 注解相较于 @Component 注解另有特殊用途，分别对应了持久层、服务层和表现层。因此，开发人员可以使用 @Component 注解自己的组件类，但是如果使用 @Repository、@Service 或 @Controller 注解来替代，那么这些类更适合由工具来处理或与切面进行关联。而且 @Repository、@Service 和 @Controller 注解也可以在将来 Spring 框架的发布中携带更多的语义。因此，对于服务层，如果在 @Component 和 @Service 注解之间进行选择，那么 @Service 注解无疑是更好的选择。同样，在持久层中，@Repository 注解已经支持作为自动异常转换的标记。

2. 元注解

Spring 提供了很多元注解。元注解就是能被应用到另一个注解上的注解。

```
@Target(ElementType.TYPE)
@Retention(RetentionPolicy.RUNTIME)
@Documented
@Component
public @interfaceService {
// …
}
```

元注解也可以被用于创建组合注解。例如，Spring Web MVC 的 @RestController 注解就是 @Controller 和 @ResponseBody 的组合。

另外，组合注解可能从元注解中重新声明任意属性来允许用户自定义。这在开发者只想暴露一个元注解的子集时会特别有用。例如，下面是一个自定义的 @Scope 注解，将作用域名称指定到 @Session 注解上，但依然允许自定义 proxyMode。

```
@Target(ElementType.TYPE)
@Retention(RetentionPolicy.RUNTIME)
@Scope("session")
public @interface SessionScope {
    ScopedProxyMode proxyMode() default ScopedProxyMode.DEFAULT;
}
```

@SessionScope 可以不声明 proxyMode 就使用，例如：

```
@Service
@SessionScope
public class SessionScopedUserService implements UserService {
    // …
```

```
}
```

或者为 proxyMode 重载一个值，例如：

```
@Service
@SessionScope(proxyMode = ScopedProxyMode.TARGET_CLASS)
publicclass SessionScopedService {
    // …
}
```

3. 自动检测类并注册bean定义

Spring 可以自动检测固有的类并在 ApplicationContext 中注册对应的 BeanDefinition。下面的两个类就是自动检测的例子。

```
@Service
public class SimpleMovieLister {
    private MovieFinder movieFinder;
    @Autowired
    public SimpleMovieLister(MovieFinder movieFinder) {
        this.movieFinder = movieFinder;
    }
}

@Repository
public class JpaMovieFinder implements MovieFinder {
    // …
}
```

要自动检测这些类并注册对应的 bean，开发人员需要添加 @ComponentScan 到自己的 @Configuration 类上，其中的 base-package 元素是这两个类的公共父类包。开发人员可以任意选择使用逗号、分号、空格分隔的列表将每个类引入父包。

```
@Configuration
@ComponentScan(basePackages = "com.waylau")
public class AppConfig {
    //…
}
```

为了更简洁，上面的示例可以使用注解的 value 属性，也就是 ComponentScan("com.waylau")。下面的示例使用 XML 配置。

```
<?xml version="1.0" encoding="UTF-8"?>
<beans xmlns="http://www.springframework.org/schema/beans"
xmlns:xsi="http://www.w3.org/2001/XMLSchema-instance"
xmlns:context="http://www.springframework.org/schema/context"
xsi:schemaLocation="http://www.springframework.org/schema/beans
http://www.springframework.org/schema/beans/spring-beans.xsd
http://www.springframework.org/schema/context
```

```
http://www.springframework.org/schema/context/spring-context.xsd">
    <context:component-scan base-package="com.waylau"/>
</beans>
```

4. 使用过滤器来自定义扫描

默认情况下，使用 @Component、@Repository、@Service、@Controller 注解，或者使用自定义的 @Component 注解的类本身仅仅用于检测候选组件。开发人员可以修改并扩展这种行为，只需应用自定义的过滤器，即在 @ComponentScan 注解中添加 include-filter 或 exclude-filter 参数即可（或者作为 component-scan 元素的 include-filter 或 exclude-filter 子元素）。每个过滤器元素需要 type 和 expression 属性。过滤器选项的描述如表 3-2 所示。

表3-2 过滤器选项及描述

过滤器选项	表达式示例	描述
annotation（默认）	com.waylau.SomeAnnotation	使用在目标组件的类级别上
assignable（分配）	com.waylau.SomeClass	目标组件分配（扩展/实现）的类（接口）
aspectj（类型表达式）	com.waylau..*Service+ AspectJ	用类型表达式来匹配目标组件
regex（正则表达式）	org.example.Default.*	用正则表达式来匹配目标组件类的名称
custom（自定义）	com.waylau.MyTypeFilter	自定义 org.springframework.core.type.TypeFilter 接口的实现类

以下示例展示了忽略所有 @Repository 注解并使用 "stub" 库来替代。

```
@Configuration
@ComponentScan(basePackages = "com.waylau",
includeFilters = @Filter(type = FilterType.REGEX, pattern =".*Stub.*Repository"),
excludeFilters = @Filter(Repository.class))
public class AppConfig {
    //…
}
```

同样地，可以使用 XML 配置，代码如下。

```
<beans>
    <context:component-scan base-package="org.example">
    <context:include-filter type="regex"
        expression=".*Stub.*Repository"/>
    <context:exclude-filter type="annotation"
        expression="org.springframework.stereotype.Repository"/>
```

```
    </context:component-scan>
</beans>
```

3.2.13 基于Java的容器配置

Spring 中新的 Java 配置支持的核心就是 @Bean 注解的方法和 @Configuration 注解的类。

（1）@Bean 注解用来指定一个方法实例，配置和初始化一个新对象交给 Spring IoC 容器管理。对于那些熟悉 Spring 配置的人来说，@Bean 注解和 <bean> 元素扮演着相同的角色。@Bean 注解的方法可以在 @Component 类中使用，但一般常用在 @Configuration 类中。

（2）@Configuration 注解的类表示其主要目的是作为 bean 定义的来源。另外，@Configuration 类允许内部 bean 依赖通过简单地调用同一类中的其他 @Bean 方法进行定义。最简单的 @Configuration 类的示例如下。

```
@Configuration
public class AppConfig {
    @Bean
    public MyService myService() {
        return newMyServiceImpl();
    }
}
```

上面基于 AppConfig 类的配置方式的效果，等同于下面基于 Spring XML 的配置方式的效果。

```
<beans>
    <bean id="myService" class="com.waylau.MyServiceImpl"/>
</beans>
```

3.2.14 环境抽象

Environment 是一个集成到容器中的特殊抽象，它针对应用的环境建立了 profile 和 properties 两个关键的概念。

profile 是包含了多个 bean 定义的一个逻辑集合，只有当指定的 profile 被激活时，其中的 bean 才会被激活。无论是通过 XML 定义还是通过注解，bean 都可以配置到 profile 中。而 Environment 对象的角色就是与 profile 相关联，然后决定激活哪一个 profile，以及哪一个 profile 为默认的 profile。

properties 在几乎所有的应用当中都有着重要的作用，当然也可能存在多个数据源，如 property 文件、JVM 系统 property、系统环境变量、JNDI、Servlet 上下文参数、ad-hoc 属性对象、Map 等。

Environment 对象与 property 相关联，然后给开发者一个方便的服务接口来配置这些数据源，并正确解析。

以下是配置 profile 的示例。

```
@Configuration
public class AppConfig {
    @Bean
    @Profile("dev")
    public DataSourcedevDataSource() {
    return newEmbeddedDatabaseBuilder()
        .setType(EmbeddedDatabaseType.HSQL)
        .addScript("classpath:com/bank/config/sql/schema.sql")
        .addScript("classpath:com/bank/config/sql/test-data.sql")
        .build();
    }
    @Bean
    @Profile("production")
    public DataSourceproductionDataSource() throwsException {
        Context ctx = newInitialContext();
        return (DataSource) ctx.lookup("java:comp/env/jdbc/datasource");
    }
}
```

从上述例子中可以看到，@Profile 注解也可以在方法级别使用，还可以声明在包含 @Bean 注解的方法中。

定义完环境抽象之后，还需要通知 Spring 要激活哪一个 profile。有多种方法来激活一个 profile，最直接的方式就是通过编程来直接调用 Environment API。ApplicationContext 中包含以下接口。

```
AnnotationConfigApplicationContext ctx = newAnnotationConfigApplication-
Context();
ctx.getEnvironment().setActiveProfiles("dev");
ctx.register(SomeConfig.class, StandaloneDataConfig.class, JndiDataConfig.
class);
ctx.refresh();
```

此外，profile 还可以通过 spring.profiles.active 中的属性来指定，可以通过系统环境变量、JVM 系统变量、Servlet 上下文中的参数，甚至是 JNDI 的一个参数来写入。在集成测试中，激活 profile 可以通过 spring-test 中的 @ActiveProfiles 来实现。以下是一个示例。

```
-Dspring.profiles.active="dev"
```

3.3 AOP

AOP（Aspect Oriented Programming，面向切面编程）并不是要代替 OOP（Object Oriented Programming，面向对象编程），而是一种对 OOP 的补充。OOP 模块化的关键单元是类，而在 AOP 中，

模块化的单元是切面（Aspect）。切面可以实现跨多个类型和对象的事务管理、日志等方面的模块化。

3.3.1 AOP概述

AOP编程的目标与OOP并没有什么不同，都是为了减少重复和专注于业务。相比之下，OOP是婉约派的选择，用继承和组合的方式，编制一套类和对象的体系；而AOP是豪放派的选择，大手一挥，凡是某包、某类、某命名方法，一并同样处理。也就是说，OOP是"绣花针"，而AOP是"砍柴刀"。

Spring框架的关键组件之一是AOP框架。虽然Spring IoC容器不依赖于AOP，但在Spring应用中，经常会使用AOP来简化编程。在Spring框架中使用AOP主要有以下优势。

（1）提供声明式企业服务，特别是可以作为EJB声明式服务的替代品。

（2）允许用户实现自定义切面。在某些不适合用OOP编程的场景中，采用AOP来补充。

（3）可以对业务逻辑的各个部分进行隔离，从而使业务逻辑各部分之间的耦合度降低，提高程序的可重用性，同时提高开发的效率。

要使用Spring AOP，需要添加spring-aop模块。

3.3.2 AOP核心概念

AOP概念并非Spring AOP所特有的，以下概念同样适用于其他AOP框架，如AspectJ。

（1）Aspect（切面）：将关注点进行模块化。某些关注点可能会横跨多个对象，如事务管理，它是Java企业级应用中一个关于横切关注点很好的例子。在Spring AOP中可以使用常规类（基于模式的方法）或@Aspect注解的常规类来实现切面。

（2）Join Point（连接点）：在程序执行过程中某个特定的点，如某方法调用或处理异常时。在Spring AOP中，一个连接点总是代表一个方法的执行。

（3）Advice（通知）：在切面的某个特定连接点上执行的动作。通知有各种类型，包括"around"、"before"和"after"等。许多AOP框架，包括Spring，都是以拦截器来实现通知模型的，并维护一个以连接点为中心的拦截器链。

（4）Pointcut（切入点）：匹配连接点的断言。通知和一个切入点表达式关联，并在满足这个切入点的连接点上运行（如当执行某个特定名称的方法时）。切入点表达式如何和连接点匹配是AOP的核心，Spring默认使用AspectJ切入点语法。

（5）Introduction（引入）：声明额外的方法或某个类型的字段。Spring允许引入新的接口（及一个对应的实现）到任何被通知的对象。例如，可以使用一个引入来使bean实现IsModified接口，以简化缓存机制。在AspectJ社区，Introduction也被称为Inter-type Declaration（内部类型声明）。

（6）Target Object（目标对象）：被一个或多个切面所通知的对象，也有人把它称为 Advised（被通知）对象。既然 Spring AOP 是通过运行时代理实现的，那这个对象永远是一个 Proxied（被代理）对象。

（7）AOP Proxy（AOP 代理）：AOP 框架创建的对象，用来实现 Aspect Contract（切面契约），包括通知方法执行等功能。在 Spring 中，AOP 代理可以是 JDK 动态代理或 CGLIB 代理。

（8）Weaving（织入）：把切面连接到其他的应用程序类型或对象上，并创建一个 Advised（被通知）对象。这些可以在编译时（如使用 AspectJ 编译器编译）、类加载时和运行时完成。

Spring 与其他纯 Java AOP 框架一样，在运行时完成织入。其中 Advice（通知）的类型主要有以下几种。

（1）Before Advice（前置通知）：在某连接点之前执行的通知，但这个通知不能阻止连接点前的执行（除非它抛出一个异常）。

（2）After Returning Advice（返回后通知）：在某连接点正常完成后执行的通知，如果一个方法没有抛出任何异常，则正常返回。

（3）After Throwing Advice（抛出异常后通知）：在方法抛出异常退出时执行的通知。

（4）After (finally) Advice（最后通知）：当某连接点退出时执行的通知（无论是正常返回还是异常退出）。

（5）Around Advice（环绕通知）：包围一个连接点的通知，如方法调用。这是最强大的一种通知类型。环绕通知可以在方法调用前后完成自定义的行为，它也会选择是否继续执行连接点，或者直接返回自己的返回值或抛出异常来结束执行。Around Advice 是最常用的一种通知类型。与 AspectJ 一样，Spring 提供所有类型的通知，但最好使用简单的通知类型来实现需要的功能。例如，如果只是需要用一个方法的返回值来更新缓存，虽然使用环绕通知也能完成同样的事情，但最好使用 After Returning 通知。用最合适的通知类型可以使编程模型变得简单，并且能够避免很多潜在的错误。例如，如果不调用 Join Point（用于 Around Advice）的 proceed() 方法，就不会有调用的问题。

在 Spring 2.0 中，所有的通知参数都是静态类型的，因此可以使用合适的类型（如一个方法执行后的返回值类型）作为通知的参数，而不是使用一个对象数组。切入点和连接点匹配的概念是 AOP 的关键，这使得 AOP 不同于其他仅仅提供拦截功能的旧技术。切入点使得通知可独立于 OO（Object Oriented，面向对象）层次。例如，一个提供声明式事务管理的 Around Advice 可以被应用到一组横跨多个对象的方法上（如服务层的所有业务操作）。

3.3.3　Spring AOP

Spring AOP 用纯 Java 实现，不需要专门的编译过程。Spring AOP 不需要控制类装载器层次，因此它适用于 Servlet 容器或应用服务器。

Spring 目前仅支持方法调用作为连接点。虽然可以在不影响 Spring AOP 核心 API 的情况下加入对成员变量拦截器的支持，但 Spring 并没有实现成员变量拦截器。如果需要通知对成员变量的访问和更新连接点，可以考虑其他语言，如 AspectJ。

Spring 实现 AOP 的方法与其他的框架不同。Spring 并不是要尝试提供最完整的 AOP 实现（尽管 Spring AOP 有这个能力），相反，它侧重于提供一种 AOP 实现和 Spring IoC 容器的整合，用于解决企业级开发中的常见问题。因此，Spring AOP 通常都和 Spring IoC 容器一起使用。Aspect 使用普通的 bean 定义语法，与其他 AOP 实现相比，这是一个显著的区别。有些任务是使用 Spring AOP 无法轻松或高效完成的，如通知一个细粒度的对象。这时，使用 AspectJ 是最好的选择。对于大多数在企业级 Java 应用中遇到的问题，Spring AOP 都能提供一个非常好的解决方案。

Spring AOP 从来没有打算通过提供一种全面的 AOP 解决方案来取代 AspectJ，它们之间的关系应该是互补的，而不是竞争的。Spring 可以无缝地整合 Spring AOP、IoC 和 AspectJ，使所有的 AOP 应用完全融入基于 Spring 的应用体系，这样的集成不会影响 Spring AOP API 或 AOP Alliance API。

Spring AOP 保留了向下兼容性，这体现了 Spring 框架的核心原则——非侵袭性，即 Spring 框架并不强迫在业务或领域模型中引入框架特定的类和接口。

3.3.4 AOP代理

Spring AOP 默认使用标准的 JDK 动态代理作为 AOP 的代理，这样任何接口（或接口的 set 方法）都可以被代理。

Spring AOP 也支持使用 CGLIB 代理，当需要代理类（而不是代理接口）时，使用 CGLIB 代理是很有必要的。如果一个业务对象并没有实现一个接口，就会默认使用 CGLIB。面向接口编程也是一个最佳实践，业务对象通常都会实现一个或多个接口。此外，在需要通知一个未在接口中声明的方法的情况下，或者需要传递一个代理对象作为一种具体类型的方法的情况下，还可以强制地使用 CGLIB。

3.3.5 实战：使用@AspectJ

@AspectJ 是用于切面的常规 Java 类注解。AspectJ 项目引入了 @AspectJ 风格作为 AspectJ 5 版本的一部分。Spring 使用与 AspectJ 5 相同的用于切入点解析和匹配的注解，但 AOP 运行时仍然是纯粹的 Spring AOP，并不依赖于 AspectJ 编译器。

1. 启用@AspectJ

可以通过 XML 或 Java 配置来启用 @AspectJ 支持。不管在任何情况下，都要确保 AspectJ 的 aspectjweaver.jar 库在应用程序的类路径中。这个库可在 AspectJ 发布的 lib 目录中或通过 Maven 的中央库得到，其配置如下。

```xml
<dependency>
    <groupId>org.springframework</groupId>
    <artifactId>spring-aspects</artifactId>
    <version>5.1.3.RELEASE</version>
</dependency>
```

2. 创建应用

下面用一个简单有趣的例子来演示 Spring AOP 的用法。此例演绎了一段"武松打虎"的故事情节——武松（Fighter）在山里等着老虎（Tiger）出现，只要发现老虎出来，就打老虎。源码可以在 aop-aspect 目录下找到。

aop-aspect 应用的 pom.xml 文件定义如下。

```xml
<?xml version="1.0" encoding="UTF-8"?>
<beans xmlns="http://www.springframework.org/schema/beans"
    xmlns:xsi="http://www.w3.org/2001/XMLSchema-instance"
    xmlns:context="http://www.springframework.org/schema/context"
    xsi:schemaLocation="
        http://www.springframework.org/schema/beans
        http://www.springframework.org/schema/beans/spring-beans.xsd
        http://www.springframework.org/schema/context
        http://www.springframework.org/schema/context/spring-context.xsd">

    <!-- 定义 bean -->
    <bean id="messageServiceImpl"
        class="com.waylau.spring.di.service.MessageServiceImpl">
        <constructor-arg name="username" value="Way Lau" />
        <constructor-arg name="age" value="30" />
    </bean>

    <bean id="messagePrinter"
        class="com.waylau.spring.di.MessagePrinter">
        <constructor-arg name="service"
            ref="messageServiceImpl" />
    </bean>
</beans>
```

3. 定义业务模型

首先定义老虎类：

```
package com.waylau.spring.aop;

public class Tiger {
    public void walk() {
        System.out.println("Tiger is walking…");
    }
}
```

老虎类只有一个 walk() 方法，只要老虎出来，就会触发这个方法。

4. 定义切面和配置

那么武松要做什么呢？他主要关注老虎的动向，等着老虎出来活动。所以在 Fighter 类中定义一个 @Pointcut。同时，在该切入点前后都可以执行相关的方法，定义 foundBefore() 和 foundAfter()：

```
package com.waylau.spring.aop;
import org.aspectj.lang.annotation.AfterReturning;
import org.aspectj.lang.annotation.Aspect;
import org.aspectj.lang.annotation.Before;
import org.aspectj.lang.annotation.Pointcut;
@Aspect
public class Fighter {
    @Pointcut("execution(* com.waylau.spring.aop.Tiger.walk())")
    public void foundTiger() {
    }

    @Before(value = "foundTiger()")
    public void foundBefore() {
        System.out.println("Fighter wait for tiger…");
    }

    @AfterReturning("foundTiger()")
    public void foundAfter() {
        System.out.println("Fighter fight with tiger…");
    }
}
```

相应的 Spring 配置如下。

```
<?xml version="1.0" encoding="UTF-8"?>
<beans xmlns="http://www.springframework.org/schema/beans"
xmlns:xsi="http://www.w3.org/2001/XMLSchema-instance"
xmlns:context="http://www.springframework.org/schema/context"
xmlns:aop="http://www.springframework.org/schema/aop"
xsi:schemaLocation="
http://www.springframework.org/schema/beans
http://www.springframework.org/schema/beans/spring-beans.xsd
http://www.springframework.org/schema/context
http://www.springframework.org/schema/context/spring-context.xsd
http://www.springframework.org/schema/aop
http://www.springframework.org/schema/aop/spring-aop.xsd">
    <!-- 启动AspectJ支持 -->
    <aop:aspectj-autoproxy />
    <!-- 定义bean -->
    <bean id="fighter" class="com.waylau.spring.aop.Fighter" />
    <bean id="tiger" class="com.waylau.spring.aop.Tiger" />
</beans>
```

5. 定义主应用

主应用定义如下。

```
package com.waylau.spring.aop;

import org.springframework.context.ApplicationContext;
import org.springframework.context.support.ClassPathXmlApplicationContext;

public class Application {
    public static void main(String[] args) {

        @SuppressWarnings("resource")
        ApplicationContext context = 
            new ClassPathXmlApplicationContext("spring.xml");

        Tiger tiger = context.getBean(Tiger.class);
        tiger.walk();
    }
}
```

6. 运行应用

运行应用，可以看到控制台最终输出如下内容。

```
Fighter wait for tiger…
Tiger is walking…
Fighter fight with tiger…
```

3.3.6 基于XML的AOP

Spring 提供基于 XML 的 AOP 支持，并提供新的 "aop" 命名空间。

在 Spring 配置中，所有的 aspect 和 advisor 元素都必须放置在元素中（应用程序上下文配置中可以有多个元素）。一个元素可以包含 pointcut、advisor 和 aspect 3 个元素（注意这些元素必须按照这个顺序声明）。

1. 声明一个pointcut

pointcut 可以在元素中声明，从而使 pointcut 定义可以在几个 aspect 和 advice 之间共享。

以下声明代表了服务层中任何业务服务都能执行的切入点的定义。

```xml
<aop:config>
<aop:pointcut id="businessService"
    expression="execution(* com.xyz.myapp.service.*.*(..))"/>
</aop:config>
```

2. 声明advice

Spring 的 advice 与 @AspectJ 风格是一致的，它们具有完全相同的语义。

以下是一个示例。

```xml
<aop:aspect id="beforeExample" ref="aBean">
<aop:before
    pointcut-ref="dataAccessOperation" method="doAccessCheck"/>
…
</aop:aspect>
```

3. 声明一个aspect

一个 aspect 就是在 Spring 应用程序上下文中定义的一个普通的 Java 对象。状态和行为被捕获到对象的字段和方法中，pointcut 和 advice 被捕获到 XML 中。

使用元素声明一个 aspect，并使用 ref 属性引用辅助 bean：

```xml
<aop:config>
<aop:aspect id="myAspect" ref="aBean">
…
</aop:aspect>
</aop:config>
<bean id="aBean" class="…">
…
</bean>
```

3.3.7 实战：基于XML的AOP

本小节基于 aop-aspect 示例进行改造，形成一个新的基于 XML 的 AOP 实战例子。新的应用源码可以在 aop-aspect-xml 目录下找到。

1. 定义业务模型

之前定义的老虎（Tiger）类保持不变。老虎类只有一个 walk() 方法，只要老虎出来，就会触发这个方法：

```java
package com.waylau.spring.aop;

public class Tiger {
    public void walk() {
        System.out.println("Tiger is walking…");
    }
}
```

之前所定义的武松（Fighter）类保持不变，稍作调整，去除注解，变成一个单纯的 POJO：

```java
package com.waylau.spring.aop;
```

```java
public class Fighter {

    public void foundBefore() {
        System.out.println("Fighter wait for tiger…");
    }

    public void foundAfter() {
        System.out.println("Fighter fight with tiger…");
    }
}
```

2. 定义切面和配置

所有 AOP 的配置都在相应的 Spring 的 XML 配置中：

```xml
<?xml version="1.0" encoding="UTF-8"?>
<beans xmlns="http://www.springframework.org/schema/beans"
    xmlns:xsi="http://www.w3.org/2001/XMLSchema-instance"
    xmlns:context="http://www.springframework.org/schema/context"
    xmlns:aop="http://www.springframework.org/schema/aop"
    xsi:schemaLocation="
        http://www.springframework.org/schema/beans
        http://www.springframework.org/schema/beans/spring-beans.xsd
        http://www.springframework.org/schema/context
        http://www.springframework.org/schema/context/spring-context.xsd
        http://www.springframework.org/schema/aop
        http://www.springframework.org/schema/aop/spring-aop.xsd">

    <!-- 启动AspectJ支持 -->
    <aop:aspectj-autoproxy />

    <!-- 定义Aspect -->
    <aop:config>
        <aop:pointcut expression="execution(* com.waylau.spring.aop.Tiger.walk())"
            id="foundTiger"/>
        <aop:aspect id="myAspect" ref="fighter">
            <aop:before method="foundBefore" pointcut-ref="foundTiger"/>
            <aop:after-returning method="foundAfter" pointcut-ref="foundTiger"/>
        </aop:aspect>
    </aop:config>

    <!-- 定义 bean -->
    <bean id="fighter" class="com.waylau.spring.aop.Fighter" />
    <bean id="tiger" class="com.waylau.spring.aop.Tiger" />
</beans>
```

3. 定义主应用

主应用保持不变，其代码如下。

```
package com.waylau.spring.aop;

import org.springframework.context.ApplicationContext;
import org.springframework.context.support.ClassPathXmlApplicationContext;

public class Application {
    public static void main(String[] args) {

        @SuppressWarnings("resource")
        ApplicationContext context =
            new ClassPathXmlApplicationContext("spring.xml");

        Tiger tiger = context.getBean(Tiger.class);
        tiger.walk();
    }
}
```

4. 运行应用

运行应用，可以看到控制台最终输出如下内容。

```
Fighter wait for tiger…
Tiger is walking…
Fighter fight with tiger…
```

3.4 资源处理

Java 的标准 java.net.URL 类和各种以 URL 为前缀的标准处理程序（如 URLDecoder、URLEncoder）并不足以满足所有对低级资源的访问。例如，想要访问需要从类路径获取的资源，或者相对于 ServletContext 的资源，标准的 URL 实现并不能支持。

Spring Resource 接口就是为了弥补上述不足。

3.4.1 常用资源接口

Spring Resource 接口是强大的用于访问低级资源的抽象，主要包含以下方法。

```
public interface Resource extends InputStreamSource {
    boolean exists();
    boolean isOpen();
    URL getURL() throws IOException;
```

```
File getFile() throws IOException;
Resource createRelative(String relativePath) throws IOException;
String getFilename();
String getDescription();
}
public interface InputStreamSource {
InputStream getInputStream() throws IOException;
}
```

(1) getInputStream()：定位并打开资源，返回一个从资源读取的 InputStream。预计每个调用都会返回一个新的 InputStream。调用方在使用完这个流后，关闭该流。

(2) exists()：返回一个布尔值，指示这个资源是否实际上以物理形式存在。

(3) isOpen()：返回一个布尔值，指示这个资源是否代表一个打开流的句柄。如果返回值为 true，则只能读取一次 InputStream，然后关闭，以避免资源泄露。除 InputStreamResource 外，对于所有的资源实现，都将返回 false。

(4) getDescription()：返回此资源的描述。这通常是完全限定的文件名或资源的实际 URL。

其他方法允许开发人员获取表示资源的实际 URL 或 File 对象（如果底层实现是兼容的，并且支持该功能）。

资源抽象在 Spring 中被广泛使用。

3.4.2 内置资源接口实现

Spring 提供了很多资源接口实现，这些实现是可以直接用的。

1. UrlResource

UrlResource 封装了一个 java.net.URL，用来访问可以通过 URL 访问的任何对象，如文件、HTTP 目标、FTP 目标等。所有的 URL 都由一个标准化的字符串表示，如使用适当的标准化前缀来表示另一个 URL 类型。

(1) file：用于访问文件系统路径。

(2) http：用于通过 HTTP 访问资源。

(3) ftp：用于通过 FTP 访问资源。

UrlResource 是由 Java 代码使用 UrlResource 构造函数显式创建的，但是当调用一个接收 String 参数的 API 方法时，通常会隐式地创建 UrlResource 来表示路径。对于后一种情况，JavaBean PropertyEditor 将最终决定创建哪种类型的资源。如果路径字符串中包含 JavaBean PropertyEditor 可以识别的前缀，如 classpath:，那么它将为该前缀创建适当的专用资源。但是，如果不能识别前缀，它会认为这只是一个标准的 URL 字符串，并会创建一个 UrlResource。

2. ClassPathResource

ClassPathResource 类代表一个应该从类路径中获得的资源，如使用线程上下文类加载器、给定的类加载器或给定的类来加载资源。如果类路径资源驻留在文件系统中，则此资源实现支持的解析为 java.io.File。

ClassPathResource 是由 Java 代码使用 ClassPathResource 构造函数显式创建的，但是当开发人员调用一个带有 String 参数的 API 方法时，通常会隐式地创建 ClassPathResource 来表示路径。对于后一种情况，JavaBean PropertyEditor 将识别字符串路径上的特殊前缀 classpath:，并在此情况下创建一个 ClassPathResource。

3. FileSystemResource

FileSystemResource 用于处理 java.io.File 资源的实现。

4. ServletContextResource

ServletContextResource 是 ServletContext 资源的实现，解释相关 Web 应用程序根目录中的相对路径。

ServletContextResource 总是支持流访问和 URL 访问，但只有在 Web 应用程序归档文件被扩展且资源位于文件系统中时才允许访问 java.io.File。不管它是否被扩展，实际上都依赖于 Servlet 容器。

5. InputStreamResource

InputStreamResource 给定 InputStream 的资源实现，只有在没有适用的资源实现的情况下才能使用。一般情况下，首选 ByteArrayResource 或任何基于文件的资源实现。

与其他 Resource 实现相比，InputStreamResource 是已打开资源的描述符，因此 isOpen() 将返回 true。如果需要将资源描述符保存在某处，或者需要多次读取流，就不要使用它。

6. ByteArrayResource

ByteArrayResource 是给定字节数组的资源实现，它为给定的字节数组创建一个 ByteArrayInputStream。

从任何给定的字节数组中加载内容都是很有用的，这样可以不必求助于一次性的 InputStreamResource。

3.4.3　ResourceLoader

ResourceLoader 接口是由可以返回（加载）Resource 实例的对象来实现的，该接口包含如下方法。

```
public interface ResourceLoader {
    Resource getResource(String location);
}
```

所有应用程序上下文都实现了 ResourceLoader 接口，因此所有的应用程序上下文都可以用来获取 Resource 实例。

当在特定的应用程序上下文中调用 getResource() 方法，并且指定的位置路径没有特定的前缀时，将返回适合该特定应用程序上下文的资源类型。例如，假设下面的代码是针对 ClassPathXml-

ApplicationContext 实例执行的。

```
Resource template = ctx.getResource("some/resource/path/myTemplate.txt");
```

上述代码将返回一个 ClassPathResource。如果对 FileSystemXmlApplicationContext 实例执行相同的方法，则会返回 FileSystemResource。对于一个 WebApplicationContext，开发人员会得到一个 ServletContextResource，以此类推。

因此，可以选择适合特定应用程序上下文的方式加载资源。

另外，也可以通过指定特殊的前缀来强制使用特定的资源，而不管应用程序的上下文类型如何。例如：

```
Resource template = ctx.getResource("classpath:some/resource/path/myTemplate.txt");
Resource template = ctx.getResource("file:///some/resource/path/myTemplate.txt");
Resource template = ctx.getResource("http://myhost.com/resource/path/myTemplate.txt");
```

3.4.4 ResourceLoaderAware

ResourceLoaderAware 接口是一个特殊的标记接口，用于标识期望通过 ResourceLoader 接口提供的对象：

```
public interface ResourceLoaderAware {
    void setResourceLoader(ResourceLoader resourceLoader);
}
```

当一个类实现了 ResourceLoaderAware 并被部署到一个应用程序上下文（作为一个 Spring 管理的 bean）时，它会被应用程序上下文识别为 ResourceLoaderAware。然后应用程序上下文将调用 setResourceLoader(ResourceLoader)，并将 ResourceLoader 自身作为参数（记住，Spring 中的所有应用程序上下文实现均使用 ResourceLoader 接口）。

由于 ApplicationContext 是一个 ResourceLoader，因此 bean 也可以实现 ApplicationContextAware 接口，并直接使用提供的应用程序上下文来加载资源。但通常情况下，最好使用专用的 ResourceLoader 接口（如果有需要）。代码只会耦合到资源加载接口，它可以被认为是一个实用接口，而不是整个 Spring ApplicationContext 接口。

从 Spring 2.5 开始，开发人员可以依靠 ResourceLoader 的自动装配来实现 ResourceLoaderAware 接口。传统的 constructor 和 byType 自动装配模式可以分别为构造函数参数和设置方法参数提供 ResourceLoader 类型的依赖关系。

3.4.5 资源作为依赖

如果 bean 本身要通过某种动态的过程来确定和提供资源路径，那么 bean 可能会使用 ResourceLoader 接口来加载资源。考虑加载某种类型的模板时，其中需要的特定资源取决于用户的角色。如果资源是静态的，那么完全清除 ResourceLoader 接口是有意义的，只要让 bean 公开其需要的 Resource 属性，并期望它们被注入其中，具体示例为：

```
<bean id="myBean" class="…">
    <property name="template" value="some/resource/path/myTemplate.txt"/>
</bean>
```

注意，资源路径没有前缀，因为应用程序上下文本身将被用作 ResourceLoader，所以根据上下文的确切类型，资源本身将根据需要通过 ClassPathResource、FileSystemResource 或 ServletContextResource 来进行加载。

如果需要强制使用特定的资源类型，则可以使用前缀。以下两个示例展示了如何强制使用 ClassPathResource 和 UrlResource（后者用于访问文件系统）。

```
<property name="template" value="classpath:some/resource/path/myTemplate.txt">
<property name="template" value="file:///some/resource/path/myTemplate.txt"/>
```

3.5 表达式语言SpEL

Spring Expression Language（SpEL）是一种强大的表达式语言，支持在运行时查询和操作对象图。其语法与 Unified EL 类似，但提供了额外的功能，特别是方法调用和基本的字符串模板功能。

虽然还有其他几种可用的 Java 表达式语言，如 OGNL、MVEL、JBoss EL 等，但 Spring 表达式语言的创建是为了向 Spring 社区提供单一支持的表达式语言，可以在所有 Spring 产品中使用 SpEL。

SpEL 的语言特性是由 Spring 项目中的需求驱动的，基于 Eclipse 的 Spring Tool Suite 中的代码自动完成。也就是说，SpEL 基于一种与技术无关的 API，允许在需要时集成其他表达式语言实现。

3.5.1 SpEL概述

SpEL 并不与 Spring 直接相关，可以独立使用。

SpEL 表达式语言支持的功能除文本表达、布尔和关系运算符、正则表达式、类表达式，以及访问属性、数组、列表、Map 外，还有调用方法、分配、调用构造函数、bean 引用、数组构建、内联列表、内联 Map、三元操作符、变量、用户定义的功能、集合投影、集合选择和模板化的表达式。

3.5.2 表达式接口

以下代码引入了 SpEL API 来评估文本字符串表达式 "Hello World"。

```
ExpressionParser parser = new SpelExpressionParser();
Expression exp = parser.parseExpression("'Hello World'");
String message = (String) exp.getValue();
```

消息变量的值只是简单的 "Hello World"。使用的 SpEL 类和接口位于 org.springframework.expression 包及其子包（如 spel.support）中。在这个例子中，表达式字符串是由单引号引起来的文本字符串，Expression 接口负责评估表达式字符串。当分别调用 parser.parseExpression 和 exp.getValue 时，可能会抛出 ParseException 和 EvaluationException 两个异常。

SpEL 支持很多功能，如调用方法、访问属性和调用构造函数。

1. 调用方法

可以在字符串上调用 concat 方法，其示例为：

```
ExpressionParser parser = new SpelExpressionParser();
Expression exp = parser.parseExpression("'Hello World'.concat('!')");
String message = (String) exp.getValue();
```

消息变量的值现在为 "Hello World!"。

2. 访问属性

可以调用 String 的 Bytes 属性，其示例为：

```
ExpressionParser parser = new SpelExpressionParser();

// 调用 'getBytes()'
Expression exp = parser.parseExpression("'Hello World'.bytes");
byte[] bytes = (byte[]) exp.getValue();
```

SpEL 还支持使用标点符号的嵌套属性，其示例为：

```
ExpressionParser parser = new SpelExpressionParser();

// 调用 'getBytes().length'
Expression exp = parser.parseExpression("'Hello World'.bytes.length");
int length = (Integer) exp.getValue();
```

3. 调用构造函数

字符串的构造函数可以被调用，其示例为：

```
ExpressionParser parser = new SpelExpressionParser();
Expression exp = parser.parseExpression("new String('hello world').toUp-
perCase()");
String message = exp.getValue(String.class);
```

3.5.3 对于bean定义的支持

SpEL 表达式可以与 XML 或基于注解的配置元数据一起使用来定义 BeanDefinitions。在这两种情况下，定义表达式的语法形式都是 #{ <expression string> }。

1. 基于XML的配置

可以使用以下表达式来设置属性或构造函数的参数值。

```
<bean id="numberGuess" class="org.spring.samples.NumberGuess">
<property name="randomNumber" value="#{ T(java.lang.Math).random()*
100.0 }"/>
<!-- … -->
</bean>
```

变量 systemProperties 是预定义的，所以可以在表达式中使用，如以下代码。

```
<bean id="taxCalculator" class="org.spring.samples.TaxCalculator">
<property name="defaultLocale" value="#{ systemProperties['user.region']
}"/>
<!-- … -->
</bean>
```

也可以通过名称引用其他 bean 属性，如以下代码。

```
<bean id="numberGuess" class="org.spring.samples.NumberGuess">
<property name="randomNumber" value="#{ T(java.lang.Math).random()*
100.0 }"/>
<!-- … -->
</bean>
<bean id="shapeGuess" class="org.spring.samples.ShapeGuess">
<property name="initialShapeSeed" value="#{ numberGuess.randomNumber
}"/>
<!-- … -->
</bean>
```

2. 基于注解的配置

@Value 注解可以放在字段、方法及构造函数参数中，以指定默认值。

以下是一个设置字段变量默认值的例子。

```
public static class FieldValueTestBean {
    @Value("#{ systemProperties['user.region'] }")
    private String defaultLocale;
```

```java
    public void setDefaultLocale(String defaultLocale) {
        this.defaultLocale = defaultLocale;
    }
    public String getDefaultLocale() {
        return this.defaultLocale;
    }
}
```

以上示例等价于在属性的 setter 方法上设置默认值。

```java
public static class PropertyValueTestBean {
    private String defaultLocale;

    @Value("#{ systemProperties['user.region'] }")
    public void setDefaultLocale(String defaultLocale) {
        this.defaultLocale = defaultLocale;
    }
    public String getDefaultLocale() {
        return this.defaultLocale;
    }
}
```

自动装配的方法和构造函数也可以使用 @Value 注解。

```java
public class SimpleMovieLister {
    private MovieFinder movieFinder;
    private String defaultLocale;

    @Autowired
    public void configure(MovieFinder movieFinder,
        @Value("#{ systemProperties['user.region'] }") String defaultLocale) {

        this.movieFinder = movieFinder;
        this.defaultLocale = defaultLocale;
    }
    // …
}
```

3.5.4　实战：使用SpEL

本小节使用 SpEL 来演示一个"商品费用结算"的例子，该例子通过 SpEL 表达式来筛选数据。

例子源码在 expression-language 目录下。

1. 自定义领域对象

创建一个新类 Item 代表商品，其代码为：

```java
package com.waylau.spring.el;

public class Item {
    private String good;
    private double weight;

    // …省略getter/setter方法

    @Override
    public String toString() {
        return "Item [good=" + good + ", weight=" + weight + "]";
    }
}
```

创建一个新类 ShopList 代表商品清单，其代码为：

```java
package com.waylau.spring.el;

import java.util.ArrayList;
import java.util.Arrays;
import java.util.List;

public class ShopList {
    private String name;
    private int count;
    private double price;

    private List<Item> items = new ArrayList<Item>();

    private Item onlyOne;

    private String[] allGood;

    // …省略getter/setter方法

    @Override
    public String toString() {
        return "ShopList [name=" + name + ", count=" + count + ", price="
                + price + ", items=" + items + ", onlyOne="
                + onlyOne + ", allGood=" + Arrays.toString(allGood) +
"]";
    }
}
```

创建一个新类 Tax 代表商品税率，其代码为：

```java
package com.waylau.spring.el;

public class Tax {
```

```
    private double ctax;
    private String name;
    
    public static String getCountry() {
        return "zh_CN";
    }
    
    public String getName() {
        return this.name;
    }
    
    public double getCtax() {
        return ctax;
    }
    
    public void setCtax(double ctax) {
        this.ctax = ctax;
    }
}
```

2. 配置文件

定义 Spring 应用的配置文件 spring.xml，这里的 SpEL 表达式是基于 XML 来定义的。

```
<?xml version="1.0" encoding="UTF-8"?>
<beans xmlns="http://www.springframework.org/schema/beans"
    xmlns:xsi="http://www.w3.org/2001/XMLSchema-instance"
    xmlns:context="http://www.springframework.org/schema/context"
    xmlns:p="http://www.springframework.org/schema/p"
    xmlns:util="http://www.springframework.org/schema/util"
    xsi:schemaLocation="
        http://www.springframework.org/schema/beans
        http://www.springframework.org/schema/beans/spring-beans.xsd
        http://www.springframework.org/schema/context
        http://www.springframework.org/schema/context/spring-context.xsd
        http://www.springframework.org/schema/util
        http://www.springframework.org/schema/util/spring-util.xsd">

    <bean id="tax" class="com.waylau.spring.el.Tax" p:ctax="10"></bean>
    
    <!-- 访问bean的属性 -->
    <bean id="list" class="com.waylau.spring.el.ShopList" p:name="shanpoo"
        p:count="2" p:price="#{tax.ctax/100 * 36.5}" />
    
    <!-- 调用bean的方法 -->
    <bean id="list2" class="com.waylau.spring.el.ShopList" p:name="shanpoo"
        p:count="2" p:price="#{tax.getCtax()/100 * 36.5}" />
```

```xml
<!-- 访问静态变量 -->
<bean id="list3" class="com.waylau.spring.el.ShopList"
    p:name="#{T(com.waylau.spring.el.Tax).country}"
    p:count="2" p:price="1" />

<!-- 访问静态方法 -->
<bean id="list4" class="com.waylau.spring.el.ShopList"
    p:name="#{T(com.waylau.spring.el.Tax).getCountry()}" p:count="2"
    p:price="1" />

<!-- 三元表达式的简化 -->
<bean id="list5" class="com.waylau.spring.el.ShopList"
    p:name="#{tax.getName()?: 'defaultTax'}"
    p:count="2" p:price="1" />

<util:list id="its">
    <bean class="com.waylau.spring.el.Item" p:good="poke" p:weight="3.34"></bean>
    <bean class="com.waylau.spring.el.Item" p:good="chicken"
        p:weight="5.66"></bean>
    <bean class="com.waylau.spring.el.Item" p:good="dark" p:weight="3.64"></bean>
    <bean class="com.waylau.spring.el.Item" p:good="egg" p:weight="2.54"></bean>
</util:list>

<!-- 展示util:list的用法 -->
<bean id="list6" class="com.waylau.spring.el.ShopList"
    p:name="#{tax.getName()?: 'defaultTax'}"
    p:count="2" p:price="1" p:items-ref="its" />

<!-- 集合筛选 -->
<bean id="list7" class="com.waylau.spring.el.ShopList"
    p:name="#{tax.getName()?: 'defaultTax'}"
    p:count="2" p:price="1" p:onlyOne="#{its[0]}" /><!-- 这里不是用ref装配 -->

<bean id="it1" class="com.waylau.spring.el.Item" p:good="poke"
    p:weight="3.34"></bean>
<bean id="it2" class="com.waylau.spring.el.Item" p:good="chicken"
    p:weight="5.66"></bean>
<util:map id="itmap">
    <entry key="poke" value-ref="it1">
    </entry>
    <entry key="chicken" value-ref="it2">
    </entry>
</util:map>
```

```xml
<!-- map集合筛选 -->
<bean id="list8" class="com.waylau.spring.el.ShopList"
    p:name="#{tax.getName()?: 'defaultTax'}"
    p:count="2" p:price="1" p:onlyOne="#{itmap['chicken']}" />

<!-- 读取.properties文件中的属性 -->
<util:properties id="itprop" location="classpath:spel.properties" />
<bean id="list9" class="com.waylau.spring.el.ShopList"
    p:name="#{itprop['username']}"
    p:price="1" />

<bean id="list10" class="com.waylau.spring.el.ShopList"
    p:items="#{its.?[weight lt 3.5]}" />
<bean id="list11" class="com.waylau.spring.el.ShopList"
    p:allGood="#{its.![good]}" />
<bean id="list12" class="com.waylau.spring.el.ShopList"
    p:allGood="#{its.?[weight gt 3.5].![good]}" />
</beans>
```

3. spel.properties文件

定义 spel.properties，用于演示读取 .properties 文件的场景。

```
username=waylau
password=123456
email=waylau521@gmail.com
```

4. 定义应用类Application

Application 类定义如下。

```java
package com.waylau.spring.el;

import org.springframework.context.ApplicationContext;
import org.springframework.context.support.ClassPathXmlApplicationContext;

public class Application {

    public static void main(String[] args) {
        @SuppressWarnings("resource")
        ApplicationContext ctx =
            new ClassPathXmlApplicationContext("spring.xml");

        ShopList list = (ShopList) ctx.getBean("list");
        System.out.println(list);

        list = (ShopList) ctx.getBean("list2");
```

```
            System.out.println(list);

            list = (ShopList) ctx.getBean("list3");
            System.out.println(list);

            list = (ShopList) ctx.getBean("list4");
            System.out.println(list);

            list = (ShopList) ctx.getBean("list5");
            System.out.println(list);

            list = (ShopList) ctx.getBean("list6");
            System.out.println(list);

            list = (ShopList) ctx.getBean("list7");
            System.out.println(list);

            list = (ShopList) ctx.getBean("list8");
            System.out.println(list);

            list = (ShopList) ctx.getBean("list9");
            System.out.println(list);

            list = (ShopList) ctx.getBean("list10");
            System.out.println(list);

            list = (ShopList) ctx.getBean("list11");
            System.out.println(list);

            list = (ShopList) ctx.getBean("list12");
            System.out.println(list);
    }
}
```

5. 运行应用

运行 Application 类,即可在控制台中看到如下信息。

```
ShopList [name=shanpoo, count=2, price=3.6500000000000004, items=[], onlyOne=null, allGood=null]
ShopList [name=shanpoo, count=2, price=3.6500000000000004, items=[], onlyOne=null, allGood=null]
ShopList [name=zh_CN, count=2, price=1.0, items=[], onlyOne=null, allGood=null]
ShopList [name=zh_CN, count=2, price=1.0, items=[], onlyOne=null, allGood=null]
ShopList [name=defaultTax, count=2, price=1.0, items=[], only-One=null, allGood=null]
```

```
ShopList [name=defaultTax, count=2, price=1.0, items=[Item
[good=poke, weight=3.34], Item [good=chicken, weight=5.66], Item
[good=dark, weight=3.64], Item [good=egg, weight=2.54]], only-
One=null, allGood=null]
ShopList [name=defaultTax, count=2, price=1.0, items=[], only-
One=Item [good=poke, weight=3.34], allGood=null]
ShopList [name=defaultTax, count=2, price=1.0, items=[], only-
One=Item [good=chicken, weight=5.66], allGood=null]
ShopList [name=waylau, count=0, price=1.0, items=[], only-
One=null, allGood=null]
ShopList [name=null, count=0, price=0.0, items=[Item [good=poke,
weight=3.34], Item [good=egg, weight=2.54]], onlyOne=null, all-
Good=null]
ShopList [name=null, count=0, price=0.0, items=[], onlyOne=null,
allGood=[poke, chicken, dark, egg]]
ShopList [name=null, count=0, price=0.0, items=[], onlyOne=null,
allGood=[chicken, dark]]
```

第4章 Spring单元测试

　　TDD（Test-Driven Development，测试驱动开发）方法要求开发人员开发功能代码之前，先编写单元测试用例代码。真正的单元测试通常运行得非常快，所以花费时间编写测试用例可以提升整个开发周期的效率。TDD是敏捷开发中的一项核心实践和技术，也是一种设计方法论。

　　Spring框架提供了Mock对象和测试支持类，有助于更轻松地进行单元测试。

4.1 Mock对象

Mock 测试就是在测试过程中，对于某些不容易构造或不容易获取的对象，用一个虚拟对象来创建以便测试的测试方法。这个虚拟对象就是 Mock 对象。Mock 对象就是真实对象在调试期间的代替品。

4.1.1 Environment

org.springframework.mock.env 包中包含 Environment 和 PropertySource 抽象的 Mock 实现。MockEnvironment 和 MockPropertySource 对于开发代码时需要依赖于特定环境属性的容器外测试很有用。

4.1.2 JNDI

org.springframework.mock.jndi 包中包含了 JNDI SPI 的实现，可以使用该实现为测试套件或独立应用程序设置简单的 JNDI 环境。例如，如果 JDBC DataSources 在测试代码中与 Java EE 容器中的 JNDI 绑定到相同的 JNDI 名称，则可以在测试场景中同时复用应用程序代码和配置，而无须进行修改。

4.1.3 Servlet API

org.springframework.mock.web 包中包含一组全面的 Servlet API Mock 对象，可用于测试 Web 上下文、控制器和过滤器。这些 Mock 对象是针对 Spring Web MVC 框架特意设计的，因此通常比动态 Mock 对象技术的 EasyMock 或替代 Servlet API Mock 对象技术的如 MockObjects 更方便使用。

Spring 5 的 Mock 对象是基于 Servlet 4.0 API 的。

4.2 测试工具类

4.2.1 测试工具

org.springframework.test.util 包中包含几个用于单元测试和集成测试的工具类。

ReflectionTestUtils 是基于反射的工具类集合。借助这个工具类的集合，开发人员可以在测试中按需更改常量值、设置非 public 字段、调用非 public 配置方法等，如以下场景。

（1）访问 ORM 框架（如 JPA 和 Hibernate 等）的 private 或 protected 字段。

（2）在 Spring 用 @Autowired、@Inject 和 @Resource 等注解的 private 或 protected 字段、setter 方法和配置方法时提供依赖注入。

（3）访问使用了 @PostConstruct 和 @PreDestroy 等注解的生命周期回调方法。

AopTestUtils 是 AOP 相关工具类的集合。这些方法可以用来获取隐藏在一个或多个 Spring 代理后面的底层目标对象的引用。

4.2.2　测试Spring Web MVC

org.springframework.test.web 包中包含 ModelAndViewAssert，可以将其与 JUnit、TestNG 或其他测试框架结合使用，来处理 Spring MVC ModelAndView 对象的单元测试。

有关 Spring Web MVC 的测试，将在第 5 章中详细介绍。

第5章 Spring集成测试

集成测试也称组装测试或联合测试。在单元测试的基础上，将所有模块按照设计要求组装成子系统或系统，进行集成测试。虽然单元测试保障了每个类、每个函数都可以正常工作，但并不能保证这些类或函数连接起来也能正常工作。

Spring框架的spring-test模块为集成测试提供了一流的支持。该Spring测试不依赖于应用程序服务器或其他部署环境。这样的测试比单元测试的运行速度要慢，但比等效的Selenium测试或依赖部署到应用程序服务器的远程测试要快得多。

5.1 集成测试概述

Spring 提供了 Spring TestContext 框架来支持注解驱动的单元测试和集成测试。TestContext 框架不受所使用的实际测试框架的影响，因此允许在各种环境中进行测试，包括 JUnit、TestNG 等。

Spring 的集成测试支持以下主要功能。

（1）在测试执行之前管理 Spring IoC 容器缓存。

（2）提供测试夹具实例的依赖注入。

（3）提供适合集成测试的事务管理。

（4）提供 Spring 的特定基类，以帮助开发人员编写集成测试用例。

5.1.1 上下文管理与缓存

在 Spring 应用中，初次启动应用往往比较耗时，不是因为 Spring 本身的开销，而是因为 Spring 容器需要时间来实例化对象。例如，具有 50~100 个 Hibernate 映射文件的项目可能需要 10~20 秒来加载映射文件。这种耗时同样体现在集成测试中。因为每个测试夹具运行测试之前产生的成本会导致整体测试运行速度变慢，从而降低开发人员的生产力，所以缓存上下文变得尤为重要。Spring TestContext 框架提供了 Spring ApplicationContexts 和 WebApplicationContexts 的一致加载及这些上下文的缓存。

默认情况下，一旦加载，就要为每个测试重新使用配置的 ApplicationContext。因此，每个测试套件只需要进行一次设置，后续的测试执行速度就会变快。在这种情况下，测试套件（Test Suite）意味着所有的测试都运行在同一个 JVM 中。使用 TestContext 框架可以查看上下文管理和上下文缓存。

5.1.2 测试夹具的依赖注入

当 TestContext 框架加载应用程序上下文时，可以通过依赖注入来选择性地配置测试类的实例。其优势是，可以在各种测试场景（如配置 Spring 管理的对象图、事务代理、数据源等）中重复使用应用程序上下文，从而避免为个别测试案例复制复杂的测试夹具设置。

5.1.3 事务管理

使用真实数据库进行测试所带来的一个常见问题是，会对持久性存储状态产生影响。即使开发人员正在使用开发数据库，对状态的更改也可能影响将来的测试。另外，许多操作（如插入或修改持久性数据）不能在事务之外执行（或验证）。

TestContext 框架解决了这个问题。默认情况下，框架会为每个测试创建并回滚事务。开发人员只需编写假定交易存在的代码即可。如果开发人员在测试中调用事务代理对象，那么这些对象将根据其配置的事务语义正确行为。另外，如果一个测试方法在测试管理的事务内部执行了删除表数据的动作，那么事务将默认回滚，数据库将在测试执行之前返回删除前的状态。应用程序上下文中定义的 PlatformTransactionManager bean 给事务管理提供了支持。

5.1.4 集成测试类

Spring TestContext 框架提供了一些支持用于简化集成测试的基础类。这些基础类为测试框架提供了定义良好的钩子方法，还有一些便利的实例变量和方法，使用户能够方便地访问。

（1）ApplicationContext：用于从整体上进行显式 bean 查找或测试上下文的状态。

（2）JdbcTemplate：用于执行 SQL 语句来查询数据库。这些查询可用于确认应用程序执行数据库相关操作前后数据库的状态，并且 Spring 保证这些查询与应用程序代码在同一个事务作用域中执行。如果需要与 ORM 工具协同使用，必须确保不会误报。

此外，开发人员可用特定的实例和方法来创建自定义的、应用程序范围内的超类。

5.1.5 JDBC

org.springframework.test.jdbc 是包含 JdbcTestUtils 的包，它是一个 JDBC 相关的工具方法集，意在简化标准数据库测试场景。JdbcTestUtils 提供了以下静态工具方法。

（1）countRowsInTable(…)：统计给定表的行数。

（2）countRowsInTableWhere(…)：使用提供的 where 语句进行筛选，统计给定表的行数。

（3）deleteFromTables(…)：删除特定表的全部数据。

（4）deleteFromTableWhere(…)：使用提供的 where 语句进行筛选并删除给定表的数据。

（5）dropTables(…)：删除指定的表。

注 意，AbstractTransactionalJUnit4SpringContextTests 和 AbstractTransactionalTestNGSpringContextTests 为前面所述的 JdbcTestUtils 提供了简便的方法。

spring-jdbc 模块提供了配置和启动嵌入式数据库的支持，可用在与数据库交互的集成测试中。

5.2 测试相关的注解

Spring 框架提供了以下特定的 Spring 注解集合，可以在单元和集成测试中结合 TestContext 框架使用。

5.2.1 @BootstrapWith

@BootstrapWith 是一个类级别的注解，用于配置如何来引导 Spring TestContext 框架启动。具体来说，@BootstrapWith 用于指定一个自定义的 TestContextBootstrapper。

5.2.2 @ContextConfiguration

@ContextConfiguration 定义了为集成测试加载和配置 ApplicationContext 的类级元数据。具体来说，@ContextConfiguration 用于声明应用程序上下文资源位置，或加载上下文的注解类。

资源位置通常位于类路径中的 XML 配置文件或 Groovy 脚本中，注解类通常是 @Configuration 类。

```
@ContextConfiguration("/test-config.xml")
public class XmlApplicationContextTests {
    // …
}

@ContextConfiguration(classes = TestConfig.class)
public class ConfigClassApplicationContextTests {
    // …
}
```

作为声明资源路径或注解类的替代方案或补充，@ContextConfiguration 可以用于声明 ApplicationContextInitializer 类。

```
@ContextConfiguration(initializers = CustomContextIntializer.class)
public class ContextInitializerTests {
    // …
}
```

@ContextConfiguration 偶尔也被用作声明 ContextLoader 策略。但要注意，通常不需要显式地配置加载器，因为默认的加载器已经支持资源路径或注解类及初始化器。

```
@ContextConfiguration(locations = "/test-context.xml",
    loader = CustomContextLoader.class)
public class CustomLoaderXmlApplicationContextTests {
    // …
}
```

5.2.3 @WebAppConfiguration

@WebAppConfiguration 是一个类级别的注解，用于声明集成测试加载的 ApplicationContext 是一个 WebApplicationContext。测试类的 @WebAppConfiguration 注解只是为了保证用于测试的 WebApplicationContext 会被加载，它使用默认值 file:src/main/webapp 作为 Web 应用的根路径（资

源基路径）。资源基路径用于在后台创建一个 MockServletContext 作为测试的 WebApplication-Context 的 ServletContext。

```
@ContextConfiguration
@WebAppConfiguration
public class WebAppTests {
    // …
}
```

可以通过隐式属性值指定不同的基本资源路径，支持 classpath: 和 file: 资源前缀。如果没有提供资源前缀，则路径被视为文件系统资源。

```
@ContextConfiguration
@WebAppConfiguration("classpath:test-web-resources")
public class WebAppTests {
    // …
}
```

注意，@WebAppConfiguration 必须与 @ContextConfiguration 一起使用。

5.2.4 @ContextHierarchy

@ContextHierarchy 是一个为集成测试定义 ApplicationContext 层次结构的类级别的注解，但也可用于测试类的层次结构。@ContextHierarchy 可声明一个或多个 @ContextConfiguration 实例列表。下面的例子展示了在同一个测试类中，@ContextHierarchy 的不同使用方法。

```
@ContextHierarchy({
@ContextConfiguration("/parent-config.xml"),
@ContextConfiguration("/child-config.xml")
})
public class ContextHierarchyTests {
    // …
}

@WebAppConfiguration
@ContextHierarchy({
@ContextConfiguration(classes = AppConfig.class),
@ContextConfiguration(classes = WebConfig.class)
})
public class WebIntegrationTests {
    // …
}
```

5.2.5 @ActiveProfiles

@ActiveProfiles 是一个类级别的注解，用于在集成测试加载 ApplicationContext 时声明哪些 bean 定义的 profiles 是应该处于激活状态的。

```
@ContextConfiguration
@ActiveProfiles("dev")
public class DeveloperTests {
    // …
}

@ContextConfiguration
@ActiveProfiles({"dev", "integration"})
public class DeveloperIntegrationTests {
    // …
}
```

5.2.6 @TestPropertySource

@TestPropertySource 是一个类级别的注解，用于配置 properties 文件的位置和内联属性，以将其添加到 Environment 的 PropertySources 集合中。测试属性源比那些从系统环境或 Java 系统属性，以及通过 @PropertySource 或编程方式声明增加的属性源具有更高的优先级。而且，内联属性比从资源路径加载的属性具有更高的优先级。

以下示例展示了如何从类路径中声明属性文件。

```
@ContextConfiguration
@TestPropertySource("/test.properties")
public class MyIntegrationTests {
    // …
}
```

以下示例展示了如何声明内联属性。

```
@ContextConfiguration
@TestPropertySource(properties = { "timezone = GMT", "port: 4242" })
public class MyIntegrationTests {
    // …
}
```

5.2.7 @DirtiesContext

@DirtiesContext 指明测试执行期间，该 Spring 应用程序上下文已经被"弄脏"，也就是被通过某种方式更改或破坏，如被更改了单例 bean 的状态。当应用程序上下文被标为"脏"（Dirty）时，它将从测试框架缓存中被移除并关闭。Spring 容器将为随后同样需要配置元数据的测试重建应用程序上下文。

@DirtiesContext 可以在同一个类或类层次结构的类级别和方法级别中使用。在这个场景下，应用程序上下文将在任意注解的方法之前或之后，以及当前测试类之前或之后被标为"脏"，这取决于配置的 methodMode 和 classMode。

以下示例解释了在多种配置场景下，什么时候上下文会被标为"脏"。

当在一个类中声明并将类模式设置为 BEFORE_CLASS 时，表示在当前测试类之前。

```
@DirtiesContext(classMode = BEFORE_CLASS)
public class FreshContextTests {
    // …
}
```

当在一个类中声明并将类模式设置为 AFTER_CLASS 或什么也不加（默认的类模式）时，表示在当前测试类之后。

```
@DirtiesContext
public class ContextDirtyingTests {
    // …
}
```

当在一个类中声明并将类模式设置为 BEFORE_EACH_TEST_METHOD 时，表示在当前测试类的每个方法之前。

```
@DirtiesContext(classMode = BEFORE_EACH_TEST_METHOD)
public class FreshContextTests {
    // …
}
```

当在一个类中声明并将类模式设置为 AFTER_EACH_TEST_METHOD 时，表示在当前测试类的每个方法之后。

```
@DirtiesContext(classMode = AFTER_EACH_TEST_METHOD)
public class ContextDirtyingTests {
    // …
}
```

当在一个方法中声明并将方法模式设置为 BEFORE_METHOD 时，表示在当前方法之前。

```
@DirtiesContext(methodMode = BEFORE_METHOD)
@Test
public void testProcessWhichRequiresFreshAppCtx() {
    // …
}
```

当在一个方法中声明并将方法模式设置为 AFTER_METHOD 或什么也不加（默认的方法模式）

时，表示在当前方法之后。

```
@DirtiesContext
@Test
public void testProcessWhichDirtiesAppCtx() {
    // …
}
```

如果 @DirtiesContext 被配置为 @ContextHierarchy 所定义的上下文层次中的一部分，则 hierarchyMode 标志可用于控制如何清除上下文缓存。默认将使用一个穷举算法来清除当前层次和与当前测试拥有共同祖先的其他上下文层次的缓存。所有拥有共同祖先的子层次应用程序上下文都会从上下文中被移除并关闭。如果穷举算法在某些场景下显得有点大材小用，那么可以指定一个更简单的当前层次算法来代替，如以下代码。

```
@ContextHierarchy({
@ContextConfiguration("/parent-config.xml"),
@ContextConfiguration("/child-config.xml")
})
public class BaseTests {
    // …
}

public class ExtendedTests extends BaseTests {
    @Test
    @DirtiesContext(hierarchyMode = CURRENT_LEVEL)
    public void test() {
        // …
    }
}
```

5.2.8　@TestExecutionListeners

@TestExecutionListeners 定义了用于配置 TestConecutionListener 实现的类级元数据，该实现应该在 TestContextManager 中注册。通常，@TestExecutionListeners 与 @ContextConfiguration 应结合使用。

```
@ContextConfiguration
@TestExecutionListeners({CustomTestExecutionListener.class,
    AnotherTestExecutionListener.class})
public class CustomTestExecutionListenerTests {
    // …
}
```

@TestExecutionListeners 默认支持继承的监听器。

5.2.9 @Commit

@Commit 表示在测试方法完成后，事务性测试方法的事务应该被提交。@Commit 可以用作 @Rollback(false) 的直接替换，以便更明确地传达代码的意图。类似于 @Rollback，@Commit 也可以被声明为类级别或方法级别的注解。

```
@Commit
@Test
public void testProcessWithoutRollback() {
    // …
}
```

5.2.10 @Rollback

@Rollback 表示测试方法完成后，是否应该回滚事务测试方法的事务。如果为 true 则事务回滚；否则事务被提交（见 @Commit）。即使未明确声明 @Rollback，Spring TestContext 框架中集成测试的回滚语义也会默认为 true。

当声明为类级注解时，@Rollback 将为测试类层次结构中的所有测试方法定义默认的回滚语义。当声明为方法级别的注解时，@Rollback 将为特定的测试方法定义回滚语义，可能会覆盖类级别的 @Rollback 或 @Commit 语义。

```
@Rollback(false)
@Test
public void testProcessWithoutRollback() {
    // …
}
```

5.2.11 @BeforeTransaction

在配置了 @Transactional 注解的事务中运行的测试方法启动事务之前，应该先执行带 @BeforeTransaction 注解的方法。该方法是一个没有返回值的 void 方法。从 Spring 4.3 开始，在基于 Java 8 的接口默认方法中声明 @BeforeTransaction，可以不使用 public。

```
@BeforeTransaction
void beforeTransaction() {
    // …
}
```

5.2.12 @AfterTransaction

当在配置了 @Transactional 注解的事务中运行的测试方法结束事务之后, 应该先执行带 @After-

Transaction 注解的方法。从 Spring 4.3 开始，在基于 Java 8 的接口默认方法中声明 @AfterTransaction，可以不使用 public。

```
@AfterTransaction
void afterTransaction() {
    // …
}
```

5.2.13 @Sql

@Sql 用于注解测试类或测试方法，以便在集成测试期间配置针对给定数据库执行的 SQL 脚本。

```
@Test
@Sql({"/test-schema.sql", "/test-user-data.sql"})
public void userTest {
    // …
}
```

5.2.14 @SqlConfig

@SqlConfig 定义用于确定如何解析和执行通过 @Sql 注解配置的 SQL 脚本。

```
@Test
@Sql(
scripts = "/test-user-data.sql",
config = @SqlConfig(commentPrefix = "'", separator = "@@")
)
public void userTest {
    // …
}
```

5.2.15 @SqlGroup

@SqlGroup 是一个集合了几个 @Sql 注解的容器注解。可以在本地使用 @SqlGroup，声明几个嵌套的 @Sql 注解，或者将其与 Java 8 对可重复注解的支持结合使用。其中 @Sql 可以简单地在相同的类或方法上多次声明，隐式地生成此容器注解。

5.2.16 Spring JUnit 4 注解

以下注解仅在与 SpringRunner、Spring 的 JUnit 4 规则或 Spring 的 JUnit 4 支持类一起使用时才受支持。

1. @IfProfileValue

@IfProfileValue 表示对特定的测试环境启用了注解测试。如果配置的 ProfileValueSource 为所提供的名称返回匹配值，则测试将启用；否则，测试将被禁用。

@IfProfileValue 可以应用于类或方法级别。@IfProfileValue 可以应用于类或方法级别，在类级别上要优先于在当前类或其子类的方法级别上。有 @IfProfileValue 注解意味着测试被隐式开启，这与 JUnit4 的 @Ignore 注解类似，区别是使用 @Ignore 注解将禁用测试。

```
@IfProfileValue(name="java.vendor", value="Oracle Corporation")
@Test
public void testProcessWhichRunsOnlyOnOracleJvm() {
    // …
}
```

或者可以配置 @IfProfileValue 的 values 列表（或语义），实现类似于 JUnit 4 环境中的 TestNG 对测试组的支持。

```
@IfProfileValue(name="test-groups", values={"unit-tests", "integra-
tiontests"})
@Test
public void testProcessWhichRunsForUnitOrIntegrationTestGroups() {
    // …
}
```

2. @ProfileValueSourceConfiguration

@ProfileValueSourceConfiguration 是类级别注解，用于当获取通过 @IfProfileValue 配置的 profile 值时，指定使用什么样的 ProfileValueSource 类型。如果一个测试没有指定 @ProfileValueSource-Configuration，那么默认使用 SystemProfileValueSource。

```
@ProfileValueSourceConfiguration(CustomProfileValueSource.class)
public class CustomProfileValueSourceTests {
    // …
}
```

3. @Timed

@Timed 用于指明被注解的测试必须在指定时限（毫秒）内结束。如果测试超过指定时限，就判定测试失败。

时限不仅包括测试方法本身所耗费的时间，还包括所有重复（见 @Repeat）及所有初始化和销毁所用的时间。

```
@Timed(millis=1000)
public void testProcessWithOneSecondTimeout() {
    // …
}
```

Spring 的 @Timed 注解与 JUnit 4 的 @Test(timeout=...) 支持相比具有不同的语义。确切地说，在 JUnit 4 中，如果一个测试方法执行时间太长，@Test(timeout=...) 将直接判定该测试失败。而 Spring 的 @Timed 注解不是直接判定测试失败，而是等待测试完成。

4. @Repeat

@Repeat 用于指明测试方法需被重复执行的次数。重复的范围不仅包括测试方法自身，还包括相应的初始化方法和销毁方法。

```
@Repeat(10)
@Test
public void testProcessRepeatedly() {
    // …
}
```

5.2.17 Spring JUnit Jupiter注解

以下注解仅在与 SpringExtension 和 JUnit Jupiter（JUnit 5 中的编程模型）一起使用时才能生效。

1. @SpringJUnitConfig

@SpringJUnitConfig 是一个组合的注解，它将 JUnit Jupiter 的 @ExtendWith(SpringExtension.class) 与 Spring TestContext 框架中的 @ContextConfiguration 结合在一起，可以在类级别用作 @ContextConfiguration 的替代实现。关于配置选项，@ContextConfiguration 和 @SpringJUnitConfig 的唯一区别是，@SpringJUnitConfig 中的 value 属性可以声明带注解的类。

```
@SpringJUnitConfig(TestConfig.class)
class ConfigurationClassJUnitJupiterSpringTests {
    // …
}
@SpringJUnitConfig(locations = "/test-config.xml")
class XmlJUnitJupiterSpringTests {
    // …
}
```

2. @SpringJUnitWebConfig

@SpringJUnitWebConfig 是一个组合的注解，它将 JUnit Jupiter 的 @ExtendWith(SpringExtension.class) 与 Spring TestContext 框架中的 @ContextConfiguration 和 @WebAppConfiguration 结合在一起，可以在类级别用作 @ContextConfiguration 和 @WebAppConfiguration 的替代实现。关于配置选项，@ContextConfiguration 和 @SpringJUnitWebConfig 的唯一区别是，@SpringJUnitWebConfig 中的 value 属性可以声明带注解的类。另外，@WebAppConfiguration 的 value 属性只能通过 @SpringJUnitWebConfig 中的 resourcePath 属性覆盖。

```
@SpringJUnitWebConfig(TestConfig.class)
class ConfigurationClassJUnitJupiterSpringWebTests {
```

```
    // …
}

@SpringJUnitWebConfig(locations = "/test-config.xml")
class XmlJUnitJupiterSpringWebTests {
    // …
}
```

3. @EnabledIf

@EnabledIf 用于表示已注解的 JUnit Jupiter 测试类或测试方法已启用,将在所提供的表达式计算结果为 true 时执行。具体来说,如果表达式的计算结果为 Boolean.TRUE 或一个等于"true"的字符串(忽略大小写),则测试将被启用。在类级别应用时,该类中的所有测试方法也会默认自动启用。

表达式可以是以下任意一种。

(1) SpEL 表达式:@EnabledIf("#{systemProperties['os.name'].toLowerCase().contains('mac')}")。

(2) Spring 环境中可用属性的占位符:@EnabledIf("${smoke.tests.enabled}")。

(3) 文本文字:@EnabledIf("true")。

注意,@EnabledIf("false") 等同于 @Disabled,@EnabledIf 可以用作元注解来创建自定义组合注解。例如,可以按以下方式创建自定义的 @EnabledOnMac 注解。

```
@Target({ElementType.TYPE, ElementType.METHOD})
@Retention(RetentionPolicy.RUNTIME)
@EnabledIf(
expression = "#{systemProperties['os.name'].toLowerCase().con-
tains('mac')}",
reason = "Enabled on Mac OS"
)
public @interface EnabledOnMac {}
```

4. @DisabledIf

@DisabledIf 用于表示已注解的 JUnit Jupiter 测试类或测试方法被禁用,并且如果提供表达式的计算结果为 true,则不执行。具体而言,如果表达式的计算结果为 Boolean.TRUE 或等于"true"的字符串(忽略大小写),则测试将被禁用。在类级别应用时,该类中的所有测试方法也会自动被禁用。

表达式可以是以下任意一种。

(1) SpEL 表达式:@DisabledIf("#{systemProperties['os.name'].toLowerCase().contains('mac')}")。

(2) Spring 环境中可用属性的占位符:@DisabledIf("${smoke.tests.disabled}")。

(3) 文本文字:@DisabledIf("true")。

注意,@DisabledIf("true") 等同于 @Disabled,@DisabledIf 可以用作元注解来创建自定义组合注解。例如,可以按以下方式创建自定义的 @DisabledOnMac 注解。

```
@Target({ElementType.TYPE, ElementType.METHOD})
@Retention(RetentionPolicy.RUNTIME)
@DisabledIf(
expression = "#{systemProperties['os.name'].toLowerCase().con-
tains('mac')}",
reason = "Disabled on Mac OS"
)
public @interface DisabledOnMac {}
```

5.3 Spring TestContext框架

Spring TestContext 框架是用于进行单元测试和集成测试的通用框架。它基于注解驱动，并且与所使用的具体测试框架无关。

5.3.1 Spring TestContext框架概述

Spring TestContext 框架在 org.springframework.test.context 包中，提供了对通用的、注解驱动的单元测试和集成测试的支持。TestContext 框架也非常重视约定大于配置，合理的默认值可以通过基于注解的配置来覆盖。

除了通用的测试基础架构外，TestContext 框架还为 JUnit 4、JUnit Jupiter（JUnit 5）和 TestNG 提供了明确的支持。对于 JUnit 4 和 TestNG，Spring 提供了抽象的支持类。此外，Spring 为 JUnit 4 提供了一个自定义的 JUnit Runner 和 JUnit 规则，以及 JUnit Jupiter 的一个自定义扩展，允许编写基于 POJO 的测试类。POJO 测试类不需要扩展特定的类层次结构，如抽象支持类等。

5.3.2 核心抽象

Spring TestContext 框架的核心由 TestContextManager 类和 TestContext、TestExecutionListener 及 SmartContextLoader 接口组成。每个测试类都会创建一个 TestContextManager，TestContextManager 反过来管理一个 TestContext 来保存当前测试的上下文，它还会在进行测试时更新 TestContext 的状态，并委托给 TestExecutionListener 实现；TestExecutionListener 实现通过提供依赖注入、管理事务等来实际执行测试；SmartContextLoader 负责为给定的测试类加载一个 ApplicationContext。

1. TestContext

TestContext 封装了执行测试的上下文，与正在使用的实际测试框架无关，并为其负责的测试实例提供上下文管理和缓存支持。如果需要，TestContext 可委托 SmartContextLoader 来加载 Application-Context。

2. TestContextManager

TestContextManager 是 Spring TestContext 框架的主要接入点，负责管理单个 TestContext，并在定义良好的测试执行点向每个注册的 TestExecutionListener 发信号通知事件。这些执行点如下。

（1）在特定测试框架的所有 before class 或 before all 方法之前。

（2）测试实例后。

（3）在特定测试框架的所有 before 或 before each 方法之前。

（4）在测试方法执行之前，但在测试设置之后。

（5）在测试方法执行之后，但在测试关闭之前。

（6）在任何一个特定测试框架的每个 after 或 after each 方法之后。

（7）在任何一个特定测试框架的每个 after class 或 after all 方法之后。

3. TestExecutionListener

TestExecutionListener 定义了 API，用于响应 TestContextManager 发布的测试执行事件，并与监听器一起注册。

4. ContextLoader

ContextLoader 是在 Spring 2.5 中引入的策略接口，主要用于在使用 Spring TestContext 框架管理集成测试时加载 ApplicationContext。

SmartContextLoader 是在 Spring 3.1 中引入的 ContextLoader 接口的扩展。SmartContextLoader SPI 取代了 Spring 2.5 中引入的 ContextLoader SPI。具体来说，SmartContextLoader 可以选择处理资源的位置，以及注解类或上下文初始值。此外，SmartContextLoader 可以在加载的上下文中设置用于激活 bean 定义的配置文件和测试属性源。

Spring 提供了以下实现。

（1）DelegatingSmartContextLoader：根据为测试类声明的配置、默认位置或默认配置类，在内部委派 AnnotationConfigContextLoader、GenericXmlContextLoader 或 GenericGroovyXmlContext Loader 中的一个。Groovy 支持仅在 Groovy 位于类路径中时才能启用。

（2）WebDelegatingSmartContextLoader：根据为测试类声明的配置、默认位置或默认配置类，在内部委派 AnnotationConfigWebContextLoader、GenericXmlWebContextLoader 或 GenericGroovyXml-WebContextLoader 中的一个。只有在测试类中存在 @WebAppConfiguration 时，才会使用 Web 的 ContextLoader。Groovy 支持仅在 Groovy 位于类路径中时才能启用。

（3）AnnotationConfigContextLoader：从注解类加载标准的 ApplicationContext。

（4）AnnotationConfigWebContextLoader：从注解类加载 WebApplicationContext。

（5）GenericGroovyXmlContextLoader：从 Groovy 脚本或 XML 配置文件的资源位置加载标准的 ApplicationContext。

（6）GenericGroovyXmlWebContextLoader：从 Groovy 脚本或 XML 配置文件的资源位置加载

WebApplicationContext。

（7）GenericXmlContextLoader：从 XML 资源位置加载标准的 ApplicationContext。

（8）GenericXmlWebContextLoader：从 XML 资源位置加载 WebApplicationContext。

（9）GenericPropertiesContextLoader：从 Java 属性文件中加载标准的 ApplicationContext。

5.3.3　引导TestContext

对于所有常见的用例来说，Spring TestContext 框架内部的默认配置都是足够的。但是，有时开发团队或第三方框架想要更改默认的 ContextLoader，实现自定义的 TestContext 或 ContextCache，增加默认的 ContextCustomizerFactory 和 TestExecutionListener 实现集等。此时，为了实现这些，Spring 提供了一个引导 TestContext 策略。

TestContextBootstrapper 定义了用于引导 TestContext 框架的 SPI。TestContextBootstrapper 被 TestContextManager 用来加载当前测试的 TestExecutionListener 实现，并构建它所管理的 TestContext。可以通过 @BootstrapWith 为测试类（或测试类层次结构）配置自定义引导策略。如果引导程序未通过 @BootstrapWith 显式配置，则将使用 DefaultTestContextBootstrapper 或 WebTestContextBootstrapper，具体取决于是否存在 @WebAppConfiguration。由于 TestContextBootstrapper SPI 未来可能会发生变化以适应新的需求，因此强烈建议开发者不要直接实现此接口，而是扩展 AbstractTestContextBootstrapper 或其具体子类之一。

5.3.4　TestExecutionListener配置

Spring 提供了以下默认注册的 TestExecutionListener 实现，顺序如下。

（1）ServletTestExecutionListener：为 WebApplicationContext 配置 Servlet API 模拟。

（2）DirtiesContextBeforeModesTestExecutionListener：处理之前模式的 @DirtiesContext 注解。

（3）DependencyInjectionTestExecutionListener：为测试实例提供依赖注入。

（4）DirtiesContextTestExecutionListener：处理 after 模式的 @DirtiesContext 注解。

（5）TransactionalTestExecutionListener：使用默认回滚语义提供事务性测试执行。

（6）SqlScriptsTestExecutionListener：执行通过 @Sql 注解配置的 SQL 脚本。

5.3.5　上下文管理

每个 TestContext 为其所负责的测试实例提供上下文管理和缓存支持。测试实例不会自动获得对配置 ApplicationContext 的访问权限。但是，如果一个测试类实现了 ApplicationContextAware 接口，那么该测试实例提供对 ApplicationContext 的引用。

提示: AbstractJUnit4SpringContextTests 和 AbstractTestNGSpringContextTests 实现了 Application-

ContextAware，因此可以自动提供对 ApplicationContext 的访问。

作为实现 ApplicationContextAware 接口的替代方法，可以通过字段或 setter 方法中的 @Autowired 注解为测试类注入应用程序上下文。例如：

```
@RunWith(SpringRunner.class)
@ContextConfiguration
public class MyTest {
@Autowired
private ApplicationContext applicationContext;
    // …
}
```

同样，如果测试配置为加载 WebApplicationContext，则可以将 Web 应用程序上下文注入自己的测试中。例如：

```
@RunWith(SpringRunner.class)
@WebAppConfiguration
@ContextConfiguration
public class MyWebAppTest {

    @Autowired
    private WebApplicationContext wac;
    // …
}
```

使用 TestContext 框架的测试类不需要扩展任何特定的类或实现特定的接口来配置其应用程序上下文，配置是通过在类级别声明 @ContextConfiguration 注解来实现的。如果测试类没有显式声明应用程序上下文资源位置或注解类，则配置的 ContextLoader 将确定如何从默认位置或默认配置类加载上下文。除了上下文资源位置和注解类以外，还可以通过应用程序上下文初始化程序来配置应用程序上下文。

5.3.6 测试夹具的依赖注入

当使用 DependencyInjectionTestExecutionListener（默认配置）时，开发人员根据测试实例的依赖关系，将 bean 注入使用 @ContextConfiguration 配置的应用程序上下文中。开发人员可以使用 setter 注入或字段注入，这取决于他选择的注解及是否将它们放置在 setter 方法或字段上。为了与 Spring 2.5 和 Spring 3.0 中引入的注解支持保持一致，可以使用 Spring 的 @Autowired 注解或 JSR-330 中的 @Inject 注解。

TestContext 框架不会检测测试实例的实例化方式，因此，对构造函数使用 @Autowired 或 @Inject 对测试类没有影响。

因为 @Autowired 被用来按类型执行自动装配，所以如果有多个相同类型的 bean 定义，那么就不能依靠这种方法来实现这些特定的 bean。在这种情况下，可以使用 @Autowired 和 @Qualifier。

从 Spring 3.0 开始，可以将 @Inject 和 @Named 一起使用。如果开发人员的测试类可以访问其 ApplicationContext，那么也可以调用 applicationContext.getBean（titleRepository）来执行显式查找。

如果不想将依赖注入应用于自己的测试实例，则不要使用 @Autowired（或 @Inject）注解字段或设置方法。或者，可以通过使用 @TestExecutionListeners 显式配置类并从监听器列表中省略 DependencyInjectionTestExecutionListener.class 来完全禁用依赖注入。

以下代码演示了在字段方法上使用 @Autowired 的方法。

```java
@RunWith(SpringRunner.class)
@ContextConfiguration("repository-config.xml")
public class HibernateTitleRepositoryTests {
    // 根据类型来注入实例
    @Autowired
    private HibernateTitleRepository titleRepository;

    @Test
    public void findById() {
        Title title = titleRepository.findById(new Long(10));
        assertNotNull(title);
    }
}
```

也可以将类配置为使用 @Autowired 进行 setter 注入。例如：

```java
@RunWith(SpringRunner.class)
@ContextConfiguration("repository-config.xml")
public class HibernateTitleRepositoryTests {
    // 根据类型来注入实例
    private HibernateTitleRepository titleRepository;

    @Autowired
    public void setTitleRepository(HibernateTitleRepository titleRepository)
    {
        this.titleRepository = titleRepository;
    }

    @Test
    public void findById() {
        Title title = titleRepository.findById(new Long(10));
        assertNotNull(title);
    }
}
```

5.3.7 如何测试request bean和session bean

很早之前，Spring 就已经支持 request 和 session scope 的 bean。从 Spring 3.2 开始，测试 request bean 和 session bean 成为一件轻而易举的事情。

以下代码展示了登录用例的 XML 配置。

```xml
<beans>
    <bean id="userService" class="com.example.SimpleUserService"
        c:loginAction-ref="loginAction"/>
    <bean id="loginAction" class="com.example.LoginAction"
        c:username="#{request.getParameter('user')}"
        c:password="#{request.getParameter('pswd')}"
        scope="request">
    <aop:scoped-proxy/>
    </bean>
</beans>
```

在 RequestScopedBeanTests 中，将 UserService 和 MockHttpServletRequest 都注入自己的测试实例中。在 requestScope() 测试方法中，可以通过在提供的 MockHttpServletRequest 中设置请求参数来设置测试工具。当在 UserService 中调用 loginUser() 方法时，可以确信用户服务访问了当前 MockHttpServletRequest 请求范围的 loginAction（刚才设置的参数）。然后，根据已知的用户名和密码对结果执行断言。

```java
@RunWith(SpringRunner.class)
@ContextConfiguration
@WebAppConfiguration
public class RequestScopedBeanTests {
    @Autowired UserService userService;
    @Autowired MockHttpServletRequest request;

    @Test
    public void requestScope() {
        request.setParameter("user", "enigma");
        request.setParameter("pswd", "$pr!ng");
        LoginResults results = userService.loginUser();
        // 断言结果
    }
}
```

session bean 的测试类似，其代码为：

```xml
<beans>
    <bean id="userService" class="com.example.SimpleUserService"
        c:userPreferences-ref="userPreferences" />
    <bean id="userPreferences" class="com.example.UserPreferences"
        c:theme="#{session.getAttribute('theme')}"
```

```
        scope="session">
    <aop:scoped-proxy/>
    </bean>
</beans>
```

```
@RunWith(SpringRunner.class)
@ContextConfiguration
@WebAppConfiguration
public class SessionScopedBeanTests {
    @Autowired UserService userService;
    @Autowired MockHttpSession session;

    @Test
    public void sessionScope() throws Exception {
        session.setAttribute("theme", "blue");
        Results results = userService.processUserPreferences();
        // …断言结果
    }
}
```

5.3.8 事务管理

在 TestContext 框架中，事务由默认配置的 TransactionalTestExecutionListener 管理，即使没有在测试类上显式声明 @TestExecutionListener。但是，为了支持事务，必须在通过 @ContextConfiguration 语义加载的 ApplicationContext 中配置一个 PlatformTransactionManager bean。另外，必须在类或方法级别为测试声明 Spring 的 @Transactional 注解。

1. 测试管理的事务

测试管理的事务是通过 TransactionalTestExecutionListener 声明式管理的事务，或者通过 TestTransaction 以编程方式进行管理的事务。这样的事务不能与 Spring 管理的事务（被加载用于测试的 ApplicationContext 内的 Spring 直接管理的事务）或应用程序管理的事务（通过测试调用的应用程序代码内的程序管理的事务）混淆。

2. 启用和禁用事务

默认情况下，测试完成后会自动回滚。如果一个测试类用 @Transactional 注解，则该类层次结构中的每个测试方法将在一个事务中运行；如果没有用 @Transactional（在类或方法级别）注解，则测试方法将不会在事务中运行。此外，使用 @Transactional 进行注解，但将传播类型设置为 NOT_SUPPORTED 的测试不会在事务中运行。

注意，AbstractTransactionalJUnit4SpringContextTests 和 AbstractTransactionalTestNGSpringContextTests 是为类级别的事务支持预配置的。

以下示例演示了为基于 Hibernate 的 UserRepository 编写的集成测试方案。正如在事务回滚和提交行为中所解释的，在执行 createUser() 方法后，不需要清理数据库，因为对数据库所做的任何更改都将由 TransactionalTestExecutionListener 自动回滚。

```
@RunWith(SpringRunner.class)
@ContextConfiguration(classes = TestConfig.class)
@Transactional
public class HibernateUserRepositoryTests {
    @Autowired
    HibernateUserRepository repository;
    @Autowired
    SessionFactory sessionFactory;
    JdbcTemplate jdbcTemplate;

    @Autowired
    public void setDataSource(DataSource dataSource) {
        this.jdbcTemplate = new JdbcTemplate(dataSource);
    }

    @Test
    public void createUser() {
        final int count = countRowsInTable("user");
        User user = new User(…);
        repository.save(user);
        sessionFactory.getCurrentSession().flush();
        assertNumUsers(count + 1);
    }

    protected int countRowsInTable(String tableName) {
        return JdbcTestUtils.countRowsInTable(this.jdbcTemplate, table-
Name);
    }

    protected void assertNumUsers(int expected) {
        assertEquals("Number of rows in the [user] table.", expected,
        countRowsInTable("user"));
    }
}
```

3. 事务回滚和提交行为

默认情况下，测试完成后会自动回滚测试。然而，事务提交和回滚行为也可以通过 @Commit 和 @Rollback 注解进行声明式配置。

4. 编程式事务管理

自 Spring 4.1 开始,可以通过 TestTransaction 的静态方法,以编程方式和测试托管的事务进行交互，例如, TestTransaction 可用于 test 、before、after 方法,以开始或结束当前测试管理的事务, 或者配置当前测试管理的事务以进行回滚或提交。无论何时启用 TransactionalTestExecutionListener, 都可以使用

TestTransaction。

以下示例演示了 TestTransaction 的一些功能。

```java
@ContextConfiguration(classes = TestConfig.class)
public class ProgrammaticTransactionManagementTests extends
    AbstractTransactionalJUnit4SpringContextTests {

    @Test
    public void transactionalTest() {
        assertNumUsers(2);
        deleteFromTables("user");
        TestTransaction.flagForCommit();
        TestTransaction.end();
        assertFalse(TestTransaction.isActive());
        assertNumUsers(0);
        TestTransaction.start();
        // …
    }

    protected void assertNumUsers(int expected) {
        assertEquals("Number of rows in the [user] table.", expected,
            countRowsInTable("user"));
    }
}
```

5. 在事务之外执行代码

有时需要在事务性测试方法之前或之后执行某些代码，并且要求这些代码是在事务性上下文之外的。例如，在执行测试之前验证初始数据库状态，或者在测试执行之后验证预期的事务性执行行为。TransactionalTestExecutionListener 完全支持这种场景，并提供了 @BeforeTransaction 和 @AfterTransaction 注解来实现这些行为。

6. 配置事务管理器

TransactionalTestExecutionListener 需要在 Spring ApplicationContext 中为测试定义一个 PlatformTransactionManager bean。如果在测试的 ApplicationContext 中有多个 PlatformTransactionManager 实例，则可以通过 @Transactional("myTxMgr") 或 @Transactional(transactionManager = "myTx- Mgr") 声明限定符，或者通过 @Configuration 类来实现 TransactionManagementConfigurer。

以下示例基于 JUnit 4 来突出显示所有与事务相关的注解。

```java
@RunWith(SpringRunner.class)
@ContextConfiguration
@Transactional(transactionManager = "txMgr")
@Commit
public class FictitiousTransactionalTest {
    @BeforeTransaction
    void verifyInitialDatabaseState() {
```

```
    // …
}
@Before
public void setUpTestDataWithinTransaction() {
    // …
}
@Test
@Rollback
public void modifyDatabaseWithinTransaction() {
    // …
}
@After
public void tearDownWithinTransaction() {
    // …
}
@AfterTransaction
void verifyFinalDatabaseState() {
    // …
}
}
```

5.3.9 执行SQL脚本

在针对关系数据库编写集成测试方案时，执行 SQL 脚本来修改数据库模式或将测试数据插入表中是很常见的。spring-jdbc 模块提供了在加载 Spring ApplicationContext 时，通过执行 SQL 脚本来初始化现有数据库的支持。当然，这些数据库也包括嵌入式的数据库。

尽管在加载 ApplicationContext 时初始化数据库进行测试是非常有用的，但有时在集成测试期间能够修改数据库也非常重要。下面介绍如何在集成测试期间以编程方式和声明方式执行 SQL 脚本。

1. 编程式执行SQL脚本

Spring 提供了以下选项，用于在集成测试方法中以编程方式执行 SQL 脚本。

- org.springframework.jdbc.datasource.init.ScriptUtils
- org.springframework.jdbc.datasource.init.ResourceDatabasePopulator
- org.springframework.test.context.junit4.AbstractTransactionalJUnit4SpringContextTests
- org.springframework.test.context.testng.AbstractTransactionalTestNGSpringContextTests

ScriptUtils 提供了一组用于处理 SQL 脚本的静态工具方法，主要用于框架内部。但是，如果需要完全控制 SQL 脚本的解析和执行方式，那么 ScriptUtils 可能会比下面介绍的其他替代方法更能满足用户需求。

ResourceDatabasePopulator 提供了一个简单的基于对象的 API，使用外部资源中定义的 SQL 脚本，以编程方式填充、初始化或清理数据库。ResourceDatabasePopulator 提供了配置分析和执行脚本时使用的字符编码、语句分隔符、注解分隔符和错误处理标志的选项，每个配置选项都有一个合理的默认值。要执行在 ResourceDatabasePopulator 中配置的脚本，可以调用 populate(Connection) 方法来针对 java.sql.Connection 执行 populator，或者调用 execute(DataSource) 方法来针对 javax.sql.DataSource 执行 populator。以下示例为测试模式和测试数据指定了 SQL 脚本，将语句分隔符设置为"@@"，然后针对数据源执行脚本。

```
@Test
public void databaseTest {
    ResourceDatabasePopulator populator = new ResourceDatabasePopula-
tor();
    populator.addScripts(
    new ClassPathResource("test-schema.sql"),
    new ClassPathResource("test-data.sql"));
    populator.setSeparator("@@");
    populator.execute(this.dataSource);
    // …
}
```

ResourceDatabasePopulator 内部其实也是使用 ScriptUtils 来解析和执行 SQL 脚本的。Abstract-TransactionalJUnit4SpringContextTests 和 AbstractTransactionalTestNGSpringContextTests 中的 executeSqlScript(..) 方法在内部使用 ResourceDatabasePopulator 来执行 SQL 脚本。

2. 声明式执行SQL脚本

Spring TestContext 框架同时也提供了声明式执行 SQL 脚本。具体而言，可以在测试类或测试方法上声明 @Sql 注解，以便将资源路径配置为在集成测试方法之前或之后，针对给定数据库执行 SQL 脚本。

注意，方法级声明会覆盖类级声明，而对 @Sql 的支持则由默认情况下启用的 SqlScriptsTestExecutionListener 提供。

每个路径资源将被解释为一个 Spring 资源。一个普通路径（如 schema.sql）将被视为与定义测试类的包相关的类路径资源。以斜杠开始的路径将被视为绝对类路径资源，如 /org/example/schema.sql。可以使用指定的资源协议来加载引用 URL 的路径，如以 classpath:、file:、http: 等为前缀的路径。

以下示例演示了如何在基于 JUnit Jupiter 的集成测试类中，在类级别和方法级别上使用 @Sql。

```
@SpringJUnitConfig
@Sql("/test-schema.sql")
class DatabaseTests {
    @Test
    void emptySchemaTest {
        // …
    }

    @Test
    @Sql({"/test-schema.sql", "/test-user-data.sql"})
    void userTest {
```

```
        // …
    }
}
```

如果没有指定 SQL 脚本，将尝试根据声明的 @Sql 的位置来自动检测默认脚本。如果无法检测到默认值，则会抛出 IllegalStateException 异常。

（1）类级别声明：如果注解的测试类为 com.example.MyTest，则相应的默认脚本为 classpath:com/example/MyTest.sql。

（2）方法级别声明：如果注解的测试方法名为 testMethod()，并且在类 com.example.MyTest 中定义，则相应的默认脚本为 classpath:com/example/MyTest.testMethod.sql。如果需要为给定的测试类或测试方法配置多组 SQL 语句，但具有不同的语法配置、不同的错误处理规则或每个集合处于不同的执行阶段，则可以声明多个 @Sql 实例。对于 Java 8，@Sql 可以用作可重复的注解；否则，@SqlGroup 注解可以用于声明多个 @Sql 实例的显式容器。

以下示例演示了如何将 @Sql 用作 Java 8 的可重复注解。在这种情况下，test-schema.sql 脚本对单行注解使用不同的语法。

```
@Test
@Sql(scripts = "/test-schema.sql", config = @SqlConfig(commentPrefix = "'"))
@Sql("/test-user-data.sql")
public void userTest {
    // …
}
```

以下示例和以上示例是一样的。不同的是，@Sql 声明在 @SqlGroup 中被组合在一起，以便与 Java 6 和 Java 7 兼容。

```
@Test
@SqlGroup({
@Sql(scripts = "/test-schema.sql", config = @SqlConfig(commentPrefix ="'")),
@Sql("/test-user-data.sql")
)}
public void userTest {
    // …
}
```

默认情况下，SQL 脚本将在相应的测试方法之前执行。但是，如果需要在测试方法之后执行特定的一组脚本（如清理数据库状态），则可以使用 @Sql 中的 executionPhase 属性，如以下示例。

```
@Test
@Sql(
scripts = "create-test-data.sql",
config = @SqlConfig(transactionMode = ISOLATED)
```

```
)
@Sql(
scripts = "delete-test-data.sql",
config = @SqlConfig(transactionMode = ISOLATED),
executionPhase = AFTER_TEST_METHOD
)
public void userTest {
    // …
}
```

其中，ISOLATED 和 AFTER_TEST_METHOD 分别从 Sql.TransactionMode 和 Sql.Execution Phase 中静态导入。

5.4　Spring MVC Test框架

Spring MVC Test 框架可以与 JUnit、TestNG 或任何其他测试框架一起使用，来测试 Spring Web MVC 代码。由于 Spring MVC Test 框架建立在 spring-test 模块的 Servlet API Mock 对象上，因此可以不依赖所运行的 Servlet 容器。它使用 DispatcherServlet 来提供完整的 Spring Web MVC 运行时的行为，并提供对使用 TestContext 框架加载实际的 Spring 配置及独立模式的支持。在独立模式下，可以手动实例化控制器并进行测试。

Spring MVC Test 还为使用 RestTemplate 的代码提供了客户端支持。客户端测试模拟服务器响应时，也不再依赖所运行的服务器。

5.4.1　服务端测试概述

使用 JUnit 或 TestNG 为 Spring MVC 控制器编写简单的单元测试，只需实例化控制器，为其注入 Mock 或 Stub 的依赖关系，并根据需要调用 MockHttpServletRequest、MockHttpServletResponse 等方法。但是，在编写这样的单元测试时，还有很多部分没有经过测试，如请求映射、数据绑定、类型转换、验证等。此外，其他控制器方法（如 @InitBinder、@ModelAttribute 和 @ExceptionHandler）也可能作为请求处理生命周期的一部分被调用。

Spring MVC Test 的目标是通过执行请求并通过实际的 DispatcherServlet 生成响应来为测试控制器提供一种有效的方法。Spring MVC Test 建立在 spring-test 模块中的 Mock 实现上，允许执行请求并生成响应，而不需要在 Servlet 容器中运行。在大多数情况下，所有操作都应该如同运行时一样工作，并且能覆盖一些单元测试无法覆盖的场景。

以下是一个基于 JUnit Jupiter 使用 Spring MVC Test 的例子。

```java
import static
    org.springframework.test.web.servlet.request.MockMvcRequestBuild-
ers.*;
import static
    org.springframework.test.web.servlet.result.MockMvcResultMatchers.*;
@SpringJUnitWebConfig(locations = "test-servlet-context.xml")
class ExampleTests {
    private MockMvc mockMvc;
    @BeforeEach
    void setup(WebApplicationContext wac) {
        this.mockMvc = MockMvcBuilders.webAppContextSetup(wac).build();
    }
    @Test
    void getAccount() throws Exception {
        this.mockMvc.perform(get("/accounts/1")
            .accept(MediaType.parseMediaType("application/json;
            charset=UTF-8")))
            .andExpect(status().isOk())
            .andExpect(content().contentType("application/json"))
            .andExpect(jsonPath("$.name").value("Lee"));
    }
}
```

以上测试依赖于 TestContext 框架的 WebApplicationContext 支持，用于从位置与测试类相同的包中的 XML 配置文件加载 Spring 配置。当然，也支持基于 Java 和 Groovy 的配置。

在该例子中，MockMvc 实例用于对 /accounts/1 执行 GET 请求，并验证结果响应的状态为 200，内容类型为 application/json，响应主体是一个属性为 name、值为 Lee 的 JSON。jsonPath 语法是通过 Jayway 的 JsonPath 项目 1 支持的。

该例子中的测试 API 是需要静态导入的，如 MockMvcRequestBuilders、MockMvcResult Matchers. 和 MockMvcBuilders。找到这些类的简单方法是搜索匹配 "MockMvc" 的类型。

5.4.2 选择测试策略

创建 MockMvc 实例有以下两种方式。第一种是通过 TestContext 框架加载 Spring MVC 配置。该框架加载 Spring 配置，并将 WebApplicationContext 注入测试中，用于构建 MockMvc 实例。

```java
@RunWith(SpringRunner.class)
@WebAppConfiguration
@ContextConfiguration("my-servlet-context.xml")
public class MyWebTests {
    @Autowired
    private WebApplicationContext wac;
    private MockMvc mockMvc;

    @Before
```

```
    public void setup() {
        this.mockMvc = MockMvcBuilders.webAppContextSetup(this.wac).
            build();
    }
    // …
}
```

第二种是简单地创建一个控制器实例,而不加载 Spring 配置。相对于 MVC JavaConfig 或 MVC 命名空间而言,默认基本的配置是自动创建的,并且可以在一定程度上进行自定义。

```
public class MyWebTests {
    private MockMvc mockMvc;
    @Before
    public void setup() {
        this.mockMvc =
        MockMvcBuilders.standaloneSetup(new AccountController()).
            build();
    }
    // …
}
```

那么,这两种方式应该如何抉择呢?

第一种方式也称 webAppContextSetup,会加载实际的 Spring MVC 配置,从而得到更完整的集成测试。由于 TestContext 框架缓存了加载的 Spring 配置,因此即使在测试套件中引入更多的测试,也可以保持测试的快速运行。此外,还可以通过 Spring 配置将 Mock 服务注入控制器中,以便专注于测试 Web 层。下面是一个用 Mockito 声明 Mock 服务的例子。

```
<bean id="accountService" class="org.mockito.Mockito" facto-
ry-method="mock">
    <constructor-arg value="com.waylau.AccountService"/>
</bean>
```

然后,可以将 Mock 服务注入测试,以便设置和验证期望值。

```
@RunWith(SpringRunner.class)
@WebAppConfiguration
@ContextConfiguration("test-servlet-context.xml")
public class AccountTests {
    @Autowired
    private WebApplicationContext wac;
    private MockMvc mockMvc;

    @Autowired
    private AccountService accountService;
        // …
}
```

第二种方式也称 standaloneSetup,其更接近于单元测试。它一次测试一个控制器,控制器可以手动注入 Mock 依赖关系，而不涉及加载 Spring 配置。这样的测试更注重风格，更容易查看哪个控制器正在被测试、是否需要特定的 Spring MVC 配置等。standaloneSetup 方式也是一种非常方便的方式——编写临时测试来验证特定行为或调试问题。

到底选择哪种方式没有绝对的答案。但是，使用 standaloneSetup 就意味着需要用额外的 webAppContextSetup 测试来验证 Spring MVC 配置，或者可以使用 webAppContextSetup 方式编写所有测试，以便始终根据实际的 Spring MVC 配置进行测试。

5.4.3 设置测试功能

无论使用哪个MockMvc 构建器所有MockMvcBuilder实现都提供了一些常用的且非常有用的功能。例如，可以为所有请求声明 accept 头，并期望所有响应中的状态为 200 及声明 contentType 头，例如：

```
MockMVc mockMvc = standaloneSetup(new MusicController())
    .defaultRequest(get("/").accept(MediaType.APPLICATION_JSON))
    .alwaysExpect(status().isOk())
    .alwaysExpect(content().contentType("application/json;char-set=UTF-8"))
    .build();
```

此外，第三方框架和应用程序可以通过MockMvcConfigurer预先打包安装指令。Spring 框架有一个内置实现，有助于跨请求保存和重用 HTTP 会话，可以使用如下代码。

```
MockMvc mockMvc = MockMvcBuilders.standaloneSetup(new TestController())
    .apply(sharedHttpSession())
    .build();
```

5.4.4 执行请求

使用任何 HTTP 方法来执行请求都是很容易的，例如：

```
mockMvc.perform(post("/hotels/{id}", 42).accept(MediaType.APPLICATION_JSON));
```

可以使用MockMultipartHttpServletRequest 来实现文件的上传请求，也可以执行内部使用请求，这样就不需要实际解析多重请求，而只需设置这些请求，例如：

```
mockMvc.perform(multipart("/doc").file("a1", "ABC".getBytes("UTF-8")));
```

可以在 URI 模板样式中指定查询参数，例如：

```
mockMvc.perform(get("/hotels?foo={foo}", "bar"));
```

或者添加表示表单参数查询的 Servlet 请求参数，例如：

```
mockMvc.perform(get("/hotels").param("foo", "bar"));
```

如果应用程序代码依赖于 Servlet 请求参数,并且不会显式检查被查询字符串(最常见的情况),那么使用哪个选项并不重要。

大多数情况下,最好从请求 URI 中省略上下文路径和 Servlet 路径。如果必须使用完整请求 URI 进行测试,则须设置相应的 contextPath 和 servletPath,以便请求映射可正常工作。

```
mockMvc.perform(get("/app/main/hotels/{id}")
    .contextPath("/app")
    .servletPath("/main"))
```

由于在每个执行请求中设置 contextPath 和 servletPath 是相当烦琐的,因此,可以通过设置通用的默认请求属性来减少设置。

```
public class MyWebTests {
    private MockMvc mockMvc;
    @Before
    public void setup() {
    mockMvc = standaloneSetup(new AccountController())
        .defaultRequest(get("/")
        .contextPath("/app").servletPath("/main")
        .accept(MediaType.APPLICATION_JSON).build();
    }
    // …
}
```

上述设置将影响通过 MockMvc 实例执行的每个请求。如果在给定的请求中也指定了相同的属性,它将覆盖默认值。

5.4.5 定义期望

期望值可以通过在执行请求后附加一个或多个 .andExpect(...) 来定义。

```
mockMvc.perform(get("/accounts/1")).andExpect(status().isOk());
```

MockMvcResultMatchers.* 提供了许多期望,其分为以下两类。

(1)断言验证响应的属性。例如,响应状态、标题和内容,这些是最重要的结果。

(2)断言检查相应结果。这些断言允许检查 Spring Web MVC 的特定方面,如是哪个控制器方法在处理请求、是否引发和处理了异常、模型的内容是什么、选择了什么视图、添加了哪些 flash 属性等。

它们还允许检查 Servlet 的特定方面,如请求和会话属性。

以下测试断言绑定或验证失败。

```
mockMvc.perform(post("/persons"))
```

```
.andExpect(status().isOk())
.andExpect(model().attributeHasErrors("person"));
```

在编写测试时,很多时候打印出执行请求的结果是很有用的。如以下示例,其中 print() 是从 MockMvcResultHandlers 静态导入的。

```
mockMvc.perform(post("/persons"))
    .andDo(print())
    .andExpect(status().isOk())
    .andExpect(model().attributeHasErrors("person"));
```

只要请求处理不会导致未处理的异常,print() 方法就会将所有可用的打印结果数据发送到 System.out。Spring Framework 4.2 引入了 log() 方法和 print() 方法的两个额外变体:一个接收 OutputStream,另一个接收 Writer。例如,调用 print(System.err) 会将打印结果数据发送到 System.err,而调用 print(myWriter) 会将结果数据打印到一个自定义写入器。如果希望将结果数据记录下来而不是打印出来,只需调用 log() 方法,该方法会将结果数据记录为 org.springframework.test.web.servlet.result 日志记录类别下的单个 DEBUG 消息。

在某些情况下,可能希望直接访问结果来验证其他方式无法验证的内容,这时可以通过在期望之后追加 .andReturn() 来实现。

```
MvcResult mvcResult = mockMvc.perform(post("/persons")).andExpect(status ()
.isOk()).andReturn();
```

如果所有测试都重复相同的期望,那么在构建 MockMvc 实例时,可以设置一个共同期望。

```
standaloneSetup(new SimpleController())
    .alwaysExpect(status().isOk())
    .alwaysExpect(content().contentType("application/json;charset=UTF-8"))
    .build()
```

当 JSON 响应内容包含使用 Spring HATEOAS 创建的超媒体链接时,可以使用 JsonPath 表达式验证生成的链接。

```
mockMvc.perform(get("/people").accept(MediaType.APPLICATION_JSON))
.andExpect(jsonPath("$.links[?(@.rel == 'self')].href")
.value("http://localhost:8080/people"));
```

当 XML 响应内容包含使用 Spring HATEOAS 创建的超媒体链接时,可以使用 XPath 表达式验证生成的链接。

```
Map<String, String> ns =
    Collections.singletonMap("ns", "http://www.w3.org/2005/Atom");
mockMvc.perform(get("/handle").accept(MediaType.APPLICATION_XML))
```

```
.andExpect(xpath("/person/ns:link[@rel='self']/@href", ns)
    .string("http://localhost:8080/people"));
```

5.4.6 注册过滤器

设置 MockMvc 实例时，可以注册一个或多个 Servlet 过滤器实例。

```
mockMvc = standaloneSetup(new PersonController())
    .addFilters(new CharacterEncodingFilter()).build();
```

已注册的过滤器将通过来自 spring-test 的 MockFilterChain 调用，最后一个过滤器将委托给 DispatcherServlet。

5.4.7 脱离容器的测试

正如之前提到的，Spring MVC Test 是建立在 spring-test 模块的 Servlet API Mock 对象之上的，并且不使用正在运行的 Servlet 容器。所以脱离容器的测试，与运行在实际客户端和服务器的完整的端到端集成测试相比，存在一些差异。

开发人员的测试往往是从一个空的 MockHttpServletRequest 开始的。无论添加什么内容到测试中，默认情况下都没有上下文路径、没有 jsessionid cookie、没有转发、没有错误或异步调度，也没有实际的 JSP 呈现。相反，"转发"和"重定向"的 URL 被保存在 MockHttpServletResponse 中，并且可以被期望所断言。这意味着如果使用的是 JSP，就可以验证请求被转发到的 JSP 页面，但不会有任何 HTML 呈现。换句话说，JSP 将不会被调用。但是要注意，所有其他不依赖于转发的呈现技术（如 Thymeleaf 和 Freemarker）都会按照预期将 HTML 呈现给响应主体。通过 @ResponseBody 方法呈现 JSON、XML 和其他格式也是如此，或者可以考虑通过 @WebIntegrationTest 从 Spring Boot 进行完整的端到端集成测试。

每种方法都有优点和缺点。Spring MVC Test 所提供的选项范围，和从经典单元测试到完整集成测试的选项范围是不同的。可以肯定的是，Spring MVC Test 中没有任何选项属于经典单元测试的范畴，只是作用与经典单元测试有点接近。例如，可以通过向控制器注入 Mock 服务来隔离 Web 层，在这种情况下，虽然只是通过 DispatcherServlet 测试 Web 层，但可以使用实际的 Spring 配置，或者一次使用专注于一个控制器的独立设置，并手动提供其所需的配置。

5.4.8 实战：服务端测试

下面新建了一个 mvc-test 应用，用于演示服务端测试。

1. 导入相关的依赖

导入与 Servlet、Spring Test、JUnit 相关的依赖，其示例为：

```xml
<dependencies>
    <dependency>
        <groupId>org.springframework</groupId>
        <artifactId>spring-context</artifactId>
        <version>${spring.version}</version>
    </dependency>
    <dependency>
        <groupId>org.springframework</groupId>
        <artifactId>spring-webmvc</artifactId>
        <version>${spring.version}</version>
    </dependency>
    <dependency>
        <groupId>javax.servlet</groupId>
        <artifactId>javax.servlet-api</artifactId>
        <version>${servlet.version}</version>
        <scope>provided</scope>
    </dependency>
    <dependency>
        <groupId>org.springframework</groupId>
        <artifactId>spring-test</artifactId>
        <version>${spring.version}</version>
        <scope>test</scope>
    </dependency>
    <dependency>
        <groupId>junit</groupId>
        <artifactId>junit</artifactId>
        <version>${junit.version}</version>
        <scope>test</scope>
    </dependency>
</dependencies>
```

2. 定义控制器

创建一个 HelloController 控制器，用于处理 HTTP 请求，其示例为：

```java
package com.waylau.spring.hello.controller;

import org.springframework.web.bind.annotation.RequestMapping;
import org.springframework.web.bind.annotation.RestController;

@RestController
public class HelloController {

    @RequestMapping("/hello")
    public String hello() {
```

```
        return "Hello World! Welcome to visit waylau.com!";
    }
}
```

当访问 /hello 接口时,应返回 "Hello World! Welcome to visit waylau.com!" 字符串。

3. 配置文件

定义 Spring 应用的配置文件 spring.xml,其示例为:

```xml
<?xml version="1.0" encoding="UTF-8"?>
<beans xmlns="http://www.springframework.org/schema/beans"
    xmlns:xsi="http://www.w3.org/2001/XMLSchema-instance"
    xmlns:context="http://www.springframework.org/schema/context"
    xmlns:mvc="http://www.springframework.org/schema/mvc"
    xsi:schemaLocation="
        http://www.springframework.org/schema/beans
        http://www.springframework.org/schema/beans/spring-beans.xsd
        http://www.springframework.org/schema/context
        http://www.springframework.org/schema/context/spring-context.xsd
        http://www.springframework.org/schema/mvc
        http://www.springframework.org/schema/mvc/spring-mvc.xsd">

    <mvc:annotation-driven/>
    <context:component-scan base-package="com.waylau.spring.*"/>

</beans>
```

上面的示例启用了 Spring MVC 的注解。

4. 编写测试类

测试类 HelloControllerTest 的代码如下。

```java
package com.waylau.spring.hello.controller;

import org.junit.Before;
import org.junit.Test;
import org.junit.runner.RunWith;
import org.springframework.beans.factory.annotation.Autowired;
import org.springframework.http.MediaType;
import org.springframework.test.context.ContextConfiguration;
import org.springframework.test.context.junit4.SpringJUnit4ClassRunner;
import org.springframework.test.context.web.WebAppConfiguration;
import org.springframework.test.web.servlet.MockMvc;
import org.springframework.test.web.servlet.setup.MockMvcBuilders;
import org.springframework.web.context.WebApplicationContext;
import static
    org.springframework.test.web.servlet.request.MockMvcRequestBuilders.get;
import static
```

```java
        org.springframework.test.web.servlet.result.MockMvcResultMatchers.content;
import static 
        org.springframework.test.web.servlet.result.MockMvcResultMatchers.status;

@RunWith(SpringJUnit4ClassRunner.class)
@ContextConfiguration("classpath:spring.xml")
@WebAppConfiguration
public class HelloControllerTest {

    private MockMvc mockMvc;

    @Autowired
    private WebApplicationContext webApplicationContext;

    @Before
    public void setUp() throws Exception {
        mockMvc = MockMvcBuilders
                .webAppContextSetup(webApplicationContext)
                .build();
    }

    @Test
    public void testHello() throws Exception {
        mockMvc.perform(get("/hello")
                .accept(MediaType
                        .parseMediaType("application/json;charset=UTF-8")))
                .andExpect(status().isOk())
                .andExpect(content().
                        contentType("application/json;charset=UTF-8"))
                .andExpect(content()
                        .string("Hello World! Welcome to visit waylau.com!"));
    }
}
```

5. 运行用例

使用 JUnit 运行 HelloControllerTest 类，能看到测试结果为绿色，代码测试成功。

第6章 Spring事务管理

在关系数据库中，一个事务可以是一条SQL语句、一组SQL语句或整个程序。事务是恢复和并发控制的基本单位。本章将详细介绍Spring的事务管理。

6.1 事务管理概述

事务具有 4 个属性：原子性、一致性、隔离性和持久性。这 4 个属性通常称为 ACID 特性。

（1）原子性（Atomicity）：指一个事务是不可分割的工作单位，其中的所有操作要么都做，要么都不做。

（2）一致性（Consistency）：指事务必须使数据库从一个一致性状态变到另一个一致性状态。一致性与原子性是密切相关的。

（3）隔离性（Isolation）：指一个事务的执行不能被其他事务干扰，即一个事务内部的操作及使用的数据与并发的其他事务是隔离的，并发执行的各个事务之间不能互相干扰。

（4）持久性（Durability）：持久性也称为永久性（Permanence），指一个事务一旦提交，其对数据库中数据的改变就是永久性的，后面的其他操作或故障不应该对其有任何影响。

6.1.1 Spring事务管理优势

Spring 框架支持全面的事务管理，它为事务管理提供了一致的抽象，具有以下优势。

（1）跨越不同事务 API 的一致编程模型，如 Java 事务 API（JTA）、JDBC、Hibernate 和 Java 持久性 API（JPA）。

（2）支持声明式事务管理。

（3）用于编程式事务管理的简单 API 比复杂事务 API（如 JTA）要简单。

（4）与 Spring 的数据访问抽象有极佳的整合能力。

6.1.2 全局事务与本地事务

Java EE 开发人员对事务管理有两种选择：全局事务或本地事务，两者都有很大的局限性。下面将讨论 Spring 框架的事务管理如何支持全局事务和本地事务模型，以及这二者的局限性。

1. 全局事务

全局事务使用户能够使用多个事务资源，通常包括关系数据库和消息队列。应用程序服务器通过 JTA 管理全局事务，而 API 的使用相当烦琐。此外，JTA 的 UserTransaction 通常需要来自 JNDI，这意味着还需要使用 JNDI 才能使用 JTA。很明显，全局事务的使用将限制应用程序代码的重用，因为 JTA 通常只在应用程序服务器环境中可用。

以前，使用全局事务的首选方式是通过 EJB CMT（容器管理事务）。CMT 是一种声明式事务管理（区别于编程式事务管理）。EJB CMT 消除了与事务相关的查找 JNDI 的过程。当然，使用 EJB 本身还是需要使用 JNDI 的，JNDI 消除了大部分（但不是全部）用于控制事务的 Java 代码。EJB CMT 的主要缺点是，CMT 与 JTA 同应用服务器环境相关联。此外，只有选择在 EJB 中实现业务逻辑时，或者至少在事务性 EJB Facade 后面才可用。一般来说，EJB 的负面影响非常大，所以这不是一个推荐的选择。

2. 本地事务

本地事务可能更容易使用，但有明显的缺点，就是它们不能在多个事务资源上工作。例如，使用 JDBC 连接管理事务的代码无法在全局 JTA 事务中运行。由于应用程序服务器不参与事务管理，因此无法确保跨多个资源的事务的正确性。还有一个缺点是，本地事务对编程模型是侵入式的。

当然，大多数应用程序使用的是单个事务资源，因此本地事务大多数情况下能够满足需求。

6.1.3 Spring事务模型

Spring 解决了全局事务和本地事务的缺点。它使应用程序开发人员能够在任何环境中使用一致的编程模型，只需编写一次代码，就能够从不同环境的不同事务管理策略中受益。Spring 框架提供了声明式和编程式事务管理。

通过编程式事务管理，开发人员可以使用 Spring 框架事务抽象在任何事务基础设施上运行。使用声明式模型，开发人员通常只用写很少的或不用写与事务管理相关的代码，因此不依赖于 Spring 框架事务 API 或任何其他事务 API。

Spring 事务抽象的核心概念是事务策略。事务策略由 org.springframework.transaction.PlatformTransactionManager 接口定义。例如：

```
public interface PlatformTransactionManager {
    TransactionStatus getTransaction(TransactionDefinition definition)
    throws TransactionException;
    void commit(TransactionStatus status) throws TransactionException;
    void rollback(TransactionStatus status) throws TransactionException;
}
```

这主要是一个服务提供者接口（SPI），虽然可以通过应用程序代码以编程方式使用。由于 PlatformTransactionManager 是一个接口，因此可以根据需要轻松进行 Mock 或 Stub，不受诸如 JNDI 等查找策略的束缚。PlatformTransactionManager 实现同 Spring 框架 IoC 容器中的任何其他对象（或 bean）一样定义。单就此优势而言，即使用户使用 JTA，Spring 框架事务也是一种有价值的抽象。Spring 的事务代码可以比直接使用 JTA 更容易测试。

PlatformTransactionManager 接口的任何方法都可以抛出未检查的 TransactionException（也就是说，它扩展了 java.lang.RuntimeException 类）。应用程序开发人员可以自行选择捕获和处理 TransactionException。

getTransaction(...) 方法根据 TransactionDefinition 参数返回一个 TransactionStatus 对象。返回的 TransactionStatus 对象可能代表一个新的事务，也可能是一个已经存在的事务（如果当前调用栈中存在匹配的事务）。后一种情况的含义是，与 Java EE 事务上下文一样，TransactionStatus 与一个执行线程相关联。

TransactionDefinition 接口指定了如下定义。

（1）隔离（Isolation）：代表了事务与其他事务的分离程度。例如，这个事务可以看到来自其他事务的未提交的写入等。

传播（Propagation）：通常在事务范围内执行的所有代码都将在该事务中运行。但是，如果在事务上下文已经存在的情况下执行事务方法，则可以指定行为。例如，代码可以在现有的事务中继续运行（常见情况），或者暂停现有事务并创建新的事务。Spring 提供了 EJB CMT 所熟悉的所有事务传播选项。要了解 Spring 中事务传播的语义，请参阅 6.3.6 小节的内容。

（2）超时（Timeout）：定义了事务超时之前能够运行多久，并由事务基础设施自动回滚。

（3）只读状态（Read-only status）：当只读取代码但不修改数据时，可以使用只读事务。在某些情况下，只读事务可以是一个有用的优化，如使用 Hibernate 时。

TransactionStatus 接口为事务代码提供了控制事务执行和查询事务状态的方法，例如：

```
public interface TransactionStatus extends SavepointManager {
    boolean isNewTransaction();
    boolean hasSavepoint();
    void setRollbackOnly();
    boolean isRollbackOnly();
    void flush();
    boolean isCompleted();
}
```

PlatformTransactionManager 能够实现多种数据源的管理，如 JDBC、JTA、Hibernate 等。

以下示例展示了如何定义本地 PlatformTransactionManager 实现，首先定义一个 JDBC 数据源：

```
<bean id="dataSource" class="org.apache.commons.dbcp.BasicDataSource"
    destroy-method="close">
    <property name="driverClassName" value="${jdbc.driverClassName}" />
    <property name="url" value="${jdbc.url}" />
    <property name="username" value="${jdbc.username}" />
    <property name="password" value="${jdbc.password}" />
</bean>
```

相关的 PlatformTransactionManager bean 定义将会有一个对 DataSource 定义的引用。其代码如下：

```
<bean id="txManager"
    class="org.springframework.jdbc.datasource.DataSourceTransaction-
Manager">
    <property name="dataSource" ref="dataSource"/>
</bean>
```

6.2 通过事务实现资源同步

通过之前的介绍，相信读者已经明白如何创建不同的事务管理器，以及它们如何链接到与事务同步的相关资源上，如 DataSourceTransactionManager 如何链接到 JDBC 数据源等。本节将介绍应用程序代码如何直接或间接使用持久化 API（如 JDBC、Hibernate 或 JPA），确保能够正确创建、重用和清理这些资源。本节还将讨论如何通过相关的 PlatformTransactionManager 来触发（可选）事务同步。

6.2.1 高级别的同步方法

高级别的同步方法是首选的方法，通常是使用 Spring 基于模板的持久性集成 API，或者使用原生的 ORM API 来管理本地的资源工厂。这些事务感知型解决方案在内部处理资源创建和重用、清理、映射等，用户无须关注这些细节，可以专注于非模板化的持久性逻辑。通常，可以使用原生的 ORM API 或 JdbcTemplate，采取模板方法进行 JDBC 访问。

6.2.2 低级别的同步方法

低级别的同步方法包括 DataSourceUtils（用于 JDBC）、EntityManagerFactoryUtils（用于 JPA）、SessionFactoryUtils（用于 Hibernate）等。当用户希望应用程序代码直接处理原生持久性 API 的资源类型时，可以使用这些类来确保获得正确的 Spring 框架管理实例、进行事务（可选）同步等。

例如，在使用 JDBC 时，不是调用 JDBC 传统 DataSource 的 getConnection() 方法，而是使用 Spring 的 org.springframework.jdbc.datasource.DataSourceUtils 类，如下所示。

```
Connection conn = DataSourceUtils.getConnection(dataSource);
```

如果现有的事务已经有一个同步（链接）到它的连接，则返回该实例。否则，方法调用会触发创建一个新的连接，该连接（可选）与所有现有事务同步，并可用于同一事务的后续重用。如前所述，所有 SQLException 都被封装在 Spring 框架的 CannotGetJdbcConnectionException 中，这是 Spring 框架未检查的 DataAccessExceptions 的层次结构之一。从这种方法中获得的信息比从 SQLException 中更多，并且这种方法具有跨数据库的可移植性，还可以在没有 Spring 事务管理的情况下工作（事务同步是可选的）。因此无论是否使用 Spring 进行事务管理，都可以使用它。

当然，一旦使用了 Spring 的 JDBC、JPA 或 Hibernate 支持，通常不会再使用 DataSourceUtils 或其他帮助类，因为通过 Spring 抽象，比直接使用相关的 API 更简便。例如，如果使用 Spring JdbcTemplate 或 jdbc.object 包来简化 JDBC 的使用，无须编写任何特殊代码，就能在后台执行正确的连接检索。

6.2.3　TransactionAwareDataSourceProxy

TransactionAwareDataSourceProxy 类是最低级别的。一般情况下，几乎不会使用这个类，而是使用上面提到的更高级别的抽象来编写新的代码。

这个类是目标 DataSource 的代理，它封装了目标 DataSource 以增加对 Spring 管理事务的感知。在这方面，它类似于由 Java EE 服务器提供的事务性 JNDI 数据源。

6.3　声明式事务管理

Spring 框架的声明式事务管理是通过 Spring AOP 实现的，它与 EJB CMT 类似，可以将事务行为指定到单个方法级别。如果需要，可以在事务上下文中调用 setRollbackOnly() 方法。这两种事务管理的区别如下。

（1）与 JTA 绑定的 EJB CMT 不同，Spring 框架的声明式事务管理适用于任何环境。通过简单地调整配置文件，可以使用 JDBC、JPA 或 Hibernate 与 JTA 事务或本地事务协同工作。

（2）可以将 Spring 框架声明式事务管理应用于任何类，而不仅仅是诸如 EJB 的特殊类。

（3）Spring 框架提供了声明式的回滚规则，这是一个与 EJB 不同的特性，它提供了回滚规则的编程式和声明式支持。

（4）Spring 框架能够通过使用 AOP 来自定义事务行为。例如，可以在事务回滚的情况下插入自定义行为。使用 EJB CMT 则不同，除 setRollbackOnly() 外，不能影响容器的事务管理。

（5）Spring 框架不支持远程调用传播事务上下文。如果需要此功能，建议使用 EJB。但是，在使用这种功能之前需要仔细考虑，因为通常情况下，远程调用事务的机会非常少。

回滚规则的概念很重要，它指定了哪些异常会导致自动回滚，可以在配置中以声明方式指定。因此，尽管可以调用 TransactionStatus 对象上的 setRollbackOnly() 来回滚当前事务，但通常都会指定 MyApplicationException 必须总是导致回滚的规则。这个规则的显著优点是，业务对象不依赖于事务基础设施。例如，通常不需要导入 Spring 事务 API 或其他 Spring API。

虽然 EJB 容器默认行为会自动回滚系统异常事务（通常是运行时的异常），但 EJB CMT 不会自动回滚应用程序异常（除 java.rmi.RemoteException 外的已检查异常）的事务。虽然声明式事务管理的 Spring 默认行为遵循 EJB 约定（仅在抛出未检查的异常时自动回滚），但指定此行为通常很有用。

6.3.1　声明式事务管理

关于 Spring 框架的声明式事务支持通过 AOP 代理来启用，并且事务性的 Advice 由元数据驱动。

AOP 与事务性元数据的结合产生了 AOP 代理，该代理使用 TransactionInterceptor 和适当的 PlatformTransactionManager 实现来驱动方法调用周围的事务。

从概念上来讲，调用事务代理的流程如图 6-1 所示。

图6-1　调用事务代理的流程

6.3.2　实战：声明式事务管理

下面将创建一个声明式事务管理的示例应用 eclarative-transaction。在这个应用中，会实现一个简单的"用户管理"功能，在执行保存用户的操作时会开启事务。同时，当遇到操作异常时，也能保证事务回滚。

1. 导入相关的依赖

声明式事务管理需要导入以下依赖。

```
<properties>
    <spring.version>5.1.5.RELEASE</spring.version>
</properties>
<dependencies>
    <dependency>
        <groupId>org.springframework</groupId>
        <artifactId>spring-context</artifactId>
        <version>${spring.version}</version>
    </dependency>
    <dependency>
        <groupId>org.springframework</groupId>
        <artifactId>spring-aspects</artifactId>
        <version>${spring.version}</version>
    </dependency>
    <dependency>
        <groupId>org.springframework</groupId>
        <artifactId>spring-jdbc</artifactId>
        <version>${spring.version}</version>
    </dependency>
    <dependency>
        <groupId>org.apache.logging.log4j</groupId>
        <artifactId>log4j-core</artifactId>
        <version>2.6.2</version>
    </dependency>
    <dependency>
        <groupId>org.apache.logging.log4j</groupId>
        <artifactId>log4j-jcl</artifactId>
        <version>2.6.2</version>
```

```xml
    </dependency>
    <dependency>
        <groupId>org.apache.logging.log4j</groupId>
        <artifactId>log4j-slf4j-impl</artifactId>
        <version>2.6.2</version>
    </dependency>
    <dependency>
        <groupId>org.apache.commons</groupId>
        <artifactId>commons-dbcp2</artifactId>
        <version>2.5.0</version>
    </dependency>
    <dependency>
        <groupId>com.h2database</groupId>
        <artifactId>h2</artifactId>
        <version>1.4.196</version>
        <scope>runtime</scope>
    </dependency>
</dependencies>
```

其中使用了 JDBC 方式来连接数据库；数据库用了 H2 内嵌数据库，方便用户进行测试；日志框架采用了 Log4j 2，主要用于打印出完整的 Spring 事务执行过程。

2. 定义领域模型

定义一个代表用户信息的 User 类，例如：

```java
package com.waylau.spring.tx.vo;

public class User {
    private String username;
    private Integer age;

    public User(String username, Integer age) {
        this.username = username;
        this.age = age;
    }

    public String getUsername() {
        return username;
    }

    public void setUsername(String username) {
        this.username = username;
    }

    public Integer getAge() {
        return age;
    }

    public void setAge(Integer age) {
```

```
        this.age = age;
    }
}
```

定义服务接口 UserService，例如：

```
package com.waylau.spring.tx.service;

import com.waylau.spring.tx.vo.User;

public interface UserService {

    void saveUser(User user);
}
```

定义服务的实现类 UserServiceImpl，例如：

```
package com.waylau.spring.tx.service;

import com.waylau.spring.tx.vo.User;

public class UserServiceImpl implements UserService {

    public void saveUser(User user) {
        throw new UnsupportedOperationException(); // 模拟异常情况
    }
}
```

在服务实现类中，并没有真的把业务数据存储到数据库中，而是抛出了一个异常，来模拟数据库操作的异常。

3. 配置文件

定义 Spring 应用的配置文件 spring.xml，例如：

```
<?xml version="1.0" encoding="UTF-8"?>
<beans xmlns="http://www.springframework.org/schema/beans"
    xmlns:xsi="http://www.w3.org/2001/XMLSchema-instance"
    xmlns:context="http://www.springframework.org/schema/context"
    xmlns:aop="http://www.springframework.org/schema/aop"
    xmlns:tx="http://www.springframework.org/schema/tx"
    xsi:schemaLocation="
        http://www.springframework.org/schema/beans
        http://www.springframework.org/schema/beans/spring-beans.xsd
        http://www.springframework.org/schema/context
        http://www.springframework.org/schema/context/spring-context.xsd
        http://www.springframework.org/schema/tx
        http://www.springframework.org/schema/tx/spring-tx.xsd
```

```xml
       http://www.springframework.org/schema/aop
       http://www.springframework.org/schema/aop/spring-aop.xsd">

    <!-- 定义Aspect -->
    <aop:config>
        <aop:pointcut id="userServiceOperation"
    expression="execution(* com.waylau.spring.tx.service.UserService.*(..))"/>
        <aop:advisor advice-ref="txAdvice"
            pointcut-ref="userServiceOperation"/>
    </aop:config>

    <!-- DataSource -->
    <bean id="dataSource" class="org.apache.commons.dbcp2.BasicDataSource"
        destroy-method="close">
        <property name="driverClassName" value="org.h2.Driver"/>
        <property name="url" value="jdbc:h2:mem:testdb"/>
        <property name="username" value="sa"/>
        <property name="password" value=""/>
    </bean>

    <!-- PlatformTransactionManager -->
    <bean id="txManager"
        class="org.springframework.jdbc.datasource.DataSourceTransactionManager">
        <property name="dataSource" ref="dataSource"/>
    </bean>

    <!-- 定义事务Advice -->
    <tx:advice id="txAdvice" transaction-manager="txManager">
        <tx:attributes>
            <!-- 所有"get"开头的都是只读 -->
            <tx:method name="get*" read-only="true"/>
            <!-- 其他方法使用默认的事务设置 -->
            <tx:method name="*"/>
        </tx:attributes>
    </tx:advice>

    <!-- 定义 bean -->
    <bean id="userService"
        class="com.waylau.spring.tx.service.UserServiceImpl" />

</beans>
```

在上述配置文件中，定义了 Advice、DataSource、PlatformTransactionManager 等事务。

4. 编写主应用类

主应用类 Application 的代码如下。

```java
package com.waylau.spring.tx;

import org.springframework.context.ApplicationContext;
import org.springframework.context.support.ClassPathXmlApplicationContext;

import com.waylau.spring.tx.service.UserService;
import com.waylau.spring.tx.vo.User;

public class Application {

    public static void main(String[] args) {
        @SuppressWarnings("resource")
        ApplicationContext context = new ClassPathXmlApplicationContext("spring.xml");
        UserService UserService = context.getBean(UserService.class);
        UserService.saveUser(new User("Way Lau", 30));
    }

}
```

在 Application 类中，会执行保存用户的操作。

5. 运行应用

运行 Application 类，能看到控制台中的打印信息如下。

```
22:50:10.217 [main] DEBUG org.springframework.context.support.ClassPathXmlApplicationContext - Refreshing org.springframework.context.support.ClassPathXmlApplicationContext@33c7e1bb
22:50:11.276 [main] DEBUG org.springframework.beans.factory.xml.XmlBeanDefinitionReader - Loaded 7 bean definitions from class path resource [spring.xml]
22:50:11.377 [main] DEBUG org.springframework.beans.factory.support.DefaultListableBeanFactory - Creating shared instance of singleton bean 'org.springframework.aop.config.internalAutoProxyCreator'
22:50:11.559 [main] DEBUG org.springframework.beans.factory.support.DefaultListableBeanFactory - Creating shared instance of singleton bean 'org.springframework.aop.support.DefaultBeanFactoryPointcutAdvisor#0'
22:50:11.597 [main] DEBUG org.springframework.beans.factory.support.DefaultListableBeanFactory - Creating shared instance of singleton bean 'dataSource'
22:50:11.982 [main] DEBUG org.springframework.beans.factory.support.DefaultListableBeanFactory - Creating shared instance of singleton bean 'txManager'
22:50:11.991 [main] DEBUG org.springframework.beans.factory.support.DefaultListableBeanFactory - Creating shared instance of singleton bean 'txAdvice'
22:50:12.031 [main] DEBUG org.springframework.transaction.interceptor.NameMatchTransactionAttributeSource - Adding transactional method [get*]
```

```
with attribute [PROPAGATION_REQUIRED,ISOLATION_DEFAULT,readOnly]
22:50:12.031 [main] DEBUG org.springframework.transaction.interceptor.
NameMatchTransactionAttributeSource - Adding transactional method [*]
with attribute [PROPAGATION_REQUIRED,ISOLATION_DEFAULT]
22:50:12.032 [main] DEBUG org.springframework.beans.factory.support.
DefaultListableBeanFactory - Creating shared instance of singleton bean
'userService'
22:50:12.139 [main] DEBUG org.springframework.jdbc.datasource.Data-
SourceTransactionManager - Creating new transaction with name [com.
waylau.spring.tx.service.UserServiceImpl.saveUser]: PROPAGATION_RE
QUIRED,ISOLATION_DEFAULT
22:50:12.390 [main] DEBUG org.springframework.jdbc.datasource.Data-
SourceTransactionManager - Acquired Connection [1893960929, URL=jdb-
c:h2:mem:testdb, UserName=SA, H2 JDBC Driver] for JDBC transaction
22:50:12.393 [main] DEBUG org.springframework.jdbc.datasource.Data-
SourceTransactionManager - Switching JDBC Connection [1893960929, URL=-
jdbc:h2:mem:testdb, UserName=SA, H2 JDBC Driver] to manual commit
22:50:12.394 [main] DEBUG org.springframework.jdbc.datasource.Data-
SourceTransactionManager - Initiating transaction rollback
22:50:12.394 [main] DEBUG org.springframework.jdbc.datasource.Data-
SourceTransactionManager - Rolling back JDBC transaction on Connection
[1893960929, URL=jdbc:h2:mem:testdb, UserName=SA, H2 JDBC Driver]
22:50:12.395 [main] DEBUG org.springframework.jdbc.datasource.Data-
SourceTransactionManager - Releasing JDBC Connection [1893960929, URL=-
jdbc:h2:mem:testdb, UserName=SA, H2 JDBC Driver] after transaction
22:50:12.395 [main] DEBUG org.springframework.jdbc.datasource.DataSour-
ceUtils - Returning JDBC Connection to DataSource
Exception in thread "main" java.lang.UnsupportedOperationException
```

从上述异常信息中，能够完整地看到整个事务的管理过程，包括创建事务、获取连接，以及遇到异常后的事务回滚、连接释放等。由此可以证明，在遇到特定的异常时，是可以进行事务回滚的。

6.3.3 事务回滚

如何告知 Spring 事务的工作将要被回滚？推荐方式是从事务上下文中正在执行的代码中抛出一个异常。Spring 框架的事务基础设施代码会捕获所有未处理的异常，因为它会唤起调用堆栈，并确定是否将事务标记为回滚。

在 Spring 框架的默认配置中，事务基础设施代码仅在运行时捕获到未检查的异常才会标记为事务回滚。换言之，如果要回滚，抛出的异常一定是 RuntimeException 的一个实例或子类。Error 默认情况下也会导致回滚，但已检查的异常不会导致在默认配置中回滚。

可以精确地配置将哪些 Exception 类型标记为回滚事务，包括已检查的异常。以下 XML 片段演示了如何配置应用程序的特定异常类型的回滚。

```xml
<tx:advice id="txAdvice" transaction-manager="txManager">
    <tx:attributes>
```

```xml
        <tx:method name="get*" read-only="true"
            rollback-for="NoProductInStockException"/>
        <tx:method name="*"/>
    </tx:attributes>
</tx:advice>
```

如果不想在抛出异常时回滚事务，还可以指定 "no-rollback-for"，例如：

```xml
<tx:advice id="txAdvice">
    <tx:attributes>
        <tx:method name="updateStock"
            no-rollback-for="InstrumentNotFoundException"/>
        <tx:method name="*"/>
    </tx:attributes>
</tx:advice>
```

当 Spring 框架的事务基础设施捕获一个异常，在检查配置的回滚规则以确定是否标记回滚事务时，最强的匹配规则将胜出。因此，在以下配置中，除 InstrumentNotFoundException 外的任何异常都会导致事务的回滚。

```xml
<tx:advice id="txAdvice">
    <tx:attributes>
        <tx:method name="*" rollback-for="Throwable"
            no-rollback-for="InstrumentNotFoundException"/>
    </tx:attributes>
</tx:advice>
```

还可以编程方式指示所需的回滚。虽然非常简单，但可将代码紧密耦合到 Spring 框架的事务基础架构上。

```java
public void resolvePosition() {
    try {
        // …
    } catch (NoProductInStockException ex) {
        // 触发回滚
        TransactionAspectSupport
            .currentTransactionStatus()
            .setRollbackOnly();
    }
}
```

如果可以，建议使用声明式方法进行回滚。

6.3.4 配置不同的事务策略

如果有多个服务层对象，并且想对它们应用一个完全不同的事务配置，那么可以通过使用不同的 pointcut 和 advice-ref 属性值定义不同的元素来实现。

假定所有服务层类都是在根 com.waylau.spring.service 包中定义的，要使所有在该包（或子包）中定义的类的实例都具有默认的事务配置，可以编写以下代码。

```xml
<?xml version="1.0" encoding="UTF-8"?>
<beans xmlns="http://www.springframework.org/schema/beans"
xmlns:xsi="http://www.w3.org/2001/XMLSchema-instance"
xmlns:aop="http://www.springframework.org/schema/aop"
xmlns:tx="http://www.springframework.org/schema/tx"
xsi:schemaLocation="
http://www.springframework.org/schema/beans
http://www.springframework.org/schema/beans/spring-beans.xsd
http://www.springframework.org/schema/tx
http://www.springframework.org/schema/tx/spring-tx.xsd
http://www.springframework.org/schema/aop
http://www.springframework.org/schema/aop/spring-aop.xsd">
    <aop:config>
    <aop:pointcut id="serviceOperation"
    expression="execution(* com.waylau.spring.service..*Service.*(..))"/>
    <aop:advisor pointcut-ref="serviceOperation" advice-ref="txAdvice"/>
    </aop:config>
    <!-- 下面两个bean将会纳入事务 -->
    <bean id="fooService"
        class="com.waylau.spring.service.DefaultFooService"/>
    <bean id="barService"
        class="com.waylau.spring.service.extras.SimpleBarService"/>
    <!-- 下面两个bean将不会纳入事务 -->
    <bean id="anotherService"
        class="org.xyz.SomeService"/><!-- 没有在指定的包中 -->
    <bean id="barManager"
        class="com.waylau.spring.service.SimpleBarManager"/><!-- 没有以Service结尾-->
    <tx:advice id="txAdvice">
    <tx:attributes>
        <tx:method name="get*" read-only="true"/>
        <tx:method name="*"/>
    </tx:attributes>
    </tx:advice>
    <!-- … -->
</beans>
```

以下示例展示了如何使用完全不同的事务配置两个不同的 bean。

```xml
<?xml version="1.0" encoding="UTF-8"?>
<beans xmlns="http://www.springframework.org/schema/beans"
xmlns:xsi="http://www.w3.org/2001/XMLSchema-instance"
xmlns:aop="http://www.springframework.org/schema/aop"
xmlns:tx="http://www.springframework.org/schema/tx"
xsi:schemaLocation="
```

```xml
   http://www.springframework.org/schema/beans
   http://www.springframework.org/schema/beans/spring-beans.xsd
   http://www.springframework.org/schema/tx
   http://www.springframework.org/schema/tx/spring-tx.xsd
   http://www.springframework.org/schema/aop
   http://www.springframework.org/schema/aop/spring-aop.xsd">
    <aop:config>
    <aop:pointcut id="defaultServiceOperation"
    expression="execution(* com.waylau.spring.service.*Service.*(..))"/>
    <aop:pointcut id="noTxServiceOperation"
    expression=
    "execution(* com.waylau.spring.service.ddl.DefaultDdlManager.*(..))"/>
    <aop:advisor pointcut-ref="defaultServiceOperation"
        advice-ref="defaultTxAdvice"/>
    <aop:advisor pointcut-ref="noTxServiceOperation"
        advice-ref="noTxAdvice"/>
    </aop:config>
    <!-- 下面两个bean将会纳入不同的事务配置 -->
    <bean id="fooService"
        class="com.waylau.spring.service.DefaultFooService"/>
    <bean id="anotherFooService"
        class="com.waylau.spring.service.ddl.DefaultDdlManager"/>
    <tx:advice id="defaultTxAdvice">
    <tx:attributes>
    <tx:method name="get*" read-only="true"/>
    <tx:method name="*"/>
    </tx:attributes>
    </tx:advice>
    <tx:advice id="noTxAdvice">
    <tx:attributes>
        <tx:method name="*" propagation="NEVER"/>
    </tx:attributes>
    </tx:advice>
    <!-- … -->
</beans>
```

6.3.5 @Transactional详解

除了基于 XML 的事务配置声明式方法外，还可以使用基于注解的方法。使用注解的好处是，声明事务的语义会使声明更接近受影响的代码，而没有太多不必要的耦合。标准的 javax.transaction.Transactional 注解也支持作为 Spring 注解的直接替代。

以下是使用 @Transactional 注解的例子。

```
@Transactional
public class DefaultFooService implements FooService {
    Foo getFoo(String fooName);
```

```
    Foo getFoo(String fooName, String barName);
    void insertFoo(Foo foo);
    void updateFoo(Foo foo);
}
```

在与上面的效果相同的情况下，如果使用基于 XML 的方式来配置，那么整体上会比较烦琐一点。

```xml
<?xml version="1.0" encoding="UTF-8"?>
<beans xmlns="http://www.springframework.org/schema/beans"
xmlns:xsi="http://www.w3.org/2001/XMLSchema-instance"
xmlns:aop="http://www.springframework.org/schema/aop"
xmlns:tx="http://www.springframework.org/schema/tx"
xsi:schemaLocation="
http://www.springframework.org/schema/beans
http://www.springframework.org/schema/beans/spring-beans.xsd
http://www.springframework.org/schema/tx
http://www.springframework.org/schema/tx/spring-tx.xsd
http://www.springframework.org/schema/aop
http://www.springframework.org/schema/aop/spring-aop.xsd">

    <bean id="fooService" class="com.waylau.spring.service.DefaultFooService"/>
    <tx:annotation-driven transaction-manager="txManager"/>
    <bean id="txManager"
    class="org.springframework.jdbc.datasource.DataSourceTransactionManager">
        <property name="dataSource" ref="dataSource"/>
    </bean>

    <!-- … -->
</beans>
```

注意，在使用代理时，应该将 @Transactional 注解仅应用于具有 public 的方法。如果使用 @Transactional 注解标注 protected、private 或包可见的方法，虽然不会引发错误，但注解的方法无法使用已配置的事务设置。

@Transactional 注解可以用于接口定义、接口方法、类定义或类的 public 方法之前。然而，仅有 @Transactional 注解是不足以激活事务行为的。@Transactional 注解只是一些元数据，既可以供具有事务的基础设施使用，也可以配置在具有事务行为的 bean 上。在前面的示例中，元素用于切换事务行为。

默认的 @Transactional 设置如下。

（1）传播设置为 PROPAGATION_REQUIRED。

（2）隔离级别为 ISOLATION_DEFAULT。

（3）事务是读—写的。

（4）事务超时默认为基础事务系统的默认超时。如果超时不受支持，则默认为无。

(5)任何 RuntimeException 都会触发回滚,并且任何已检查的异常都不会触发回滚。

6.3.6 事务传播机制

本小节将详细介绍 Spring 事务传播机制,其类型定义在 Propagation 枚举类中。

```
public enum Propagation {
REQUIRED(TransactionDefinition.PROPAGATION_REQUIRED),
SUPPORTS(TransactionDefinition.PROPAGATION_SUPPORTS),
MANDATORY(TransactionDefinition.PROPAGATION_MANDATORY),
REQUIRES_NEW(TransactionDefinition.PROPAGATION_REQUIRES_NEW),
NOT_SUPPORTED(TransactionDefinition.PROPAGATION_NOT_SUPPORTED),
NEVER(TransactionDefinition.PROPAGATION_NEVER),
NESTED(TransactionDefinition.PROPAGATION_NESTED);
// …
}
```

下面主要对常用的 PROPAGATION_REQUIRED、PROPAGATION_REQUIRES_NEW 和 PROPAGATION_NESTED 做详细介绍。

1. PROPAGATION_REQUIRED

PROPAGATION_REQUIRED 表示假如当前正要执行的事务不在另外一个事务中,那么就开启一个新的事务。

例如,ServiceB.methodB() 的事务级别定义为 PROPAGATION_REQUIRED,那么由于执行 ServiceA.methodA() 时,ServiceA.methodA() 已经开启了事务,因此这时调用 ServiceB.methodB(),ServiceB.methodB() 会看到自己已经运行在 ServiceA.methodA() 的事务内部,就不再开启新的事务。而假如 ServiceA.methodA() 运行时发现自己没有在事务中,它就会为自己分配一个事务。

这样,在 ServiceA.methodA() 或 ServiceB.methodB() 内的任何地方出现异常,事务都会被回滚。即使 ServiceB.methodB() 的事务已经被提交,但是 ServiceA.methodA() 出现异常了要回滚,那么 ServiceB.methodB() 也会回滚。

图 6-2 所示为 PROPAGATION_REQUIRED 类型的事务处理流程。

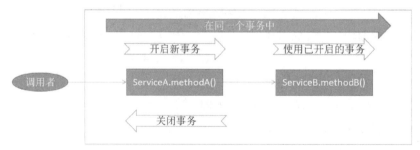

图6-2 PROPAGATION_REQUIRED类型的事务处理流程

2. PROPAGATION_REQUIRES_NEW

例如，定义 ServiceA.methodA() 的事务级别为 PROPAGATION_REQUIRED，ServiceB.methodB() 的事务级别为 PROPAGATION_REQUIRES_NEW，那么当执行到 ServiceB.methodB() 的时候，ServiceA.methodA() 所在的事务就会挂起，ServiceB.methodB() 会开启一个新的事务。等 ServiceB.methodB 的事务完成以后，ServiceA.methodA() 才继续执行。PROPAGATION_REQUIRES_NEW 与 PROPAGATION_REQUIRED 的区别在于，事务的回滚程度不同。因为 ServiceB.methodB() 已经开启了一个新事务，所以就会存在两个不同的事务。如果 ServiceB.methodB() 已经提交，那么 ServiceA.methodA() 回滚失败，ServiceB.methodB() 是不会回滚的。即使 ServiceB.methodB() 回滚失败，它抛出的异常被 ServiceA.methodA() 捕获，ServiceA.methodA() 事务仍然可以提交。

图 6-3 所示为 PROPAGATION_REQUIRES_NEW 类型的事务处理流程。

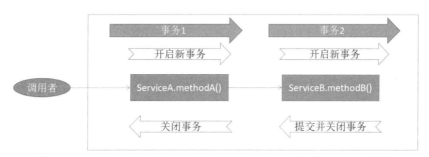

图6-3　PROPAGATION_REQUIRES_NEW 类型的事务处理流程

3. PROPAGATION_NESTED

PROPAGATION_NESTED 使用具有可回滚到多个保存点的单个物理事务。PROPAGATION_NESTED 与 PROPAGATION_REQUIRES_NEW 的区别是，PROPAGATION_REQUIRES_NEW 另开启一个事务，将会与其父事务相互独立，而 PROPAGATION_NESTED 的事务和其父事务是相依的，它的提交要和父事务一起。也就是说，如果父事务最后回滚，它也要回滚。如果子事务回滚或提交，不会导致父事务回滚或提交，但父事务回滚将导致子事务回滚。

图 6-4 所示为 PROPAGATION_NESTED 类型的事务处理流程。

图6-4　PROPAGATION_NESTED 类型的事务处理流程

6.4 编程式事务管理

Spring 框架提供了两种编程式事务管理方式,使用 TransactionTemplate 和 PlatformTransactionManager 实现。

6.4.1 TransactionTemplate

TransactionTemplate 使用一种回调方法,使应用程序代码可以获取和释放事务资源,这样可以让开发人员更加专注于业务逻辑的编写。

以下是使用 TransactionTemplate 的示例。

```
public class SimpleService implements Service {
    private final TransactionTemplate transactionTemplate;

    public SimpleService(PlatformTransactionManager transactionManager)
    {
        Assert.notNull(transactionManager,
            "The 'transactionManager'argument must not be null.");
        this.transactionTemplate =
            new TransactionTemplate(transactionManager);
    }

    public Object someServiceMethod() {
        return transactionTemplate.execute(new TransactionCallback() {
            // the code in this method executes in a transactional context
            public Object doInTransaction(TransactionStatus status) {
                updateOperation1();
                return resultOfUpdateOperation2();
            }
        });
    }
}
```

以下是 bean 配置的示例。

```
<bean id="sharedTransactionTemplate"
    class="org.springframework.transaction.support.TransactionTemplate">
    <property name="isolationLevelName" value="ISOLATION_READ_UNCOMMITTED"/>
    <property name="timeout" value="30"/>
</bean>
```

6.4.2　PlatformTransactionManager

也可以直接使用 org.springframework.transaction.PlatformTransactionManager 来管理事务。只需通过 bean 引用将正在使用的 PlatformTransactionManager 的实现传递给 bean。然后，使用 TransactionDefinition 和 TransactionStatus 对象来启动、回滚和提交事务。

以下是使用 PlatformTransactionManager 的示例。

```
DefaultTransactionDefinition def = new DefaultTransactionDefinition();
def.setName("SomeTxName");
def.setPropagationBehavior(TransactionDefinition.PROPAGATION_REQUIRED);

TransactionStatus status = txManager.getTransaction(def);

try {
    // …
}
catch (MyException ex) {
    txManager.rollback(status);
    throw ex;
}
txManager.commit(status);
```

6.4.3　声明式事务管理和编程式事务管理

如果应用中只有少量的事务操作，编程式事务管理通常是一个很好的选择。例如，如果 Web 应用程序只需要某些更新操作的事务，则可能不想使用 Spring 或其他技术来设置事务代理。在这种情况下，使用 TransactionTemplate 可能是一个好方法。因为它能够很明确地设置事务，并与具体的业务逻辑代码靠得更近。

如果应用程序有大量的事务操作，则声明式事务管理通常是更好的选择。它使事务管理不受业务逻辑的影响，而且在配置上也很简单。当使用 Spring 框架而不使用 EJB CMT 时，声明式事务管理的配置成本往往很低。

第7章 Spring Web MVC

Spring Web MVC也称Spring MVC,它实现了Web开发中常见的MVC模式,是Spring技术栈中应用最为广泛的框架之一。本章将详细介绍Spring Web MVC的原理及用法。

7.1 Spring Web MVC 概述

Spring Web MVC 简称 Spring MVC，实现了 Web 开发中经典的 MVC（Model-View-Controller）模式，同类产品还包括 Structs 等。

MVC 由以下三部分组成。

（1）模型（Model）：应用程序的核心功能，管理模块中用到的数据和值。

（2）视图（View）：提供模型的展示，是应用程序的外观。

（3）控制器（Controller）：对用户的输入做出反应，管理用户和视图的交互，是连接模型和视图的枢纽。

Spring Web MVC 是基于 Servlet API 构建的，自 Spring 框架诞生之日起，就包含在 Spring 中。要使用 Spring Web MVC 框架的功能，需要添加 spring-webmvc 模块。

7.2 DispatcherServlet

在 Java Web 企业级应用中，Servlet 是业务处理的核心。像许多其他 Web 框架一样，Spring Web MVC 围绕前端控制器模式进行设计，其中 DispatcherServlet 为所有的请求处理提供调度，并将实际工作交由可配置、可委托的组件执行。该模型非常灵活，支持多种工作流程。

7.2.1 DispatcherServlet概述

DispatcherServlet 需要根据 Servlet 规范，使用 Java 配置或在 web.xml 中进行声明和映射。DispatcherServlet 依次使用 Spring 配置来发现其在请求映射、查看解析、异常处理等方面所需的委托组件。

以下是注册和初始化 DispatcherServlet 的 Java 配置示例。该类将由 Servlet 容器自动检测。

```
public class MyWebApplicationInitializer
    implements WebApplicationInitializer {

    @Override
    public void onStartup(ServletContext servletCxt) {
        // 加载 Spring Web用于配置
        AnnotationConfigWebApplicationContext ac =
            newAnnotationConfigWebApplicationContext();
        ac.register(AppConfig.class);
        ac.refresh();

        // 创建并注册DispatcherServlet
```

```
        DispatcherServlet servlet = newDispatcherServlet(ac);
        ServletRegistration.Dynamic registration =
            servletCxt.addServlet("app", servlet);
        registration.setLoadOnStartup(1);
        registration.addMapping("/app/*");
    }
}
```

以下是在 web.xml 中进行声明和映射的示例。

```
<web-app>
<listener>
    <listener-class>
    org.springframework.web.context.ContextLoaderListener
    </listener-class>
</listener>
<context-param>
    <param-name>contextConfigLocation</param-name>
    <param-value>/WEB-INF/app-context.xml</param-value>
</context-param>
<servlet>
    <servlet-name>app</servlet-name>
    <servlet-class>
        org.springframework.web.servlet.DispatcherServlet
    </servlet-class>
    <init-param>
        <param-name>contextConfigLocation</param-name>
        <param-value></param-value>
    </init-param>
    <load-on-startup>1</load-on-startup>
</servlet>
<servlet-mapping>
    <servlet-name>app</servlet-name>
    <url-pattern>/app/*</url-pattern>
</servlet-mapping>
</web-app>
```

7.2.2 上下文层次结构

DispatcherServlet 需要一个 WebApplicationContext（普通 ApplicationContext 的扩展）用于其配置。WebApplicationContext 有一个指向它所关联的 ServletContext 和 Servlet 的链接，也可以绑定 ServletContext，以便应用程序可以使用 RequestContextUtils 上的静态方法来查找 WebApplicationContext 是否需要访问它。

但对大多数应用来说，单个 WebApplicationContext 可以使应用看上去更加简单，也可以支持上下文层次结构中包含一个根 WebApplicationContext，以及多个 DispatcherServlet（或其他 Servlet）实例，这些 DispatcherServlet 实例都有自己的子 WebApplicationContext 配置，但同时又共享根 WebApplicationContext 的配置。

根 WebApplicationContext 通常包含需要跨多个 Servlet 实例共享的基础架构 bean，如数据存储库和业务服务。这些 bean 被有效地继承，并且可以在 Servlet 的子 WebApplicationContext 中重写，子 WebApplicationContext 通常包含给定 Servlet 本地的 bean，如图 7-1 所示。

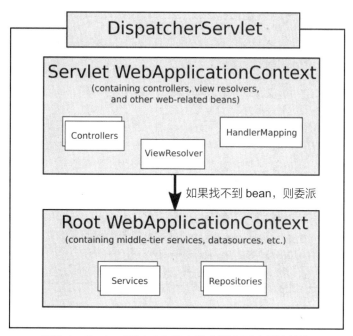

图7-1　Spring Web MVC中的典型上下文层次结构

以下是使用 WcbApplicationContext 层次结构的配置示例。

```
public class MyWebAppInitializer
    extends AbstractAnnotationConfigDispatcherServletInitializer {
    @Override
    protected Class<?>[] getRootConfigClasses() {
        return newClass<?[] { RootConfig.class };
    }
    @Override
    protected Class<?>[] getServletConfigClasses() {
        return newClass<?[] { App1Config.class };
    }
    @Override
    protected String[] getServletMappings() {
        return newString[] { "/app1/*" };
    }
}
```

相当于以下在 web.xml 中进行配置的示例。

```xml
<web-app>
<listener>
    <listener-class>
        org.springframework.web.context.ContextLoaderListener
    </listener-class>
</listener>
<context-param>
    <param-name>contextConfigLocation</param-name>
    <param-value>/WEB-INF/root-context.xml</param-value>
</context-param>
<servlet>
    <servlet-name>app1</servlet-name>
    <servlet-class>
        org.springframework.web.servlet.DispatcherServlet
    </servlet-class>
<init-param>
    <param-name>contextConfigLocation</param-name>
    <param-value>/WEB-INF/app1-context.xml</param-value>
</init-param>
<load-on-startup>1</load-on-startup>
</servlet>
<servlet-mapping>
    <servlet-name>app1</servlet-name>
    <url-pattern>/app1/*</url-pattern>
</servlet-mapping>
</web-app>
```

7.2.3 处理流程

DispatcherServlet 按如下方式处理请求。

（1）在请求中搜索并绑定 WebApplicationContext，作为控制器和进程中其他元素可以使用的属性。WebApplicationContext 在 DispatcherServlet.WEB_APPLICATION_CONTEXT_ATTRIBUTE 关键字下被默认为绑定。

（2）将语言环境解析程序绑定到请求，以启用进程中的元素来解析处理请求（呈现视图、准备数据等）时要使用的语言环境。如果不需要区域解析，则可以跳过该步骤。

（3）主题解析器必须等待请求来决定使用哪个主题。如果不使用主题，则可以忽略。

（4）如果指定了 multipart 文件解析器，则将检查请求中的 multipart；如果找到 multipart，则将请求封装在 MultipartHttpServletRequest 中，以供进程中的其他元素进一步处理。

（5）搜索适当的处理程序。如果找到处理程序，则执行与处理程序（预处理程序、后处理程序和控制器）关联的执行链，以便准备模型或渲染。对于注解控制器，响应可以直接呈现（在 HandlerAdapter 内）而无须返回视图。

（6）如果返回模型，则会呈现视图。如果没有返回模型（可能是由于预处理程序或后处理程序拦截了请求，也可能出于其他安全原因），则不会呈现视图，因为请求可能已经被满足。

WebApplicationContext 中声明的 HandlerExceptionResolver bean 属于异常解析器，用于解决请求处理期间抛出的异常，这些异常解析器允许定制逻辑来解决异常。Spring DispatcherServlet 还支持返回最后修改日期。确定特定请求的最后修改日期的过程为：DispatcherServlet 查找适当的处理程序映射并测试找到的处理程序是否实现了 LastModified 接口；如果实现了，则将 LastModified 接口的 long getLastModified(request) 方法的值返回给客户端。

开发者可以通过将 Servlet 的初始化参数（init-param 元素）添加到 web.xml 文件中的 Servlet 声明中，来自定义单个 DispatcherServlet 实例。

7.2.4 拦截

所有 HandlerMapping 实现都支持处理程序拦截器，如果想要将特定功能应用于某些请求（如检查委托人），该拦截器非常有用。拦截器必须实现 org.springframework.web.servlet 包中的 HandlerInterceptor，并提供以下 3 种方法来执行各种预处理和后处理。

（1）preHandle(...)：在执行实际处理程序之前。

（2）postHandle(...)：在处理程序执行后。

（3）afterCompletion(...)：在完成请求后。

其中，preHandle(...) 方法返回一个布尔值，可以使用此方法来中断或继续处理执行链。当此方法返回 true 时，继续处理程序执行链；当返回 false 时，DispatcherServlet 假定拦截器本身已经处理了请求（如已经呈现适当的视图），并且中断了执行链中的其他拦截器和实际处理器。

7.3 过滤器

spring-web 模块提供了很多有用的过滤器。Spring 过滤器的实现依赖于 Servlet 容器。在实现上基于函数回调，可以对所有请求进行过滤。但其缺点是，一个过滤器实例只能在容器初始化时调用一次。

使用过滤器的目的是做一些过滤操作，获取想要的数据。例如，在过滤器中修改字符编码或修改 HttpServletRequest 的一些参数，来过滤低俗字眼、危险字符等。

7.3.1 HTTP PUT 表单

浏览器只能通过 HTTP GET 或 HTTP POST 提交表单数据，而非浏览器客户端则可以使用 HTTP PUT 和 PATCH。Servlet API 规定 ServletRequest.getParameter*() 方法只能用于 HTTP POST 请

求。那么，如果用户想使用 HTTP PUT 来请求表单该怎么办呢？

可以通过 spring-web 模块提供的 HttpPutFormContentFilter 拦截内容类型为 "application/x-www-form-urlencoded" 的 HTTP PUT 和 PATCH 请求。从请求主体读取表单数据，并封装为 ServletRequest，以使表单数据可以通过 ServletRequest.getParameter*() 方法访问到。

7.3.2 转发头

当请求经过负载均衡器等代理时，主机、端口等信息可能会发生改变，这对于需要创建资源链接的应用程序提出了挑战，因为链接应反映原始请求的主机、端口等客户视角。RFC 7239 规范[1]定义了代理如何 "Forwarded"（转发）HTTP 头，转发时需要提供有关原始请求的信息。还有其他一些非标准转发的使用，如 "X-Forwarded-Host" "X-Forwarded-Port" 和 "X-Forwarded-Proto" 等。

ForwardedHeaderFilter 检测、提取并使用来自 "Forwarded" 头或 "X-Forwarded-Host" "X-Forwarded-Port" 和 "X-Forwarded-Proto" 的信息。它包装请求以覆盖主机、端口，并 "隐藏" 转发的头以供后续处理。

注意，使用转发头时存在一定的安全隐患，因为在应用程序级别很难确定转发头是否可信。这就是为什么应该正确配置网络（尤其是从外部过滤不可信）的转发头。没有代理并且不需要使用转发头的应用程序可以配置 ForwardedHeaderFilter，以删除并忽略这些头。

7.3.3 ShallowEtagHeaderFilter

ShallowEtagHeaderFilter 是 Spring 提供的支持 ETag 的一个过滤器。ETag 指被请求变量的实体值，是一个可以与 Web 资源关联的记号。Web 资源可以是一个 Web 页，也可以是 JSON 或 XML 文档，服务器单独负责判断记号是什么及其含义，并在 HTTP 响应头中将其传送到客户端。以下是服务器端返回的格式：

```
ETag:"D41D8CD98F00B204E9800998ECF8427E"
```

客户端的查询更新格式为：

```
If-None-Match:"D41D8CD98F00B204E9800998ECF8427E"
```

如果 ETag 无变化，则返回状态 304，这与 Last-Modified 一样。

ShallowEtagHeaderFilter 将 JSP 等的内容缓存生成 MD5 的 key，然后将响应中作为头的 ETag 返回给客户端。下次客户端对相同的资源（或相同的 URL）发出请求时，会将之前生成的 key 作为 If-None-Match 的值发送到服务器。Filter 将客户端传来的值和服务器上的值做比较，如果相同则

---

[1] 规范可见 https://tools.ietf.org/html/rfc7239。

返回 304，否则将发送新的内容到客户端。

7.3.4 CORS

Spring Web MVC 通过控制器上的注解为 CORS 配置提供细粒度的支持。但是，当与 Spring Security 一起使用时，建议依靠内置的 CorsFilter，且必须排在 Spring Security 的过滤器链之前。

有关 CORS 方面的详细内容，将在 7.6 节中继续讲解。有关 Spring Security 的内容，会在第 8 章详细讲解。

7.4 控制器

@Controller 和 @RestController 是 Spring Web MVC 中实现控制器的常用注解。这些注解可以用来表示请求映射、请求输入及异常处理等。带注解的控制器具有灵活的方法签名，不需要扩展基类，也不需要实现特定的接口。

以下是使用 @Controller 注解的示例。

```
@Controller
public class HelloController {

    @GetMapping("/hello")
    public String handle(Model model) {
    model.addAttribute("message", "Hello World!");
        return "index";
    }
}
```

7.4.1 声明控制器

开发者可以使用 Servlet 的 WebApplicationContext 中的标准 Spring bean 来定义控制器 bean。@Controller 的原型允许自动检测，可以被 Spring 自动注册。

如果要启用 @Controller bean 的自动检测，可以将组件扫描添加到 Java 配置中，其代码如下：

```
@Configuration
@ComponentScan("com.waylau.spring")
public class WebConfig {
    // …
}
```

上述配置相当于以下基于 XML 的配置。

```xml
<?xml version="1.0" encoding="UTF-8"?>
<beans xmlns="http://www.springframework.org/schema/beans"
    xmlns:xsi="http://www.w3.org/2001/XMLSchema-instance"
    xmlns:p="http://www.springframework.org/schema/p"
    xmlns:context="http://www.springframework.org/schema/context"
    xsi:schemaLocation="
    http://www.springframework.org/schema/beans
    http://www.springframework.org/schema/beans/spring-beans.xsd
    http://www.springframework.org/schema/context
    http://www.springframework.org/schema/context/spring-context.xsd">

    <context:component-scan base-package="com.waylau.spring"/>
    <!-- … -->
</beans>
```

@RestController 相当于 @Controller 与 @ResponseBody 的组合，主要用于返回 RESTful 应用常用的 JSON 格式数据，如下所示。

```
@RestController = @Controller + @ResponseBody
```

其中，@ResponseBody 注解指示方法返回值应绑定到 Web 响应的正文；@RestController 注解暗示用户这是一个支持 REST 的控制器。

7.4.2 请求映射

@RequestMapping 注解用于将请求映射到控制器方法上。它具有与 URL、HTTP 方法、请求参数、头和媒体类型进行匹配的各种属性。可以在类级使用它来表示共享映射，或者在方法级使用，以将映射范围缩小到特定的端点映射。

@RequestMapping 还有一些基于特定 HTTP 方法的快捷方式变体，包括 @GetMapping、@PostMapping、@PutMapping、@DeleteMapping 和 @PatchMapping。

在类级别仍需要用 @RequestMapping 来表示共享映射。

以下是类级别和方法级别映射的示例。

```
@RestController
@RequestMapping("/persons")
class PersonController {

    @GetMapping("/{id}")
    public Person getPerson(@PathVariable Long id) {
        // …
    }
    @PostMapping
    @ResponseStatus(HttpStatus.CREATED)
    public void add(@RequestBody Person person) {
```

```
        // …
    }
}
```

7.4.3 处理器方法

以下是 Spring Web MVC 中常用的处理器方法。

1. @RequestParam

@RequestParam 将 Servlet 请求参数（查询参数或表单数据）绑定到控制器中的方法参数上，以下代码展示了此用法。

```
@Controller
@RequestMapping("/pets")
public class EditPetForm {
    // …
    @GetMapping
    public String setupForm(@RequestParam("petId") int petId,
        Model model) {

        Pet pet = this.clinic.loadPet(petId);
        model.addAttribute("pet", pet);
        return "petForm";
    }
    // …
}
```

2. @RequestHeader

@RequestHeader 将请求头绑定到控制器中的方法参数上。

以下示例所示的是获取请求头上的 Accept-Encoding 和 Keep-Alive 的值。

```
@GetMapping("/demo")
public void handle(@RequestHeader("Accept-Encoding") String encoding,
    @RequestHeader("Keep-Alive") long keepAlive) {
    //…
}
```

3. @CookieValue

@CookieValue 将 HTTP cookie 的值绑定到控制器中的方法参数上。

以下示例演示了如何获取 cookie 值。

```
@GetMapping("/demo")
public void handle(@CookieValue("JSESSIONID") String cookie) {
    //…
}
```

如果目标方法的参数类型不是字符串，则自动将应用类型转换为字符串。

4. @ModelAttribute

在方法参数上使用 @ModelAttribute 来访问模型中的属性，如果不存在，则将其实例化。模型属性还覆盖了名称与字段名称匹配的 HTTP Servlet 请求参数的值，它无须解析和转换单个查询参数或表单字段，例如：

```
@PostMapping("/owners/{ownerId}/pets/{petId}/edit")
public String processSubmit(@ModelAttribute Pet pet) { }
```

5. @RequestAttribute

可以使用 @RequestAttribute 注解来访问先前创建的请求属性，如通过 Servlet 过滤器或 HandlerInterceptor。

```
@GetMapping("/")
public String handle(@RequestAttribute Client client) {
    // …
}
```

6. 重定向属性

默认情况下，所有模型属性都在重定向 URL 中作为 URI 模板变量公开，例如：

```
@PostMapping("/files/{path}")
public String upload(…) {
    // …
    return "redirect:files/{path}";
}
```

7. @RequestBody

使用 @RequestBody 通过 HttpMessageConverter 读取请求体并将其反序列化为一个 Object，下面是一个带有 @RequestBody 参数的例子。

```
@PostMapping("/accounts")
public void handle(@RequestBody Account account) {
    // …
}
```

8. HttpEntity

HttpEntity 与 @RequestBody 很相似，只是前者是基于容器对象来公开请求头和正文的，例如：

```
@PostMapping("/accounts")
public void handle(HttpEntity<Account> entity) {
    // …
}
```

9. @ResponseBody

在一个方法上使用 @ResponseBody 注解,将通过 HttpMessageConverter 返回经过序列化的响应主体,例如:

```
@GetMapping("/accounts/{id}")
@ResponseBody
public Account handle() {
    // …
}
```

10. ResponseEntity

ResponseEntity 与 @ResponseBody 很相似,只是前者是基于容器对象来指定请求头和正文的,例如:

```
@PostMapping("/something")
public ResponseEntity<String> handle() {
    // …
    URI location = …
    return new ResponseEntity.created(location).build();
}
```

11. Jackson JSON

Spring Web MVC 为 Jackson 的序列化视图提供了内置的支持。以下是在 @ResponseBody 或 ResponseEntity 控制器方法上使用 Jackson 的 @JsonView 注解来激活序列化视图类的例子。

```
@RestController
public class UserController {
    @GetMapping("/user")
    @JsonView(User.WithoutPasswordView.class)
    public User getUser() {
        return new User("eric", "7!jd#h23");
    }
}

public class User {
    public interface WithoutPasswordView {};
    public interface WithPasswordView extends WithoutPasswordView {};
    private String username;
    private String password;
    public User() {
    }

    public User(String username, String password) {
        this.username = username;
        this.password = password;
    }
```

```
@JsonView(WithoutPasswordView.class)
public String getUsername() {
    return this.username;
}

@JsonView(WithPasswordView.class)
public String getPassword() {
    return this.password;
}
}
```

7.4.4 模型方法

可以在 @RequestMapping 方法参数上使用 @ModelAttribute 来创建或访问模型中的 Object，并将其绑定到请求中。@ModelAttribute 也可以用于控制器的方法级注解，其目的不是处理请求，而是在请求处理之前添加常用模型属性。

控制器可以有任意数量的 @ModelAttribute 方法。所有这些方法在相同控制器中的 @RequestMapping 方法之前被调用。@ModelAttribute 方法也可以通过 @ControllerAdvice 在控制器之间共享。

@ModelAttribute 方法具有灵活的方法签名，除了其本身或任何与请求主体相关的内容，还支持许多与 @RequestMapping 方法相同的参数。以下是使用 @ModelAttribute 方法的示例。

```
@ModelAttribute
public void populateModel(@RequestParam String number, Model model) {
    model.addAttribute(accountRepository.findAccount(number));
    // …
}
```

7.4.5 绑定器方法

@Controller 或 @ControllerAdvice 类中的 @InitBinder 方法可用于将请求参数绑定到 @ModelAttribute 参数（命令对象）上，也可用于类型转换。

除了 @ModelAttribute(command object) 参数外，@InitBinder 方法还支持许多与 @RequestMapping 方法相同的参数，例如：

```
@Controller
public class FormController {
    @InitBinder
    public void initBinder(WebDataBinder binder) {
        SimpleDateFormat dateFormat = new SimpleDateFormat("yyyy-MM-dd");
        dateFormat.setLenient(false);
        binder.registerCustomEditor(Date.class,
```

```
            new CustomDateEditor(dateFormat, false));
    }
    // …
}
```

7.5 异常处理

如果在请求映射期间发生异常或请求处理程序（如 @Controller）抛出异常，则 DispatcherServlet 将委托 HandlerExceptionResolver bean 来解决异常并提供替代处理。这个处理通常是一个错误响应。

下面列出了可用的 HandlerExceptionResolver 实现。

（1）SimpleMappingExceptionResolver：处理异常类名称和错误视图名称之间的映射，用于在浏览器应用程序中呈现错误页面。

（2）DefaultHandlerExceptionResolver：用于解决 Spring Web MVC 引发的异常并将其映射到 HTTP 状态代码。

（3）ResponseStatusExceptionResolver：使用 @ResponseStatus 注解来解决异常，并根据注解中的值将其映射到 HTTP 状态代码。

（4）ExceptionHandlerExceptionResolver：通过在 @Controller 或 @ControllerAdvice 类中调用 @ExceptionHandler 方法来解决异常（见 @ExceptionHandler 方法）。

7.5.1 @ExceptionHandler

@Controller 和 @ControllerAdvice 类可以用 @ExceptionHandler 方法处理来自控制器方法的异常，例如：

```
@Controller
public class SimpleController {
    // …
    @ExceptionHandler
    public ResponseEntity<String> handle(IOException ex) {
        // …
    }
}
```

@ExceptionHandler 注解可以列出要匹配的异常类型，或者简单地将目标异常声明为方法参数。当多个异常方法同时匹配时，根异常匹配通常优先于引发异常匹配。准确地说，ExceptionDepth-Comparator 根据抛出异常类型的深度对异常进行排序。

在 Spring Web MVC 中，@ExceptionHandler 方法建立在 DispatcherServlet 级别的 Handler-

ExceptionResolver 机制上。

7.5.2 框架异常处理

只需在 Spring 配置中声明多个 HandlerExceptionResolver bean，并根据需要设置其顺序属性，就可以形成一个异常解析链。order 属性越高，异常解析器的定位就越靠后。

HandlerExceptionResolver 可以指定返回的内容：指向错误视图的 ModelAndView；如果在解析器中处理了异常，则为 Empty ModelAndView；如果异常未解决，则返回 null，供后续解析器尝试使用；如果异常仍然存在，则允许冒泡到 Servlet 容器。

MVC Config 内置了多种解析器，用于默认的 Spring MVC 异常声明、@ResponseStatus 注解的异常声明，以及 @ExceptionHandler 方法。也可以自定义这些解析器的列表或将其替换掉。

7.5.3 REST API异常

REST 服务的一个常见要求是，在响应正文中包含错误详细信息。Spring 框架不会自动执行此操作，只有特定应用程序的响应正文中包含错误详细信息。但是，@RestController 可以使用带有 ResponseEntity 返回值的 @ExceptionHandler 方法来设置响应的状态和主体。这些方法也可以在 @ControllerAdvice 类中声明以全局应用。

如果想要实现自定义错误信息的全局异常处理，那么应用程序应该扩展 ResponseEntityExceptionHandler，它提供了对 Spring MVC 引发的异常处理及钩子方法，并可以定制响应主体。如果要使用它，需要创建一个 ResponseEntityExceptionHandler 的子类，使用 @ControllerAdvice 注解覆盖必要的方法，并将其声明为 Spring bean。

7.5.4 注解异常

带有 @ResponseStatus 注解的异常类会被 ResponseStatusExceptionResolver 解析。可以实现自定义的一些异常，同时在页面上进行显示。具体的使用方法如下。

定义一个异常类：

```
@ResponseStatus(value = HttpStatus.FORBIDDEN,
    reason = "用户名和密码不匹配！")
public class UserNameNotMatchPasswordException
    extends RuntimeException{
}
```

抛出异常：

```
@RequestMapping("/testResponseStatusExceptionResolver")
public String testResponseStatusExceptionResolver(@RequestParam("i")
```

```
int i){
    if (i==13){
        throw new UserNameNotMatchPasswordException();
    }
    return "success";
}
```

7.5.5 容器错误页面

如果异常未被 HandlerExceptionResolver 处理，或者响应状态设置为错误状态（4xx、5xx），则 Servlet 容器可能会在 HTML 中呈现默认错误页面。默认错误页面可以在 web.xml 中声明，例如：

```
<error-page>
<location>/error</location>
</error-page>
```

鉴于上述情况，当异常冒泡时或响应具有错误状态时，Servlet 会在容器内将 ERROR 分配到配置的 URL（如 "/error"），然后由 DispatcherServlet 进行处理，可能会将其映射到一个 @Controller。该实现可以通过模型返回错误视图名称或呈现 JSON 响应，例如：

```
@RestController
public class ErrorController {
    @RequestMapping(path = "/error")
    public Map<String, Object> handle(HttpServletRequest request) {
        Map<String, Object> map = new HashMap<String, Object>();
        map.put("status", request.getAttribute("javax.servlet.error.
           status_code"));
        map.put("reason", request.getAttribute("javax.servlet.error.
           message"));
        return map;
    }
}
```

注意，因为 Servlet API 不提供在 Java 中创建错误页面映射的方法，所以需要同时使用 WebApplicationInitializer 和 web.xml 来实现。

7.6 CORS处理

出于安全原因，浏览器禁止对当前源以外的资源进行 AJAX 调用。CORS（Cross-Origin Resource Sharing，跨域资源共享）是一个 W3C 标准。它允许浏览器向跨源服务器发出 XMLHttpRequest 请求，从而克服了 AJAX 只能同源使用的限制。

Spring MVC HandlerMapping 提供了对 CORS 的内置支持。在成功将请求映射到处理程序后，HandlerMapping 会检查给定请求和处理程序的 CORS 配置并采取进一步的操作。预检请求被直接处理掉，而简单和实际的 CORS 请求会被拦截，验证是否需要设置 CORS 响应头。

为了实现跨域请求（Origin 头域存在且与请求的主机不同），需要有一些明确声明的 CORS 配置。如果找不到匹配的 CORS 配置，则会拒绝预检请求。如果没有将 CORS 头添加到简单和实际的 CORS 请求的响应中，则会被浏览器拒绝。

每个 HandlerMapping 可以单独配置基于 URL 模式的 CorsConfiguration 映射。在大多数情况下，应用程序将使用 MVC 配置来实现全局映射。HandlerMapping 级别的全局 CORS 配置可以与更细粒度的处理器级 CORS 配置相结合。例如，带注解的控制器可以使用类级别或方法级别的 @CrossOrigin 注解。

7.6.1 @CrossOrigin

@CrossOrigin 注解用于在带注解的控制器方法上启用跨域请求，例如：

```
@RestController
@RequestMapping("/account")
public class AccountController {
    @CrossOrigin
    @GetMapping("/{id}")
    public Account retrieve(@PathVariable Long id) {
        // …
    }
    @DeleteMapping("/{id}")
    public void remove(@PathVariable Long id) {
        // …
    }
}
```

@CrossOrigin 默认启用以下内容。

（1）所有的源。

（2）所有的头。

（3）控制器方法所映射到的所有 HTTP 方法。

（4）allowCredentials 默认情况下未启用，因为它建立了一个信任级别，用于公开敏感的用户特定信息，如 Cookie 和 CSRF 令牌，并且只能在适当的情况下使用。

（5）maxAge 默认设置为 30 分钟。

@CrossOrigin 也在类级别上得到支持，并由所有方法继承，例如：

```
@RestController
@RequestMapping("/account")
public class AccountController {
```

```
@CrossOrigin
@GetMapping("/{id}")
public Account retrieve(@PathVariable Long id) {
    // …
}
@DeleteMapping("/{id}")
public void remove(@PathVariable Long id) {
    // …
}
}
```

@CrossOrigin 可以在类级别和方法级别使用，例如：

```
@CrossOrigin(maxAge = 3600)
@RestController
@RequestMapping("/account")
public class AccountController {
    @CrossOrigin("http://domain2.com")
    @GetMapping("/{id}")
    public Account retrieve(@PathVariable Long id) {
        // …
    }
    @DeleteMapping("/{id}")
    public void remove(@PathVariable Long id) {
        // …
    }
}
```

7.6.2 全局CORS配置

除了细粒度的控制器方法级配置外，可能还需要定义一些全局 CORS 配置。可以在任何 HandlerMapping 上分别设置基于 URL 的 CorsConfiguration 映射。但是，大多数应用程序将使用 MVC 的 Java 配置或 XML 配置来完成此操作，默认情况下，全局配置启用以下内容。

（1）所有的源。

（2）所有的头。

（3）GET、HEAD 和 POST 方法。

（4）allowCredentials 默认情况下未启用，因为它建立了一个信任级别，用于公开敏感的用户特定信息，如 Cookie 和 CSRF 令牌，并且只能在适当的情况下使用。

（5）maxAge 默认设置为 30 分钟。

7.6.3 自定义

可以通过基于 Java 或 XML 的配置来自定义 CORS。

1. Java配置

如果要在 MVC 的 Java 配置中启用 CORS，则使用 CorsRegistry 回调，例如：

```
@Configuration
@EnableWebMvc
public class WebConfig implements WebMvcConfigurer {
    @Override
    public void addCorsMappings(CorsRegistry registry) {
        registry.addMapping("/api/**")

        .allowedOrigins("http://domain2.com")
        .allowedMethods("PUT", "DELETE")
        .allowedHeaders("header1", "header2", "header3")
        .exposedHeaders("header1", "header2")
        .allowCredentials(true).maxAge(3600);
        // …
    }
}
```

2. XML配置

如果要在 XML 命名空间中启用 CORS，则使用元素，例如：

```
<mvc:cors>
    <mvc:mapping path="/api/**"
        allowed-origins="http://domain1.com, http://domain2.com"
        allowed-methods="GET, PUT"
        allowed-headers="header1, header2, header3"
        exposed-headers="header1, header2" allow-credentials="true"
        max-age="123" />
    <mvc:mapping path="/resources/**"
        allowed-origins="http://domain1.com" />
</mvc:cors>
```

7.6.4 CORS 过滤器

开发者可以通过内置的 CorsFilter 来应用 CORS 支持，配置过滤器将 CorsConfigurationSource 传递给其构造函数，例如：

```
CorsConfiguration config = new CorsConfiguration();
Config.applyPermitDefaultValues()
config.setAllowCredentials(true);
config.addAllowedOrigin("http://domain1.com");
config.addAllowedHeader("");
```

```
config.addAllowedMethod("");

UrlBasedCorsConfigurationSource source =
    new UrlBasedCorsConfigurationSource();
source.registerCorsConfiguration("/**", config);

CorsFilter filter = new CorsFilter(source);
```

7.7 HTTP缓存

Cache-Control 用于指定所有缓存机制在整个请求/响应链中必须服从的指令，这些指令用于阻止缓存对请求或响应造成干扰，它们通常会覆盖默认缓存算法。缓存指令是单向的，即请求中存在一个指令，并不意味着响应中也存在同一个指令。

Last-Modified 字段值通常用于验证缓存是否失效。简单来说，如果实体值在 Last-Modified 值之后没有被更改，则认为该缓存条目有效。

ETag 是一个 HTTP 响应头，由 HTTP/1.1 兼容的 Web 服务器返回，用于确定给定 URL 中的内容是否已经被更改。它可以被认为是 Last-Modified 头更复杂的后继者。当服务器返回带有 ETag 头的表示时，客户端可以在随后的 GET 中的 If-None-Match 头中使用此头。如果内容未更改，则服务器返回 "304: Not Modified"。

7.7.1 缓存控制

Spring Web MVC 支持许多缓存策略，并提供了为应用程序配置 Cache-Control 头的方法。

Spring Web MVC 提供了约定的 setCachePeriod(int seconds) 方法，该方法有不同的返回值，分别如下。

（1）值为 -1：不会生成 Cache-Control 响应头。

（2）值为 0：使用 "Cache-Control: no-store" 指令时，将阻止缓存。

（3）值 n>0：使用 "Cache-Control: max-age=n" 指令时，将给定响应缓存 n 秒。

CacheControl 构建器类简单地描述了可用的 Cache-Control 指令，使构建自己的 HTTP 缓存策略变得更加容易。一旦构建完成，一个 CacheControl 实例可以被接收为 Spring Web MVC API 中的一个参数。

```
// 缓存1个小时 - "Cache-Control: max-age=3600"
CacheControl ccCacheOneHour = CacheControl.maxAge(1, TimeUnit.HOURS);
// 阻止缓存 - "Cache-Control: no-store"
CacheControl ccNoStore = CacheControl.noStore();
```

```
// 在公共和私人缓存中缓存10天
// 公共缓存不应该转换响应
// "Cache-Control: max-age=864000, public, no-transform"
CacheControl ccCustom =
    CacheControl.maxAge(10, TimeUnit.DAYS).noTransform().cachePublic();
```

7.7.2 静态资源

应该为静态资源提供适当的 Cache-Control 和头,以获得最佳性能,以下是一个配置示例。

```
@Configuration
@EnableWebMvc
public class WebConfig implements WebMvcConfigurer {
    @Override
    public void addResourceHandlers(ResourceHandlerRegistry registry) {
        registry.addResourceHandler("/resources/**")
            .addResourceLocations("/public-resources/")
            .setCacheControl(CacheControl.maxAge(1, TimeUnit.HOURS).
            cachePublic());
    }
}
```

如果是基于 XML,则上述配置相当于:

```
<mvc:resources mapping="/resources/**" location="/public-resources/">
<mvc:cache-control max-age="3600" cache-public="true"/>
</mvc:resources>
```

7.7.3 控制器缓存

Spring Web MVC 控制器可以支持 Cache-Control、ETag 和 If-Modified-Since 等 HTTP 请求。控制器可以使用 HttpEntity 类型与请求 / 响应进行交互,返回 ResponseEntity 的控制器可以包含 HTTP 缓存信息,例如:

```
@GetMapping("/book/{id}")
public ResponseEntity<Book> showBook(@PathVariable Long id) {
    Book book = findBook(id);
    String version = book.getVersion();
    return ResponseEntity
        .ok()
        .cacheControl(CacheControl.maxAge(30, TimeUnit.DAYS))
        .eTag(version)
        .body(book);
}
```

@RequestMapping 方法也可以支持相同的行为,其实现如下。

```
@RequestMapping
public String myHandleMethod(WebRequest webRequest, Model model) {
    long lastModified = // 1.特定于应用程序的计算
    if (request.checkNotModified(lastModified)) {
        // 2.快速退出,不需要进一步处理
        return null;
    }

    // 3.或者另外请求处理
    model.addAttribute(…);
    return "myViewName";
}
```

7.8 MVC配置

Spring Web MVC 提供了基于 Java 和 XML 的配置,其默认的配置值可以满足大多数的应用场景。当然,Spring Web MVC 也提供了 API,以方便开发人员来自定义配置。

接下来将详细介绍 Spring Web MVC 的配置。

7.8.1 启用MVC配置

在基于 Java 的配置中,启用 MVC 配置时使用 @EnableWebMvc 注解,其用法如下。

```
@Configuration
@EnableWebMvc
public class WebConfig {
}
```

如果是使用基于 XML 的配置,则需要使用 <mvc:annotation-driven> 元素,其用法如下。

```
<?xml version="1.0" encoding="UTF-8"?>
<beans xmlns="http://www.springframework.org/schema/beans"
    xmlns:mvc="http://www.springframework.org/schema/mvc"
    xmlns:xsi="http://www.w3.org/2001/XMLSchema-instance"
    xsi:schemaLocation="
        http://www.springframework.org/schema/beans
        http://www.springframework.org/schema/beans/spring-beans.xsd
        http://www.springframework.org/schema/mvc
        http://www.springframework.org/schema/mvc/spring-mvc.xsd">

    <mvc:annotation-driven/>

</beans>
```

7.8.2 类型转换

默认情况下，Number 和 Date 类型的格式化程序已安装，并且支持 @NumberFormat 和 @DateTimeFormat 注解。如果 Joda 类库存在于类路径中，还会安装对 Joda 时间格式库的全面支持。

在 Java 配置中注册自定义格式化器和转换器的实现如下。

```java
@Configuration
@EnableWebMvc
public class WebConfig implements WebMvcConfigurer {

    @Override
    public void addFormatters(FormatterRegistry registry) {
        // …
    }
}
```

如果是使用基于 XML 的配置，则用法如下。

```xml
<?xml version="1.0" encoding="UTF-8"?>
<beans xmlns="http://www.springframework.org/schema/beans"
    xmlns:mvc="http://www.springframework.org/schema/mvc"
    xmlns:xsi="http://www.w3.org/2001/XMLSchema-instance"
    xsi:schemaLocation="
        http://www.springframework.org/schema/beans
        http://www.springframework.org/schema/beans/spring-beans.xsd
        http://www.springframework.org/schema/mvc
        http://www.springframework.org/schema/mvc/spring-mvc.xsd">

    <mvc:annotation-driven conversion-service="conversionService"/>

    <bean id="conversionService"
            class="org.springframework.format.support.FormattingConversionServiceFactoryBean">
        <property name="converters">
            <set>
                <bean class="org.example.MyConverter"/>
            </set>
        </property>
        <property name="formatters">
            <set>
                <bean class="org.example.MyFormatter"/>
                <bean class="org.example.MyAnnotationFormatterFactory"/>
            </set>
        </property>
        <property name="formatterRegistrars">
            <set>
                <bean class="org.example.MyFormatterRegistrar"/>
            </set>
```

```
        </property>
    </bean>
</beans>
```

7.8.3 验证

默认情况下，如果 bean 验证存在于类路径中，如 Hibernate Validator、LocalValidatorFactoryBean 被注册为全局验证器，则可用于加了 @Valid 和 Validated 的控制器方法参数的验证。

在 Java 配置中，可以自定义全局的 Validator 实例，例如：

```
@Configuration
@EnableWebMvc
public class WebConfig implements WebMvcConfigurer {

    @Override
    public Validator getValidator(); {
        // …
    }
}
```

如果是使用基于 XML 的配置，则用法如下。

```
<?xml version="1.0" encoding="UTF-8"?>
<beans xmlns="http://www.springframework.org/schema/beans"
    xmlns:mvc="http://www.springframework.org/schema/mvc"
    xmlns:xsi="http://www.w3.org/2001/XMLSchema-instance"
    xsi:schemaLocation="
        http://www.springframework.org/schema/beans
        http://www.springframework.org/schema/beans/spring-beans.xsd
        http://www.springframework.org/schema/mvc
        http://www.springframework.org/schema/mvc/spring-mvc.xsd">

    <mvc:annotation-driven validator="globalValidator"/>

</beans>
```

7.8.4 拦截器

在 Java 配置的应用中，注册拦截器用于传入请求，例如：

```
@Configuration
@EnableWebMvc
public class WebConfig implements WebMvcConfigurer {

    @Override
```

```java
    public void addInterceptors(InterceptorRegistry registry) {
        registry.addInterceptor(new LocaleInterceptor());
        registry.addInterceptor(new ThemeInterceptor())
            .addPathPatterns("/**").excludePathPatterns("/admin/**");
        registry.addInterceptor(new SecurityInterceptor())
            .addPathPatterns("/secure/*");
    }
}
```

如果是使用基于 XML 的配置，则用法如下。

```xml
<mvc:interceptors>
    <bean class="org.springframework.web.servlet.i18n.LocaleChangeInterceptor"/>
    <mvc:interceptor>
        <mvc:mapping path="/**"/>
        <mvc:exclude-mapping path="/admin/**"/>
        <bean class="org.springframework.web.servlet.theme.ThemeChangeInterceptor"/>
    </mvc:interceptor>
    <mvc:interceptor>
        <mvc:mapping path="/secure/*"/>
        <bean class="org.example.SecurityInterceptor"/>
    </mvc:interceptor>
</mvc:interceptors>
```

7.8.5 内容类型

根据请求的媒体类型，可以配置 Spring MVC，如 Accept 头、URL 路径扩展、查询参数等。

默认情况下，首先根据类路径依赖关系将 json、xml、rss 和 atom 注册为已知扩展，检查 URL 路径扩展，然后检查 Accept 头。如果将这些默认值仅更改为 Accept header，并且使用基于 URL 的内容类型解析，则需要考虑路径扩展中的查询参数策略。更多详细信息，请参见后缀匹配和 RFD。

在 Java 配置中，自定义请求的内容类型示例如下。

```java
@Configuration
@EnableWebMvc
public class WebConfig implements WebMvcConfigurer {

    @Override
    public void configureContentNegotiation(ContentNegotiationConfigurer configurer) {
        configurer.mediaType("json", MediaType.APPLICATION_JSON);
    }
}
```

如果是使用基于 XML 的配置，则用法如下。

```xml
<mvc:annotation-driven content-negotiation-manager="contentNegotiation-
Manager"/>

<bean id="contentNegotiationManager"
    class="org.springframework.web.accept.ContentNegotiationManagerFac-
toryBean">
    <property name="mediaTypes">
        <value>
            json=application/json
            xml=application/xml
        </value>
    </property>
</bean>
```

7.8.6 消息转换器

自定义 HttpMessageConverter 可以在 Java 配置中通过覆盖 configureMessageConverters() 方法来实现，如果想要替换由 Spring MVC 创建的默认转换器，或者只想定制默认转换器或将其他转换器添加到默认转换器，则可以重写 extendMessageConverters() 方法。

以下示例展示了如何添加 Jackson JSON 和 XML 转换器的自定义 ObjectMapper。

```java
@Configuration
@EnableWebMvc
public class WebConfiguration implements WebMvcConfigurer {

    @Override
    public void configureMessageConverters(List<HttpMessageConverter<?>>
converters) {
        Jackson2ObjectMapperBuilder builder = new Jackson2ObjectMapper-
Builder()
                .indentOutput(true)
                .dateFormat(new SimpleDateFormat("yyyy-MM-dd"))
                .modulesToInstall(new ParameterNamesModule());
        converters.add(new MappingJackson2HttpMessageConverter(builder.
build()));
        converters.add(new MappingJackson2XmlHttpMessageConverter(build-
er.xml().build()));
    }
}
```

如果是使用基于 XML 的配置，则用法如下。

```xml
<mvc:annotation-driven>
    <mvc:message-converters>
        <bean class="org.springframework.http.converter.json.Mapping-
Jackson2HttpMessageConverter">
            <property name="objectMapper" ref="objectMapper"/>
```

```xml
        </bean>
        <bean class="org.springframework.http.converter.xml.Mapping-
Jackson2XmlHttpMessageConverter">
            <property name="objectMapper" ref="xmlMapper"/>
        </bean>
    </mvc:message-converters>
</mvc:annotation-driven>

<bean id="objectMapper" class="org.springframework.http.converter.json.
Jackson2ObjectMapperFactoryBean"
      p:indentOutput="true"
      p:simpleDateFormat="yyyy-MM-dd"
      p:modulesToInstall="com.fasterxml.jackson.module.paramnames.
ParameterNamesModule"/>

<bean id="xmlMapper" parent="objectMapper" p:createXmlMapper="true"/>
```

7.8.7 视图控制器

视图控制器是定义一个 ParameterizableViewController 的快捷方式，它可以在被调用时立即将请求转发到视图。如果在视图生成响应之前没有执行 Java 控制器逻辑，则在静态情况下使用。

以下是在 Java 中将 "/" 请求转发到名为 "home" 的视图的示例。

```java
@Configuration
@EnableWebMvc
public class WebConfig implements WebMvcConfigurer {

    @Override
    public void addViewControllers(ViewControllerRegistry registry) {
        registry.addViewController("/").setViewName("home");
    }
}
```

如果是使用基于 XML 的配置，则用法如下。

```xml
<mvc:view-controller path="/" view-name="home"/>
```

7.8.8 视图解析器

MVC 配置简化了视图解析器的注册。

以下是 Java 配置示例，它使用 FreeMarker HTML 模板和 Jackson 作为 JSON 呈现的视图解析器。

```java
@Configuration
@EnableWebMvc
public class WebConfig implements WebMvcConfigurer {
```

```
    @Override
    public void configureViewResolvers(ViewResolverRegistry registry) {
        registry.enableContentNegotiation(new MappingJackson2Json-
View());
        registry.jsp();
    }
}
```

如果使用基于 XML 的配置，则用法如下。

```xml
<mvc:view-resolvers>
    <mvc:content-negotiation>
        <mvc:default-views>
            <bean class="org.springframework.web.servlet.view.json.MappingJackson2JsonView"/>
        </mvc:default-views>
    </mvc:content-negotiation>
    <mvc:jsp/>
</mvc:view-resolvers>
```

7.8.9 静态资源

静态资源选项提供了一种便捷的方式来从基于资源的位置列表中提供静态资源。

在下面的示例中，如果请求以"/resources"开头，则会使用相对路径查找并提供 Web 应用程序根目录"/public"或"/static"下的类路径的静态资源，这些资源将在未来 1 年内到期，以确保最大限度地利用浏览器缓存并减少浏览器发出的 HTTP 请求。Last-Modified 头也被评估，如果存在，则返回 304 状态码。

```java
@Configuration
@EnableWebMvc
public class WebConfig implements WebMvcConfigurer {

    @Override
    public void addResourceHandlers(ResourceHandlerRegistry registry) {
        registry.addResourceHandler("/resources/**")
            .addResourceLocations("/public", "classpath:/static/")
            .setCachePeriod(31556926);
    }
}
```

如果使用基于 XML 的配置，则用法如下。

```xml
<mvc:resources mapping="/resources/**"
    location="/public, classpath:/static/"
    cache-period="31556926" />
```

7.8.10 DefaultServletHttpRequestHandler

DefaultServletHttpRequestHandler 允许将 DispatcherServlet 映射到 "/"（从而覆盖容器默认的 Servlet 映射），同时仍允许静态资源请求由容器的默认 Servlet 处理。它使用 "/**" 的 URL 映射和相对于其他 URL 映射的最低优先级来配置 DefaultServletHttpRequestHandler。

该处理程序将把所有请求转发给默认的 Servlet。重要的是，它会保持最后的所有其他 URL HandlerMapping 的顺序。如果开发者使用或者设置自定义 HandlerMapping 实例，则要确保将其顺序属性设置为低于 DefaultServletHttpRequestHandler 的值（Integer.MAX_VALUE）。如果要启用该处理程序，则使用以下代码。

```
@Configuration
@EnableWebMvc
public class WebConfig implements WebMvcConfigurer {

    @Override
    public void configureDefaultServletHandling(DefaultServletHandler-
Configurer configurer) {
        configurer.enable();
    }
}
```

如果使用基于 XML 的配置，则用法如下。

```
<mvc:default-servlet-handler/>
```

7.8.11 路径匹配

路径匹配允许自定义与 URL 匹配和 URL 处理相关的选项。

在 Java 中配置的示例如下。

```
@Configuration
@EnableWebMvc
public class WebConfig implements WebMvcConfigurer {

    @Override
    public void configurePathMatch(PathMatchConfigurer configurer) {
        configurer
            .setUseSuffixPatternMatch(true)
            .setUseTrailingSlashMatch(false)
            .setUseRegisteredSuffixPatternMatch(true)
            .setPathMatcher(antPathMatcher())
            .setUrlPathHelper(urlPathHelper());
    }

    @Bean
```

```
    public UrlPathHelper urlPathHelper() {
        //…
    }

    @Bean
    public PathMatcher antPathMatcher() {
        //…
    }
}
```

如果使用基于 XML 的配置，则用法如下。

```
<mvc:annotation-driven>
    <mvc:path-matching
        suffix-pattern="true"
        trailing-slash="false"
        registered-suffixes-only="true"
        path-helper="pathHelper"
        path-matcher="pathMatcher"/>
</mvc:annotation-driven>

<bean id="pathHelper" class="org.example.app.MyPathHelper"/>
<bean id="pathMatcher" class="org.example.app.MyPathMatcher"/>
```

7.9 实战：基于Spring Web MVC的REST接口

本节将基于 Spring Web MVC 技术来实现 REST 接口。

示例程序名为"mvc-rest"，其作用是实现简单的 REST 样式的 API。

7.9.1 接口设计

本小节将在系统中实现两个接口，分别为 GET http://localhost:8080/hello 和 GET http://localhost:8080/hello/way。

其中，第一个接口"/hello"将会返回"Hello World!"字符串；第二个接口"/hello/way"将会返回一个包含用户信息的 JSON 字符串。

7.9.2 系统配置

需要在应用中添加如下依赖。

```xml
<properties>
    <spring.version>5.1.5.RELEASE</spring.version>
    <jetty.version>9.4.14.v20181114</jetty.version>
    <jackson.version>2.9.7</jackson.version>
</properties>
<dependencies>
    <dependency>
        <groupId>org.springframework</groupId>
        <artifactId>spring-webmvc</artifactId>
        <version>${spring.version}</version>
    </dependency>
    <dependency>
        <groupId>org.eclipse.jetty</groupId>
        <artifactId>jetty-servlet</artifactId>
        <version>${jetty.version}</version>
        <scope>provided</scope>
    </dependency>
    <dependency>
        <groupId>com.fasterxml.jackson.core</groupId>
        <artifactId>jackson-core</artifactId>
        <version>${jackson.version}</version>
    </dependency>
    <dependency>
        <groupId>com.fasterxml.jackson.core</groupId>
        <artifactId>jackson-databind</artifactId>
        <version>${jackson.version}</version>
    </dependency>
</dependencies>
```

其中，spring-webmvc 是为了使用 Spring MVC 的功能；jetty-servlet 是为了提供内嵌的 Servlet 容器，这样就无须依赖外部容器，可以直接运行应用；jackson-core 和 jackson-databind 为应用提供 JSON 序列化的功能。

7.9.3 后台编码实现

后台编码实现如下。

1. 领域模型

创建一个 User 类，代表用户信息，其代码如下。

```java
public class User {
    private String username;
    private Integer age;

    public User(String username, Integer age) {
        this.username = username;
        this.age = age;
```

```
    }

    public String getUsername() {
        return username;
    }

    public void setUsername(String username) {
        this.username = username;
    }

    public Integer getAge() {
        return age;
    }

    public void setAge(Integer age) {
        this.age = age;
    }
}
```

2. 控制器

创建 HelloController，用于处理用户的请求，其代码如下。

```
@RestController
public class HelloController {

    @RequestMapping("/hello")
    public String hello() {
        return "Hello World! Welcome to visit waylau.com!";
    }

    @RequestMapping("/hello/way")
    public User helloWay() {
        return new User("Way Lau", 30);
    }
}
```

其中，映射到"/hello"的方法将会返回"Hello World!"字符串；而映射到"/hello/way"将会返回一个包含用户信息的 JSON 字符串。

7.9.4 应用配置

在本应用中，采用基于 Java 注解的配置。

AppConfiguration 主应用配置如下。

```
import org.springframework.context.annotation.ComponentScan;
import org.springframework.context.annotation.Configuration;
```

```java
import org.springframework.context.annotation.Import;

@Configuration
@ComponentScan(basePackages = { "com.waylau.spring" })
@Import({ MvcConfiguration.class })
public class AppConfiguration {

}
```

上述配置中，AppConfiguration 会扫描 "com.waylau.spring" 包下的文件，并自动将相关的 bean 进行注册。

AppConfiguration 同时又引入了 MVC 的配置类 MvcConfiguration，其配置如下。

```java
@EnableWebMvc
@Configuration
public class MvcConfiguration implements WebMvcConfigurer {

    public void extendMessageConverters(List<HttpMessageConverter<?>> converters) {
        converters.add(new MappingJackson2HttpMessageConverter());
    }
}
```

MvcConfiguration 配置类一方面启用了 MVC 的功能，另一方面添加了 Jackson JSON 的转换器。

最后，需要引入 Jetty 服务器 JettyServer，其配置如下。

```java
import org.eclipse.jetty.server.Server;
import org.eclipse.jetty.servlet.ServletContextHandler;
import org.eclipse.jetty.servlet.ServletHolder;
import org.springframework.web.context.ContextLoaderListener;
import org.springframework.web.context.WebApplicationContext;
import org.springframework.web.context.support.AnnotationConfigWebApplicationContext;
import org.springframework.web.servlet.DispatcherServlet;
import com.waylau.spring.mvc.configuration.AppConfiguration;

public class JettyServer {
    private static final int DEFAULT_PORT = 8080;
    private static final String CONTEXT_PATH = "/";
    private static final String MAPPING_URL = "/*";

    public void run() throws Exception {
        Server server = new Server(DEFAULT_PORT);
        server.setHandler(servletContextHandler(webApplicationContext()));
        server.start();
        server.join();
    }
```

```
    private ServletContextHandler servletContextHandler(WebApplication-
Context context) {
        ServletContextHandler handler = new ServletContextHandler();
        handler.setContextPath(CONTEXT_PATH);
        handler.addServlet(new ServletHolder(new DispatcherServlet(con-
text)),
            MAPPING_URL);
        handler.addEventListener(new ContextLoaderListener(context));
        return handler;
    }

    private WebApplicationContext webApplicationContext() {
        AnnotationConfigWebApplicationContext context =
            new AnnotationConfigWebApplicationContext();
        context.register(AppConfiguration.class);
        return context;
    }
}
```

JettyServer 将会在 Application 类中进行启动，其代码如下。

```
public class Application {

    public static void main(String[] args) throws Exception {
        new JettyServer().run();;
    }

}
```

7.9.5 运行应用

在编辑器中，直接运行 Application 类即可。启动后，应能看到如下控制台信息。

```
2018-12-12 19:56:29.030:INFO::main: Logging initialized @2488ms to org.
eclipse.jetty.util.log.StdErrLog
2018-12-12 19:56:30.101:INFO:oejs.Server:main: jetty-9.4.14.v20181114;
built: 2018-11-14T21:20:31.478Z; git: c4550056e785fb5665914545889f21d-
c136ad9e6; jvm 1.8.0_162-b12
2018-12-12 19:56:30.340:INFO:oejshC.ROOT:main: Initializing Spring root
WebApplicationContext
十二月 12, 2018 7:56:30 下午 org.springframework.web.context.ContextLoad-
er initWebApplicationContext
信息: Root WebApplicationContext: initialization started
十二月 12, 2018 7:56:34 下午 org.springframework.web.context.ContextLoad-
er initWebApplicationContext
信息: Root WebApplicationContext initialized in 3736 ms
2018-12-12 19:56:34.154:INFO:oejshC.ROOT:main: Initializing Spring
```

```
DispatcherServlet 'org.springframework.web.servlet.DispatcherServ-
let-4e04a765'
十二月 12, 2018 7:56:34 下午 org.springframework.web.servlet.Framework-
Servlet initServletBean
信息: Initializing Servlet 'org.springframework.web.servlet.Dispatch-
erServlet-4e04a765'
十二月 12, 2018 7:56:34 下午 org.springframework.web.servlet.Framework-
Servlet initServletBean
信息: Completed initialization in 7 ms
2018-12-12 19:56:34.163:INFO:oejsh.ContextHandler:main: Started
o.e.j.s.ServletContextHandler@60df60da{/,null,AVAILABLE}
2018-12-12 19:56:34.955:INFO:oejs.AbstractConnector:main: Started
ServerConnector@2c07545f{HTTP/1.1,[http/1.1]}{0.0.0.0:8080}
2018-12-12 19:56:34.956:INFO:oejs.Server:main: Started @8536ms
```

在浏览器中分别访问 http://localhost:8080/hello 和 http://localhost:8080/hello/way 地址进行测试，能看到图 7-2 和图 7-3 所示的响应效果。

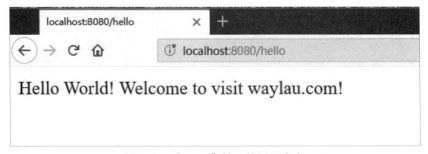

图7-2　"/hello"接口的返回内容

图7-3　"/hello/way"接口的返回内容

第8章 Spring Security

在任何应用中都不可忽视安全的重要性。Spring Security为基于Java EE的企业级软件应用程序提供全面的安全服务。特别是使用Spring框架构建的项目，可以更好地使用Spring Security来加快构建的速度。

8.1 基于角色的权限管理

本节将讨论权限管理中角色的概念,以及基于角色的机制来进行权限管理。

8.1.1 角色的概念

当说到程序的权限管理时,人们往往会想到"角色"这一概念。角色是代表一系列行为或责任的实体,用于限定在软件系统中能做什么、不能做什么。用户账号往往与角色相关联,因此,一个用户在软件系统中能做什么,取决于与之关联的角色什么。

例如,一个用户以关联了"项目管理员"角色的账号登录系统,那这个用户就可以做项目管理员能做的所有事情,如列出项目中的应用、管理项目组成员、生成项目报表等。

从这个意义上来说,角色更多的是一种行为的概念,表示用户能在系统中进行的操作。

8.1.2 基于角色的访问控制

既然角色代表了可执行的操作,那么一个合乎逻辑的做法是,在软件开发中使用角色来控制对软件功能和数据的访问。这种权限控制方法就称为基于角色的访问控制(Role Based Access Control,RBAC。)

有两种已经在使用的 RBAC 方式:隐式访问控制和显式访问控制。

1. 隐式访问控制(ImplicitAccessControl)

前面提到,角色代表一系列可执行的操作。但如何知道一个角色到底关联了哪些可执行的操作呢?答案是,目前大多数的应用中,用户并不能明确地知道一个角色到底关联了哪些可执行操作。可能用户心里是清楚的(例如,一个有"管理员"角色的用户可以锁定用户账号、进行系统配置;一个关联了"消费者"这一角色的用户可在网站上选购商品),但这些应用系统并没有明确定义一个角色到底包含哪些可执行的操作。

以"项目管理员"角色来说,系统中并没有对其能进行什么样的操作进行明确定义,它仅是一个字符串名词。开发人员通常将这个名词写在程序里以进行访问控制。例如,判断一个用户是否能查看项目报表,程序员可能会进行如下编码。

```
if (user.hasRole("Project Manager") ) {
    // 显示报表按钮
} else {
    // 不显示按钮
}
```

在上面的示例中,开发人员判断用户是否有项目管理员角色来决定是否显示查看项目报表的按钮。注意,上面的代码中并没有明确的语句来定义项目管理员这一角色到底包含哪些可执行的操作。

它只是假设一个关联了项目管理员角色的用户可查看项目报表,而开发人员也是基于这一假设来写 if/else 语句的。这种方式就是基于角色的隐式访问控制。

然而,这种权限访问控制是非常脆弱的。一个极小的权限方面的需求变更都可能导致上面的代码需要重新修改。

举例来说,假如某一天这个开发团队被告知,需要让一个"部门管理员"角色也可以查看项目报表,那么之前的隐式访问控制的代码就要修改为:

```
if (user.hasRole("Project Manager") || user.hasRole("Department Manager")) {
    // 显示报表按钮
} else {
    // 不显示按钮
}
```

随后,开发人员需要更新测试用例、重新编译系统,还可能需要重走软件质量控制(QA)流程,然后再重新部署上线。这一切仅仅是因为一个极小的权限方面的需求变更。后面如果需求方又说让另一个角色也可查看报表,或是前面关于"部门管理员可查看报表"的需求不再需要了,该怎么办?如果需求方要求动态地创建、删除角色,以便自行配置角色,又该如何应对呢?

像上面的情况,这种隐式的(静态字符串)基于角色的访问控制难以满足需求。理想的情况是,如果权限需求发生变更,不需要修改任何代码。

2. 显式访问控制(ExplicitAccessControl)

怎样才能达到上面所说的理想情况呢?实际上,可以通过显式地(明确地)界定在应用中能执行的操作来实现。回顾上面隐式的权限控制的例子,思考一下这些代码最终的目的及最终要做什么样的控制。从根本上来说,这些代码最终是在保护资源(项目报表),是要界定一个用户能对这些资源执行什么样的操作(查看/修改)。当将权限访问控制分解到这种最原始的层次,就可以用一种更细粒度、更富有弹性的方式来表达权限控制策略。

可以修改上面的代码块,以基于资源的语义来更有效地进行权限访问控制。

```
if (user.isPermitted("projectReport:view:12345")) {
    // 显示报表按钮
} else {
    // 不显示按钮
}
```

上面的例子中,可明确地看到是在控制什么。不要太在意冒号分隔的语法,这仅是一个例子,重点是上面的语句明确地表示了"如果当前用户允许查看编号为 12345 的项目报表,则显示项目报表按钮"。也就是说,明确地说明了一个用户账号可对一个资源实例进行的具体操作。

8.1.3 哪种方式更好

显式访问控制的代码块与隐式访问控制的代码的主要区别在于，显式访问控制的代码块是基于什么是受保护的，而不是谁可能有能力做什么。二者的区别看似简单，但后者对系统开发及部署有着深刻的影响。显式访问控制与隐式访问控制相比，具有以下优势。

（1）更少的代码重构：因为是基于系统的功能（系统的资源及对资源的操作）来进行权限控制的，所以系统的功能需求一旦确定下来，一段时间内对它的改动是比较少的。只有当系统的功能需求改变时，才会涉及权限代码的改变。例如，上面提到的查看项目报表的功能，显式的权限控制不会像传统隐式的权限控制那样，因不同的用户/角色要进行某个操作就需要重构代码。只要这个功能存在，显式的权限控制代码就不需要改变。

（2）资源和操作更直观：保护资源对象、控制对资源对象的操作，这样的权限控制方式更符合人们的思想习惯。正因为符合这种直观的思维方式，所以面向对象的编辑思想及 REST 通信模型被广泛采用。

（3）安全模型更有弹性：隐式的示例代码中没有确定哪些用户/组/角色可对资源进行什么操作，这意味着它可支持任何安全模型的设计。例如，可以将操作（权限）直接分配给用户，或者将操作（权限）分配到一个角色，然后再将角色与用户关联，或者将多个角色关联到组（Group）上，等等。完全可以根据应用的特点定制权限模型。

（4）外部安全管理更高效：由于源代码只反映资源/行为，而不是用户/组/角色，因此资源/行为与用户/组/角色的关联可以通过外部的模块/专用工具/管理控制台来完成。这意味着在权限需求变化时，开发人员并不需要花费时间来修改代码，业务分析师甚至最终用户就可以通过相应的管理工具来修改权限策略配置。

（5）运行时可做修改：因为基于资源的权限控制代码并不依赖于行为的主体（如组/角色/用户），开发者并没有将行为主体的字符名词写在代码中，所以开发者甚至可以在程序运行时通过修改主体对资源进行配置的方式来应对权限方面的需求变更，再也不需要像隐式的访问控制那样重构代码。显式访问控制更适合当前的软件应用。

不管是隐式访问控制还是显式访问控制，都有其合适的场景。庆幸的是，在 Java 平台有很多现成的现代权限管理框架可供选择，如 Apache Shiro（http://shiro.apache.org/）和 Spring Security（http://projects.spring.io/spring-security/），一个是以简洁好用而被业界广泛应用，另一个则以功能强大而著称。

关于这两个框架的用法，读者可以参考笔者的另外两本开源书《Apache Shiro 1.2.x 参考手册》（https://github.com/waylau/apache-shiro-1.2.x-reference）及《Spring Security 教程》（https://github.com/waylau/spring-security-tutorial）。

8.2 Spring Security基础

Spring Security 的出现有很多原因，但主要是由于 Java EE 的 Servlet 规范和 EJB 规范缺乏对企业应用的安全性方面的支持。Spring Security 能够解决安全性方面的问题，并拥有数十个其他有用的可自定义的安全功能。

在 Java 领域，另一个值得关注的安全框架是 Apache Shiro。但与 Apache Shiro 相比，Spring Security 的功能更加强大，与 Spring 的兼容也更好。

8.2.1　Spring Security的认证模型

应用程序安全性主要包括认证（Authentication）与授权（Authorization）两个方面。

（1）认证：指建立主体（Principal）的过程。主体通常是指可以在应用程序中执行操作的用户、设备或其他系统。

（2）授权：或称为访问控制（Access-Control），是指决定是否允许主体在应用程序中执行操作。为了到达需要授权的点，认证过程已经建立了主体的身份。

在认证方面，Spring Security 支持各种各样的认证模型。这些认证模型大多是由第三方提供的，或者由诸如互联网工程任务组的相关标准机构开发。此外，Spring Security 提供了一组认证功能。具体来说，Spring Security 目前支持以下所有技术的身份验证集成。

- HTTP BASIC 认证头（基于 IETF RFC 的标准）
- HTTP Digest 认证头（基于 IETF RFC 的标准）
- HTTP X.509 客户端证书交换（基于 IETF RFC 的标准）
- LDAP（一种常见的跨平台身份验证需求，特别是在大型环境中比较常用）
- 基于表单的身份验证（用于简单的用户界面需求）
- OpenID 身份验证
- 基于预先建立的请求头的验证（如 Computer Associates Siteminder）
- Jasig Central Authentication Service（CAS，一个流行的开源单点登录系统）
- 远程方法调用（RMI）和 HttpInvoker（Spring 远程协议）的透明认证上下文传播
- 自动 "remember-me" 身份验证（可以选中一个框，以避免在预定时间段内重新验证）
- 匿名身份验证（允许每个未经身份验证的调用自动承担特定的安全身份）
- Run-as 身份验证（一个调用应使用不同的安全身份继续运行，这是有用的）
- Java 认证和授权服务（Java Authentication and Authorization Service，JAAS）
- Java EE 容器认证（如果需要，仍然可以使用容器管理身份验证）
- Kerberos

第三方公司或者社区也贡献了诸多特性，可以方便地集成到 Spring Security 应用中，具体如下。

- Java Open Source Single Sign-On（JOSSO）
- OpenNMS Network Management Platform
- AppFuse
- AndroMDA
- Mule ESB
- Direct Web Request（DWR）
- Grails
- Tapestry
- JTrac
- Jasypt
- Roller
- Elastic Path
- Atlassian Crowd

许多独立软件供应商（ISV）选择采用 Spring Security，就是出于这种灵活的认证模型。这样可以快速地将解决方案与最终客户需求进行组合，从而避免进行大量的工作或需求变更。如果上述认证机制都不符合需求，那么也可以基于 Spring Security 来实现自己的认证机制。

Spring Security 还提供了一组深层次的授权功能，其包括以下三个主要方面。

（1）对 Web 请求进行授权。

（2）授权某个方法是否可以被调用。

（3）授权访问单个领域对象实例。

8.2.2　Spring Security的安装

Spring Security 的安装非常简单，以下展示了如何采用 Maven 和 Gradle 两种方式来安装。

1. 使用Maven

使用 Maven 的最少依赖如下所示。

```
<dependencyManagement>
    <dependencies>
        <!-- …省略其他依赖 -->
        <dependency>
            <groupId>org.springframework.security</groupId>
            <artifactId>spring-security-bom</artifactId>
```

```xml
            <version>5.2.0.BUILD-SNAPSHOT</version>
            <type>pom</type>
            <scope>import</scope>
        </dependency>
    </dependencies>
</dependencyManagement>


<dependencies>
    <!-- …省略其他依赖 -->
    <dependency>
        <groupId>org.springframework.security</groupId>
        <artifactId>spring-security-web</artifactId>
    </dependency>
    <dependency>
        <groupId>org.springframework.security</groupId>
        <artifactId>spring-security-config</artifactId>
    </dependency>
</dependencies>
```

2. 使用Gradle

使用 Gradle 的最少依赖如下所示。

```
plugins {
    id "io.spring.dependency-management" version "1.0.6.RELEASE"
}

dependencyManagement {
    imports {
        mavenBom 'org.springframework.security:spring-securi-
ty-bom:5.2.0.BUILD-SNAPSHOT'
    }
}

// …省略其他依赖

dependencies {
    compile "org.springframework.security:spring-security-web"
    compile "org.springframework.security:spring-security-config"
}
```

8.2.3 模块

自 Spring 3 开始，Spring Security 将代码划分到不同的 jar 中，这使不同的功能模块和第三方依赖显得更加清晰。Spring Security 主要包括以下 9 个核心模块。

1. Core-spring-security-core.jar

该 jar 包含核心 authentication 和 authorization 的类和接口、远程支持和基础配置 API。任何使用 Spring Security 的应用都需要引入这个 jar。它支持本地应用、远程客户端、方法级别的安全配置和 JDBC 用户配置，其主要包含的顶级包如下。

（1）org.springframework.security.core：核心。

（2）org.springframework.security.access：访问。

（3）org.springframework.security.authentication：认证。

（4）org.springframework.security.provisioning：配置。

2. Remoting-spring-security-remoting.jar

该 jar 提供与 Spring Remoting 整合的支持，开发者并不需要这个，除非需要使用 Spring Remoting 写一个远程客户端，其主包为 org.springframework.security.remoting。

3. Web-spring-security-web.jar

该 jar 包含 Filter 和关于 Web 安全的基础代码，如果使用 Spring Security 进行 Web 安全认证和基于 URL 的访问控制，则需要它。其主包为 org.springframework.security.web。

4. Config-spring-security-config.jar

该 jar 包含安全命名空间解析代码和 Java 配置代码。如果使用 Spring Security XML 命名空间进行配置或 Spring Security 的 Java 配置支持，则需要它，其主包为 org.springframework.security.config。注意，不要在代码中直接使用这个 jar 中的类。

5. LDAP-spring-security-ldap.jar

该 jar 用于进行 LDAP 认证和配置代码。如果进行 LDAP 认证或管理 LDAP 用户实体，则需要它，其顶级包为 org.springframework.security.ldap。

6. ACL-spring-security-acl.jar

该 jar 用于进行特定领域对象的 ACL（访问控制列表）实现。使用它可以对特定对象的实例进行一些安全配置，其顶级包为 org.springframework.security.acls。

7. CAS-spring-security-cas.jar

该 jar 用于进行 Spring Security CAS 客户端集成。如果使用一个单点登录服务器进行 Spring Security Web 安全认证，则需要引入它，其顶级包为 org.springframework.security.cas。

8. OpenID-spring-security-openid.jar

该 jar 用于进行 OpenID Web 认证支持。如果基于一个外部 OpenID 服务器对用户进行验证，则需要使用它，其顶级包为 org.springframework.security.openid。

一般情况下，spring-security-core 和 spring-security-config 都会引入该 jar，在 Web 开发中，通常还会引入 spring-security-web。

9. Test-spring-security-test.jar

该 jar 用于测试 Spring Security，在开发环境中通常需要添加该包。

8.2.4　Spring Security 5的新特性及高级功能

本书案例采用 Spring Security 5 进行编写。Spring Security 5 相对于之前的版本，主要新增了如下特性。

（1）为 OAuth 2.0 登录添加了支持。

（2）支持初始响应式编程。有关响应式编程方面的内容，可以参阅笔者所著的《Spring 5 开发大全》。

针对 Web 方面的开发，Spring Security 提供的高级功能如下。

1. Remember-Me认证

Remember-Me 身份验证是指网站能够记住身份之间的会话，这通常是通过发送 cookie 到浏览器，cookie 在未来会话中被检测到，并导致自动登录发生。Spring Security 为这些操作提供了必要的钩子，并且有两个具体的实现。

（1）使用散列来保存基于 cookie 的令牌的安全性。

（2）使用数据库或其他持久存储机制来存储生成的令牌。

在本书的案例中，将会通过散列的方式来实现 Remember-Me 认证。

2. 使用HTTPS

可以使用 requires-channel 属性直接支持 URL 采用 HTTPS 协议。

```
<http>
    <intercept-url pattern="/secure/**" access="ROLE_USER"
        requires-channel="https"/>
    <intercept-url pattern="/**" access="ROLE_USER"
        requires-channel="any"/>
    …
</http>
```

注意，当用户尝试使用 HTTP 访问与"/secure/**"模式匹配的任何内容时，都会首先将其重定向到 HTTPS 的 URL 上。

如果开发者的应用程序想使用 HTTP/HTTPS 的非标准端口，则可以指定端口映射列表，如下所示。

```
<http>
    …
    <port-mappings>
        <port-mapping http="9080" https="9443"/>
    </port-mappings>
</http>
```

注意，安全起见，应用程序应该始终采用 HTTPS 在整个过程中进行安全连接，以避免中间人发起攻击。

3. 会话管理

在会话管理方面，Spring Security 提供了诸如检测超时、控制并发会话、防御会话固定攻击等功能。

（1）检测超时。开发者可以配置 Spring Security，以检测提交的无效会话 ID，并将用户重定向到适当的 URL。这是通过 session-management 元素实现的。

```
<http>
    …
    <session-management invalid-session-url="/invalidSession.htm" />
</http>
```

注意，使用此机制来检测超时，如果用户注销，然后在不关闭浏览器的情况下重新登录，则可能会报告错误。这是因为当会话 cookie 无效时，会话 cookie 不会被清除，即使用户已经注销，也将被重新提交。开发者可能需要在注销时显式删除 JSESSIONID cookie，例如，在注销处理程序中使用以下语法。

```
<http>
    <logout delete-cookies="JSESSIONID" />
</http>
```

但这种用法并不是每个 Servlet 容器都支持，所以开发者需要在自己的环境中测试它。

（2）控制并发会话。如果开发者希望对单个用户登录应用程序的能力加以限制，Spring Security 可支持添加以下功能。首先，需要将以下监听器添加到 web.xml 文件中，以使 Spring Security 更新有关会话的生命周期事件。

```
<listener>
    <listener-class>
    org.springframework.security.web.session.HttpSessionEventPublisher
    </listener-class>
</listener>
```

然后将以下内容添加到应用程序上下文中。

```
<http>
    …
    <session-management>
    <concurrency-control max-sessions="1" />
    </session-management>
</http>
```

这将阻止用户多次登录（第二次登录将导致第一次登录无效）。通常，开发者更希望防止第二次登录，在这种情况下可以使用：

```
<http>
    …
    <session-management>
    <concurrency-control max-sessions="1"
        error-if-maximum-exceeded="true" />
    </session-management>
</http>
```

第二次登录将被拒绝。如果正在使用基于表单的登录,用户被拒绝登录的信息将被发送到 authentication-failure-url。如果第二次认证是通过另一个非交互机制进行的,如 Remember-Me,则会向客户端发送 unauthorized(401)错误。如果想要使用错误页面,则可以将 session-authentication-error-url 属性添加到 session-management 元素中。

(3)防御会话固定攻击。会话固定攻击(Session Fixation Attacks)是潜在的风险,恶意攻击者可能会通过访问站点来创建会话,然后通过这个会话进行攻击(拥有会话在一定程度上表明攻击者通过了认证)。而 Spring Security 通过创建新会话或在用户登录时以其他方式更改会话 ID 来自动防范此问题。如果不需要此保护或此保护与其他要求冲突,则可以在元素中使用 session-fixation-protection 属性来进行设置。

4. 支持OpenID

Spring Security 命名空间支持 OpenID 登录,例如:

```
<http>
    <intercept-url pattern="/**" access="ROLE_USER" />
    <openid-login />
</http>
```

在 OpenID 中进行注册,并将用户信息添加到内存中。

```
<user name="http://jimi.hendrix.myopenid.com/"
    authorities="ROLE_USER"/>
```

开发者可以登录 myopenid.com 网站进行身份验证,也可以通过在 openid-login 元素上设置 user-service-ref 属性来选择特定的 UserDetailsService bean 并使用 OpenID。

Spring Security 也支持 OpenID 的属性交换。例如,以下配置将尝试从 OpenID 提供程序中检索电子邮件和属性,供应用程序使用。

```
<openid-login>
    <attribute-exchange>
    <openid-attribute name="email"
        type="http://axschema.org/contact/email" required="true"/>
    <openid-attribute name="name" type="http://axschema.org/namePerson"/>
    </attribute-exchange>
</openid-login>
```

5. 自定义过滤器

如果开发者以前使用过 Spring Security，就会知道该框架维护一连串的过滤器。开发者可能希望在特定位置将自己的过滤器添加到堆栈中，或使用当前没有命名空间配置选项（如 CAS）的 Spring Security 过滤器，也可能希望使用自定义版本的标准命名空间过滤器，如由元素创建的UsernamePasswordAuthenticationFilter。

使用命名空间时，应始终严格执行过滤器的顺序。创建应用程序上下文时，过滤器 bean 将通过命名空间处理代码进行排序，并且标准 Spring Security 过滤器在命名空间中具有别名。

8.3 实战：基于Spring Security的安全认证功能

本节将演示基于 Spring Security 的安全认证功能，该应用代码可以在 security-basic 目录下找到。

8.3.1 添加依赖

添加 Spring Security 的依赖时，由于使用的是 Snapshot 版本，因此要配置 Spring Snapshots 仓库，其配置如下。

```xml
<!-- Spring Snapshots仓库 -->
<repositories>
    <repository>
        <id>spring-snapshots</id>
        <name>Spring Snapshots</name>
        <url>https://repo.spring.io/snapshot</url>
    </repository>
</repositories>

<properties>
    <spring.version>5.1.5.RELEASE</spring.version>
    <jetty.version>9.4.14.v20181114</jetty.version>
    <jackson.version>2.9.7</jackson.version>
    <spring-security.version>5.2.0.BUILD-SNAPSHOT</spring-security.version>
</properties>

<dependencies>
    <dependency>
        <groupId>org.springframework</groupId>
        <artifactId>spring-webmvc</artifactId>
        <version>${spring.version}</version>
    </dependency>
    <dependency>
```

```xml
    <groupId>org.eclipse.jetty</groupId>
    <artifactId>jetty-servlet</artifactId>
    <version>${jetty.version}</version>
    <scope>provided</scope>
</dependency>
<dependency>
    <groupId>com.fasterxml.jackson.core</groupId>
    <artifactId>jackson-core</artifactId>
    <version>${jackson.version}</version>
</dependency>
<dependency>
    <groupId>com.fasterxml.jackson.core</groupId>
    <artifactId>jackson-databind</artifactId>
    <version>${jackson.version}</version>
</dependency>

<!-- 安全相关的依赖 -->
<dependency>
    <groupId>org.springframework.security</groupId>
    <artifactId>spring-security-web</artifactId>
    <version>${spring-security.version}</version>
</dependency>
<dependency>
    <groupId>org.springframework.security</groupId>
    <artifactId>spring-security-config</artifactId>
    <version>${spring-security.version}</version>
</dependency>
</dependencies>
```

8.3.2 添加业务代码

业务代码中包含模型和控制器。

1. User模型

User 类的代码如下。

```java
package com.waylau.spring.mvc.vo;

public class User {
    private String username;
    private Integer age;

    public User(String username, Integer age) {
        this.username = username;
        this.age = age;
    }
```

```
    //…省略getter/setter方法
}
```

2. 控制器

控制器 HelloController 的代码如下。

```
package com.waylau.spring.mvc.controller;

import org.springframework.web.bind.annotation.RequestMapping;
import org.springframework.web.bind.annotation.RestController;
import com.waylau.spring.mvc.vo.User;

@RestController
public class HelloController {

    @RequestMapping("/hello")
    public String hello() {
        return "Hello World! Welcome to visit waylau.com!";
    }

    @RequestMapping("/hello/way")
    public User helloWay() {
        return new User("Way Lau", 30);
    }
}
```

上述控制器的逻辑非常简单，当访问"/hello"时，会响应一段文本。当访问"/hello/way"时，会返回一个 POJO 对象，该 POJO 对象可以根据消息转换器的设置生成不同格式的消息。

8.3.3 配置消息转换器

添加 Spring Web MVC 的配置类 MvcConfiguration，并在该配置中启用消息转换器。

```
package com.waylau.spring.mvc.configuration;

import java.util.List;

import org.springframework.context.annotation.Configuration;
import org.springframework.http.converter.HttpMessageConverter;
import org.springframework.http.converter.json.MappingJackson2HttpMessageConverter;
import org.springframework.web.servlet.config.annotation.EnableWebMvc;
import org.springframework.web.servlet.config.annotation.WebMvcConfigurer;
```

```
@EnableWebMvc // 启用MVC
@Configuration
public class MvcConfiguration implements WebMvcConfigurer {

    public void extendMessageConverters(List<HttpMessageConverter<?>> cs) {
        // 使用Jackson JSON来进行消息转换
        cs.add(new MappingJackson2HttpMessageConverter());
    }
}
```

由于预先在 pom.xml 中添加了 Jackson JSON 的依赖，因此可以使用 Jackson JSON 来进行消息转换，将响应消息体转为 JSON 格式。

8.3.4 配置Spring Security

以下是针对 Spring Security 的配置。

```
package com.waylau.spring.mvc.configuration;

import org.springframework.context.annotation.Bean;
import org.springframework.security.config.annotation.web.builders.HttpSecurity;
import org.springframework.security.config.annotation.web.configuration.EnableWebSecurity;
import org.springframework.security.config.annotation.web.configuration.WebSecurityConfigurerAdapter;
import org.springframework.security.core.userdetails.User;
import org.springframework.security.core.userdetails.UserDetailsService;
import org.springframework.security.provisioning.InMemoryUserDetailsManager;

@EnableWebSecurity // 启用Spring Security功能
public class WebSecurityConfig
    extends WebSecurityConfigurerAdapter {

    /**
     * 自定义配置
     */
    @Override
    protected void configure(HttpSecurity http) throws Exception {
        http.authorizeRequests().anyRequest().authenticated()//所有请求都需认证
            .and()
            .formLogin() // 使用form表单登录
            .and()
            .httpBasic(); // HTTP基本认证
    }
```

```java
@SuppressWarnings("deprecation")
@Bean
public UserDetailsService userDetailsService() {
    InMemoryUserDetailsManager manager = 
            new InMemoryUserDetailsManager();
    
    manager.createUser(
            User.withDefaultPasswordEncoder()   // 密码编码器
                .username("waylau")    // 用户名
                .password("123")       // 密码
                .roles("USER")         // 角色
                .build()
            );
    return manager;
}
```

在上述配置中，要启动 Spring Security 功能，需要在配置类上添加 @EnableWebSecurity 注解。

安全配置类 WebSecurityConfig 继承自 WebSecurityConfigurerAdapter。WebSecurityConfigurer-Adapter 提供用于创建一个 Websecurityconfigurer 实例的基类，允许重写其方法。这里重写了 configure 方法：authorizeRequests().anyRequest().authenticated() 方法意味着所有请求都需要认证；formLogin() 方法表明这是个基于表单的身份验证；httpBasic() 方法表明该认证是一个 HTTP 基本认证。

UserDetailsService 用于提供身份认证信息。本例中使用了基于内存的信息管理器 InMemory-UserDetailsManager，同时初始化了一个用户名为 "waylau"、密码为 "123"、角色为 "USER" 的身份信息。withDefaultPasswordEncoder() 方法指定了该用户身份信息使用默认的密码编码器。

8.3.5 创建应用配置类

AppConfiguration 是整个应用的配置类，用于导入 Spring Web MVC 及 Spring Security 的配置信息，其代码如下。

```java
import org.springframework.context.annotation.ComponentScan;
import org.springframework.context.annotation.Configuration;
import org.springframework.context.annotation.Import;

@Configuration
@ComponentScan(basePackages = { "com.waylau.spring" })
@Import({ WebSecurityConfig.class, MvcConfiguration.class})
public class AppConfiguration {

}
```

8.3.6 创建内嵌Jetty的服务器

创建内嵌了 Jetty 的服务器，其代码如下。

```java
package com.waylau.spring.mvc;

import java.util.EnumSet;
import javax.servlet.DispatcherType;
import org.eclipse.jetty.server.Server;
import org.eclipse.jetty.servlet.FilterHolder;
import org.eclipse.jetty.servlet.ServletContextHandler;
import org.eclipse.jetty.servlet.ServletHolder;
import org.springframework.web.context.ContextLoaderListener;
import org.springframework.web.context.WebApplicationContext;
import org.springframework.web.context.support.AnnotationConfigWebApplicationContext;
import org.springframework.web.filter.DelegatingFilterProxy;
import org.springframework.web.servlet.DispatcherServlet;
import com.waylau.spring.mvc.configuration.AppConfiguration;

public class JettyServer {
    private static final int DEFAULT_PORT = 8080;
    private static final String CONTEXT_PATH = "/";
    private static final String MAPPING_URL = "/*";

    public void run() throws Exception {
        Server server = new Server(DEFAULT_PORT);
        server.setHandler(servletContextHandler(webApplicationContext()));
        server.start();
        server.join();
    }

    private ServletContextHandler servletContextHandler(WebApplicationContext ct) {
        // 启用Session管理器
        ServletContextHandler handler = 
                new ServletContextHandler(ServletContextHandler.SESSIONS);

        handler.setContextPath(CONTEXT_PATH);
        handler.addServlet(new ServletHolder(new DispatcherServlet(ct)),
                MAPPING_URL);
        handler.addEventListener(new ContextLoaderListener(ct));

        // 添加Spring Security过滤器
        FilterHolder filterHolder=new FilterHolder(DelegatingFilterProxy.class);
        filterHolder.setName("springSecurityFilterChain");
```

```
        handler.addFilter(filterHolder, MAPPING_URL,
                EnumSet.of(DispatcherType.REQUEST));

        return handler;
    }

    private WebApplicationContext webApplicationContext() {
        AnnotationConfigWebApplicationContext context =
                new AnnotationConfigWebApplicationContext();
        context.register(AppConfiguration.class);
        return context;
    }
}
```

JettyServer 将 Spring 的上下文 Servlet、监听器、过滤器等信息都传给了 Jetty 服务。

8.3.7 应用启动器

创建应用启动类 Application，其代码如下。

```
package com.waylau.spring.mvc;

public class Application {

    public static void main(String[] args) throws Exception {
        new JettyServer().run();
    }

}
```

8.3.8 运行应用

右键运行 Application 类即可启动应用。

如果访问 http://localhost:8080/hello/way 时跳转到了登录界面，则意味着被安全认证拦截了。正如 WebSecurityConfig 所配置的那样，登录界面是一个 form 表单，如图 8-1 所示。

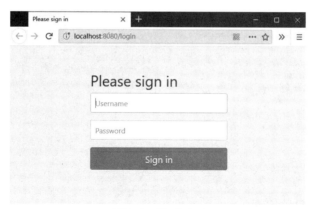

图8-1　登录界面

尝试输入一个错误的用户名和密码，可以看到图 8-2 所示的提示信息。

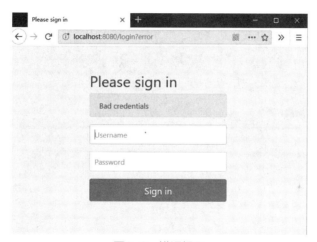

图8-2　错误提示

用初始化的用户名、密码进行登录，登录成功后可以看到能够正常访问应用的 API 了，如图 8-3 所示。

图8-3　成功登录

第9章 MyBatis 基础

大部分应用都会将数据存储于关系型数据库中。在Java领域，MyBatis是非常流行的处理Java对象与数据库关系之间的映射的框架，在互联网公司被广泛采用。本章主要介绍MyBatis的基础知识。

9.1 MyBatis 概述

MyBatis 是一款优秀的持久层框架,它支持定制化 SQL、存储过程及高级映射。MyBatis 避免了几乎所有的 JDBC 代码和手动设置参数及获取结果集。MyBatis 可以使用简单的 XML 或注解来配置和映射原生信息,将接口和 Java POJO(Plain Ordinary Java Objects,普通的 Java 对象)映射成数据库中的记录。

要使用 MyBatis,只需将 mybatis-3.4.6.jar 文件置于 classpath 中即可。

如果使用 Maven 来构建项目,则将下面的 MyBatis 的依赖配置于 pom.xml 文件中即可。

```
<dependency>
  <groupId>org.mybatis</groupId>
  <artifactId>mybatis</artifactId>
  <version>3.4.6</version>
</dependency>
```

9.2 与Hibernate对比

在 Java 领域,另外一款比较著名的 ORM 框架是 Hibernate,接下来就这两款框架从下面 3 个方面进行比较。

9.2.1 框架复杂度

Hibernate 功能强大,数据库无关性好,O/R 映射能力强,是 JPA 规范的实现之一。当然,Hibernate 的代码量也相对比较庞大,框架复杂度比较高,算是重量级的 ORM 解决方案。

MyBatis 则相对简单,它并没有试图成为一个完整的 ORM 框架。MyBatis 力图简化传统 JDBC 开发过程中需要编写很多样板代码的烦琐过程。MyBatis 可以使用简单的 XML 或注解方式来配置和映射原生信息,将接口和 Java POJO 映射成数据库中的记录。因此,MyBatis 整体框架的复杂度会比较低,开发者也能够轻易读懂源码。

9.2.2 学习成本

就上手程度而言,真正掌握 Hibernate 要比掌握 MyBatis 难一些,MyBatis 框架相对容易上手,与 JDBC 开发方式接近。

然而,不能只考虑两者的特性及性能,更要根据项目需求去考虑究竟哪一个更适合项目开发。

例如，一个项目中用到的复杂查询基本没有，只是简单的增、删、改、查，这时选择 Hibernate 效率就很高，因为基本的 SQL 语句已经被封装好了，不需要再写 SQL 语句，这就节省了大量的时间。但是对于一个大型项目而言，其复杂语句较多，选择 MyBatis 就能高效许多，而且语句的管理也比较方便。

由于互联网公司的产品需求是经常变化的，无法形成稳定的编程模型，因此 MyBatis 的开发方式更加适合互联网公司。

9.2.3 性能

首先，由于 Hibernate 比 MyBatis 抽象封装的程度更高，因此理论上 Hibernate 的单个语句的执行效率会低一点。所有的框架都一样，排除算法上的差异，越是底层，执行效率越高。

其次，MyBatis 可以最大化利用原生 SQL 性能，如可以按需指定查询的字段、自由关联表结构、使用原生数据库的执行计划等。而 Hibernate 是面向对象的，查询时会将表中的所有字段查询出来，这一点会有性能消耗。

最后，Hibernate 要考虑跨数据库，只能使用通用的数据库查询方案，很多数据库原生的特性也就无法支持。

综上所述，MyBatis 拥有比 Hibernate 更好的性能，这也是互联网公司选择 MyBatis 的重要原因之一。

有关 Hibernate 方面的知识，可以参阅笔者所著的《Spring Boot 企业级应用开发实战》。

9.3 四大核心概念

本节将讲解 MyBatis 的四大核心概念，即 SqlSessionFactoryBuilder、SqlSessionFactory、SqlSession 和 Mapper。

9.3.1 SqlSessionFactoryBuilder

从命名上可以看出，这个是一个采用 Builder 模式、用于创建 SqlSessionFactory 的类。SqlSessionFactoryBuilder 根据配置来构造 SqlSessionFactory。

其中配置方式主要有两种。

1. XML文件方式

XML 文件方式是常用的一种，下面的代码演示了 XML 文件配置方式。

```
String resource = "org/mybatis/example/mybatis-config.xml";
```

```
InputStream inputStream = Resources.getResourceAsStream(resource);

SqlSessionFactory sqlSessionFactory = new SqlSessionFactoryBuilder().
build(inputStream);
```

mybatis-config.xml 就是配置文件，其内容如下。

```xml
<?xml version="1.0" encoding="UTF-8" ?>
<!DOCTYPE configuration
PUBLIC "-//mybatis.org//DTD Config 3.0//EN"
"http://mybatis.org/dtd/mybatis-3-config.dtd">
<configuration>
    <environments default="development">
        <environment id="development">
            <transactionManager type="JDBC"/>
            <dataSource type="POOLED">
                <property name="driver" value="${driver}"/>
                <property name="url" value="${url}"/>
                <property name="username" value="${username}"/>
                <property name="password" value="${password}"/>
            </dataSource>
        </environment>
    </environments>
    <mappers>
        <mapper resource="org/mybatis/example/BlogMapper.xml"/>
    </mappers>
</configuration>
```

2. Java Config

第二种配置方式是通过 Java 代码来配置，用法如下。

```
DataSource dataSource = BlogDataSourceFactory.getBlogDataSource();

TransactionFactory transactionFactory = new JdbcTransactionFactory();

Environment environment = new Environment("development", transaction-
Factory, dataSource);

Configuration configuration = new Configuration(environment);
configuration.addMapper(BlogMapper.class);

SqlSessionFactory sqlSessionFactory = new SqlSessionFactoryBuilder().
build(configuration);
```

Java Config 相较于 XML 文件的方式而言，会有一些限制，如修改了配置文件需要重新编译、注解方式没有 XML 配置项多等。所以，业界大多数情况下是选择 XML 文件的方式。但到底选择哪种方式，取决于自己团队的需要。例如，项目的 SQL 语句不复杂，也不需要一些高级的 SQL 特性，那么 Java Config 则会更加简洁一点；反之，则可以选择 XML 文件的方式。

9.3.2　SqlSessionFactory

SqlSessionFactory，顾名思义，是用于生产 SqlSession 的工厂。

通过如下方式获取 SqlSession 实例。

```
SqlSession session = sqlSessionFactory.openSession();
```

9.3.3　SqlSession

SqlSession 包含执行 SQL 的所有方法，可以通过 SqlSession 实例来直接执行已映射的 SQL 语句，例如：

```
SqlSession session = sqlSessionFactory.openSession();
try {
Blog blog = session.selectOne(
"org.mybatis.example.BlogMapper.selectBlog", 101);
} finally {
session.close();
}
```

下面的方式可以做到类型安全。

```
SqlSession session = sqlSessionFactory.openSession();
try {
BlogMapper mapper = session.getMapper(BlogMapper.class);
Blog blog = mapper.selectBlog(101);
} finally {
session.close();
}
```

9.3.4　Mapper

Mapper 用于 Java 与 SQL 之间的映射，包括 Java 映射为 SQL 语句，以及 SQL 返回结果映射为 Java。

下面是一个常见的 Mapper 接口映射文件。

```
<?xml version="1.0" encoding="UTF-8" ?>
<!DOCTYPE mapper
PUBLIC "-//mybatis.org//DTD Mapper 3.0//EN"
"http://mybatis.org/dtd/mybatis-3-mapper.dtd">
<mapper namespace="org.mybatis.example.BlogMapper">
<select id="selectBlog" resultType="Blog">
```

```
select * from Blog where id = #{id}
</select>
</mapper>
```

其中 "org.mybatis.example.BlogMapper" 是要映射的接口，"selectBlog" 是 BlogMapper 上的方法，负责具体执行 "select * from Blog where id = #{id}" 这个 SQL 语句。

这样，就能通过

```
Blog blog = session.selectOne(
"org.mybatis.example.BlogMapper.selectBlog", 101);
```

或者

```
BlogMapper mapper = session.getMapper(BlogMapper.class);
Blog blog = mapper.selectBlog(101);
```

来获取执行的结果。

当然，如果采用注解的方式，则可以省去 XML 映射文件。以下是采用注解方式的示例。

```
public interface BlogMapper {
@Select("SELECT * FROM blog WHERE id = #{id}")
Blog selectBlog(int id);
}
```

9.4 生命周期及作用域

理解 MyBatis 的不同作用域和生命周期是至关重要的，因为错误使用 MyBatis 的作用域和生命周期将会导致非常严重的并发问题。

9.4.1 SqlSessionFactoryBuilder

SqlSessionFactoryBuilder 类可以被实例化、使用和丢弃，一旦创建了 SqlSessionFactory，就不再需要它了。因此，SqlSessionFactoryBuilder 实例的最佳作用域是方法作用域（也就是局部方法变量）。可以重用 SqlSessionFactoryBuilder 来创建多个 SqlSessionFactory 实例，但最好还是不要让其一直存在，以释放资源给其他更重要的事情。

9.4.2 SqlSessionFactory

SqlSessionFactory 一旦被创建，就应该在应用运行期间一直存在，没有任何理由对其进行清除或重建。使用 SqlSessionFactory 的最佳实践是在应用运行期间不要重复创建多次，多次重建 SqlSessionFactory 被视为一种代码"坏味道"（Bad Smell）。因此，SqlSessionFactory 的最佳作用域是应用作用域。有很多方法可以实现最佳实践，最简单的就是使用单例模式或者静态单例模式。

9.4.3 SqlSession

每个线程都应该有自己的 SqlSession 实例。SqlSession 实例不是线程安全的，因此是不能被共享的，所以它的最佳作用域是请求或方法作用域。绝对不能将 SqlSession 实例的引用放在一个类的静态域，也绝不能将 SqlSession 实例的引用放在任何类型的管理作用域。如果正在使用一种 Web 框架，则要考虑将 SqlSession 放在一个与 HTTP 请求对象相似的作用域中。换句话说，每次收到 HTTP 请求，就打开一个 SqlSession；返回一个响应，就关闭 SqlSession。这个关闭操作是很重要的，应该把该操作放到 finally 块中，以确保每次都能执行关闭。下面的示例就是确保 SqlSession 关闭的标准模式。

```
SqlSession session = sqlSessionFactory.openSession();
try {
  // 业务逻辑…
} finally {
  session.close();
}
```

确保所有的代码中都采用这种模式，以保证所有数据库资源都能被正确地关闭。

9.4.4 Mapper实例

Mapper 是一个用来绑定映射语句的接口。Mapper 接口的实例是从 SqlSession 中获得的。因此从技术层面来讲，任何 Mapper 实例的最大作用域都和请求它们的 SqlSession 相同。尽管如此，Mapper 实例的最佳作用域还是方法作用域。也就是说，Mapper 实例应该在调用它们的方法中被请求，用过之后即可废弃。并不需要显式地关闭 Mapper 实例，尽管在整个请求作用域（Request Scope）中保持 Mapper 实例也不会有什么问题，但在这个作用域中管理太多资源的话会难以控制。所以要保持简单，最好把 Mapper 放在方法作用域（Method Scope）中。下面的示例就展示了这个实践。

```
SqlSession session = sqlSessionFactory.openSession();
try {
  BlogMapper mapper = session.getMapper(BlogMapper.class);
  // 业务逻辑…
} finally {
  session.close();
}
```

第10章 MyBatis高级应用

前面已经介绍了MyBatis的基础知识,本章将继续探讨MyBatis的高级应用,包括如何配置MyBatis、如何编写映射文件等。

10.1 配置文件

MyBatis 配置是定义在 mybatis-config.xml 文件中的,该配置文件主要包含如下配置内容。

10.1.1 properties

MyBatis 的 properties(属性)都是可外部配置且可动态替换的,既可以在典型的 Java 属性文件中配置,也可以通过 properties 元素的子元素来传递,例如:

```
<properties resource="org/mybatis/example/config.properties">
  <property name="username" value="dev_user"/>
  <property name="password" value="F2Fa3!33TYyg"/>
</properties>
```

其中的属性就可以在整个配置文件中被用来替换需要动态配置的属性值,例如:

```
<dataSource type="POOLED">
  <property name="driver" value="${driver}"/>
  <property name="url" value="${url}"/>
  <property name="username" value="${username}"/>
  <property name="password" value="${password}"/>
</dataSource>
```

在以上代码中,username 和 password 将由 properties 元素中设置的相应值来替换。driver 和 url 属性将由 config.properties 文件中对应的值来替换。这样就为配置提供了诸多灵活的选择。

属性也可以被传递到 SqlSessionFactoryBuilder.build() 方法中,例如:

```
SqlSessionFactory factory = new SqlSessionFactoryBuilder().build(reader, props);
// …
```

或者

```
SqlSessionFactory factory = new SqlSessionFactoryBuilder().build(reader, environment, props);
// …
```

如果属性在多个地方进行了配置,那么 MyBatis 将按照下面的顺序来加载。

(1)首先读取 properties 元素中指定的属性。

(2)其次根据 properties 元素中的 resource 属性读取类路径下的属性文件,或者根据 url 属性指定的路径读取属性文件,并覆盖已读取的同名属性。

(3)最后读取作为方法参数传递的属性,并覆盖已读取的同名属性。

因此,通过方法参数传递的属性具有最高优先级,resource/url 属性中指定的配置文件次之,优先级最低的是 properties 属性中指定的属性。

从 MyBatis 3.4.2 开始，可以为占位符指定一个默认值，例如：

```xml
<dataSource type="POOLED">
  <!-- … -->
  <property name="username"
    value="${username:ut_user}"/> <!-- 如果'username'属性不存在  则默认值是 'ut_user' -->
</dataSource>
```

这个特性默认是关闭的。若想为占位符指定一个默认值，则需开启这个特性，例如：

```xml
<properties resource="org/mybatis/example/config.properties">
  <!-- … -->
  <property name="org.apache.ibatis.parsing.PropertyParser.enable-default-value"
    value="true"/> <!-- 开启特性 -->
</properties>
```

可以通过增加一个指定的属性来改变分隔符和默认值的字符，例如：

```xml
<properties resource="org/mybatis/example/config.properties">
  <!-- … -->
  <property name="org.apache.ibatis.parsing.PropertyParser.default-value-separator"
    value="?:"/> <!-- 改变分隔符 -->
</properties>
<dataSource type="POOLED">
  <!-- … -->
  <property name="username" value="${db:username?:ut_user}"/>
</dataSource>
```

10.1.2 settings

settings 会改变 MyBatis 运行时的行为。

一个配置完整的 settings 元素的示例代码如下。

```xml
<settings>
  <setting name="cacheEnabled" value="true"/>
  <setting name="lazyLoadingEnabled" value="true"/>
  <setting name="multipleResultSetsEnabled" value="true"/>
  <setting name="useColumnLabel" value="true"/>
  <setting name="useGeneratedKeys" value="false"/>
  <setting name="autoMappingBehavior" value="PARTIAL"/>
  <setting name="autoMappingUnknownColumnBehavior" value="WARNING"/>
  <setting name="defaultExecutorType" value="SIMPLE"/>
  <setting name="defaultStatementTimeout" value="25"/>
  <setting name="defaultFetchSize" value="100"/>
  <setting name="safeRowBoundsEnabled" value="false"/>
```

```
<setting name="mapUnderscoreToCamelCase" value="false"/>
<setting name="localCacheScope" value="SESSION"/>
<setting name="jdbcTypeForNull" value="OTHER"/>
<setting name="lazyLoadTriggerMethods" value="equals,clone,hashCode,-toString"/>
</settings>
```

表10-1描述了上述设置中各个项目的含义及默认值。

表10-1 各个项目含义及默认值

设置名	描述	有效值	默认值
cacheEnabled	全局开启或关闭配置文件中的所有映射器已经配置的缓存	true | false	true
lazyLoadingEnabled	延迟加载的全局开关。当开启时，所有关联对象都会延迟加载。在特定关联关系中可通过设置fetchType属性来覆盖该项的开关状态	true | false	false
aggressiveLazy-Loading	当开启时，任何方法的调用都会加载该对象的所有属性。否则，每个属性都会按需加载（参考lazyLoadTriggerMethods）	true | false	false（在MyBatis 3.4.1及之前的版本中，默认值为true）
multipleResultSetsEnabled	是否允许单一语句返回多结果集。如果允许，则需要驱动支持	true | false	true
useColumnLabel	使用列标签代替列名。不同的驱动在这方面会有不同的表现，具体可参考相关驱动文档或通过测试这两种不同的模式来观察所用驱动的结果	true | false	true
useGeneratedKeys	允许JDBC支持自动生成主键，需要驱动支持。若设置为true，则此强制支持自动生成主键，尽管一些驱动不能支持，但仍可正常工作（如Derby）	true | false	false
autoMappingBe-havior	指定MyBatis如何自动映射列到字段或属性。NONE表示取消自动映射；PARTIAL只会自动映射没有定义嵌套结果集映射的结果集。FULL会自动映射任意复杂的结果集（无论是否嵌套）	NONE | PARTIAL | FULL	PARTIAL

续表

设置名	描述	有效值	默认值
autoMappingUnknownColumnBehavior	指定发现自动映射目标未知列（或者未知属性类型）的行为。NONE: 不做任何反应 WARNING: 输出提醒日志 (org.apache.ibatis.session.AutoMappingUnknownColumnBehavior的日志等级必须设置为WARNING) FAILING: 映射失败 (抛出SqlSessionException)	NONE | WARNING | FAILING	NONE
defaultExecutorType	配置默认的执行器。SIMPLE 就是普通的执行器；REUSE 执行器会重用预处理语句（prepared statements）；BATCH 执行器将重用语句并执行批量更新	SIMPLE | REUSE | BATCH	SIMPLE
defaultStatementTimeout	设置超时时间，决定驱动等待数据库响应的秒数	任意正整数	未设置 (null)
defaultFetchSize	为驱动的结果集获取数量（fetchSize）设置一个提示值。此参数只可以在查询设置中被覆盖	任意正整数	未设置 (null)
safeRowBoundsEnabled	允许在嵌套语句中使用分页（RowBounds）。若允许使用则设置为 false	true | false	false
safeResultHandlerEnabled	允许在嵌套语句中使用分页（ResultHandler）。若允许使用则设置为 false	true | false	true
mapUnderscoreToCamelCase	是否开启自动驼峰命名规则（camel case）映射，即从经典数据库列名 A_COLUMN 到经典 Java 属性名 aColumn 的类似映射	true | false	false
localCacheScope	MyBatis 利用本地缓存机制（Local Cache）防止循环引用（circular references）和加速重复嵌套查询。默认值为 SESSION，这种情况下会缓存一个会话中执行的所有查询。若设置值为 STATEMENT，则本地会话仅用在语句执行上，对相同 SqlSession 的不同调用将不会共享数据	SESSION | STATEMENT	SESSION

213

续表

设置名	描述	有效值	默认值
jdbcTypeForNull	当没有为参数提供特定的 JDBC 类型时，为空值指定 JDBC 类型。某些驱动需要指定列的 JDBC 类型，多数情况下直接用一般类型即可，如 NULL、VARCHAR 或 OTHER	JdbcType 常量，常用值：NULL、VARCHAR 或 OTHER	OTHER
lazyLoadTrigger-Methods	指定哪个对象的方法触发一次延迟加载	用逗号分隔的方法列表	equals,clone,hashCode,toString
defaultScriptingLanguage	指定动态 SQL 生成的默认语言	一个类型别名或完全限定类名	org.apache.ibatis.scripting.xmltags.XMLLanguageDriver
defaultEnumType-Handler	指定 Enum 使用的默认 TypeHandler（新增于 MyBatis3.4.5）	一个类型别名或完全限定类名	org.apache.ibatis.type.EnumTypeHandler
callSettersOnNulls	指定当结果集中的值为 null 时，是否调用映射对象的 setter（map 对象时为 put）方法，这在依赖于 Map.keySet() 或 null 值初始化时比较有用。注意，基本类型（int、boolean 等）是不能设置为 null 的	true | false	false
returnInstance-ForEmptyRow	当返回行的所有列都是空时，MyBatis默认返回 null。当开启这个设置时，MyBatis会返回一个空实例。注意，它也适用于嵌套的结果集（如集合或关联，新增于 MyBatis 3.4.2）	true | false	false
logPrefix	指定 MyBatis 增加到日志名称的前缀	任何字符串	未设置
logImpl	指定 MyBatis 所用日志的具体实现，未指定时将自动查找	SLF4J | LOG4J | LOG4J2 | JDK_LOGGING | COMMONS_LOGGING | STDOUT_LOGGING | NO_LOGGING	未设置
proxyFactory	指定 Mybatis 创建的具有延迟加载能力的对象所用到的代理工具	CGLIB | JAVASSIST	JAVASSIST（MyBatis 3.3 以上）

续表

设置名	描述	有效值	默认值
vfsImpl	指定 VFS 的实现	自定义 VFS 实现的类的全限定名，以逗号分隔	未设置
useActualParamName	允许使用方法签名中的名称作为语句参数名称。为了使用该特性，项目必须采用 Java 8 编译，并且加上 parameters 选项（新增于 MyBatis 3.4.1）	true | false	true
configurationFactory	指定一个提供 Configuration 实例的类。这个被返回的 Configuration 实例用来加载被反序列化对象的延迟加载属性值。这个类必须包含一个签名为 static Configuration getConfiguration() 的方法（新增于 MyBatis 3.2.3）	类型别名或全类名	未设置

10.1.3　typeAliases

typeAliases 用于给 Java 类型起一个别名，让类型看起来更加简洁，示例如下。

```xml
<typeAliases>
  <typeAlias alias="Author" type="domain.blog.Author"/>
  <typeAlias alias="Blog" type="domain.blog.Blog"/>
  <typeAlias alias="Comment" type="domain.blog.Comment"/>
  <typeAlias alias="Post" type="domain.blog.Post"/>
  <typeAlias alias="Section" type="domain.blog.Section"/>
  <typeAlias alias="Tag" type="domain.blog.Tag"/>
</typeAliases>
```

通过上述配置，Blog 就可以用在任何使用 domain.blog.Blog 的地方。

也可以指定一个包名，MyBatis 会在包名下面搜索需要的 Java Bean，比如：

```xml
<typeAliases>
  <package name="domain.blog"/>
</typeAliases>
```

每一个在 domain.blog 包中的 Java Bean，在没有注解的情况下，会将 Bean 的非限定类名的首字母小写来作为其别名，如 domain.blog.Author 的别名为 author。

若存在注解，则别名为其注解值，示例如下。

```java
@Alias("author")
public class Author {
```

```
    //…
}
```

MyBatis 为常用的 Java 类型设置了别名，具体如表 10-2 所示。这些别名都是不区分大小写的。

表10-2　常用Java类型的别名

别名	映射的类型	别名	映射的类型
_byte	byte	double	Double
_long	long	float	Float
_short	short	boolean	Boolean
_int	int	date	Date
_integer	int	decimal	BigDecimal
_double	double	bigdecimal	BigDecimal
_float	float	object	Object
_boolean	boolean	map	Map
string	String	hashmap	HashMap
byte	Byte	list	List
long	Long	arraylist	ArrayList
short	Short	collection	Collection
int	Integer	iterator	Iterator
integer	Integer	—	—

10.1.4　typeHandlers

MyBatis 无论是在预处理语句（PreparedStatement）中设置一个参数时，还是从结果集中取出一个值时，都会用类型处理器将获取的值以合适的方式转换为 Java 类型。

表 10-3 描述了 MyBatis 内置的默认类型处理器。

表10-3　MyBatis内置的默认类型处理器

类型处理器	Java 类型	JDBC 类型
BooleanTypeHandler	java.lang.Boolean, boolean	数据库兼容的 BOOLEAN
ByteTypeHandler	java.lang.Byte, byte	数据库兼容的 NUMERIC 或 BYTE

续表

类型处理器	Java 类型	JDBC 类型
ShortTypeHandler	java.lang.Short, short	数据库兼容的 NUMERIC 或 SMALLINT
IntegerTypeHandler	java.lang.Integer, int	数据库兼容的 NUMERIC 或 INTEGER
LongTypeHandler	java.lang.Long, long	数据库兼容的 NUMERIC 或 BIGINT
FloatTypeHandler	java.lang.Float, float	数据库兼容的 NUMERIC 或 FLOAT
DoubleTypeHandler	java.lang.Double, double	数据库兼容的 NUMERIC 或 DOUBLE
BigDecimalTypeHandler	java.math.BigDecimal	数据库兼容的 NUMERIC 或 DECIMAL
StringTypeHandler	java.lang.String	CHAR, VARCHAR
ClobReaderTypeHandler	java.io.Reader	—
ClobTypeHandler	java.lang.String	CLOB, LONGVARCHAR
NStringTypeHandler	java.lang.String	NVARCHAR, NCHAR
NClobTypeHandler	java.lang.String	NCLOB
BlobInputStreamTypeHandler	java.io.InputStream	—
ByteArrayTypeHandler	byte[]	数据库兼容的字节流类型
BlobTypeHandler	byte[]	BLOB, LONGVARBINARY
DateTypeHandler	java.util.Date	TIMESTAMP
DateOnlyTypeHandler	java.util.Date	DATE
TimeOnlyTypeHandler	java.util.Date	TIME
SqlTimestampTypeHandler	java.sql.Timestamp	TIMESTAMP
SqlDateTypeHandler	java.sql.Date	DATE
SqlTimeTypeHandler	java.sql.Time	TIME
ObjectTypeHandler	Any	OTHER 或未指定类型
EnumTypeHandler	Enumeration Type	VARCHAR 或任何兼容的字符串类型，用以存储枚举的名称（而不是索引值）

续表

类型处理器	Java 类型	JDBC 类型
EnumOrdinalTypeHandler	Enumeration Type	任何兼容的NUMERIC或DOUBLE类型，存储枚举的序数值（而不是名称）
SqlxmlTypeHandler	java.lang.String	SQLXML
InstantTypeHandler	java.time.Instant	TIMESTAMP
LocalDateTimeTypeHandler	java.time.LocalDateTime	TIMESTAMP
LocalDateTypeHandler	java.time.LocalDate	DATE
LocalTimeTypeHandler	java.time.LocalTime	TIME
OffsetDateTimeTypeHandler	java.time.OffsetDateTime	TIMESTAMP
OffsetTimeTypeHandler	java.time.OffsetTime	TIME
ZonedDateTimeTypeHandler	java.time.ZonedDateTime	TIMESTAMP
YearTypeHandler	java.time.Year	INTEGER
MonthTypeHandler	java.time.Month	INTEGER
YearMonthTypeHandler	java.time.YearMonth	VARCHAR 或 LONGVARCHAR
JapaneseDateTypeHandler	java.time.chrono.JapaneseDate	DATE

提示：从 MyBatis 3.4.5 开始，MyBatis 默认支持 JSR-310（日期和时间 API）。

可以重写类型处理器或创建类型处理器来处理不支持的或非标准的类型。

具体做法可分为以下两步。

1. 实现TypeHandler接口或继承BaseTypeHandler类

要创建类型处理器，第一步是实现 org.apache.ibatis.type.TypeHandler 接口，或者继承一个很便利的 org.apache.ibatis.type.BaseTypeHandler 类，其代码如下：

```java
@MappedJdbcTypes(JdbcType.VARCHAR)
public class ExampleTypeHandler extends BaseTypeHandler<String> {

  @Override
  public void setNonNullParameter(PreparedStatement ps, int i, String parameter, JdbcType jdbcType) throws SQLException {
    ps.setString(i, parameter);
  }

  @Override
  public String getNullableResult(ResultSet rs, String columnName) throws SQLException {
```

```
    return rs.getString(columnName);
  }

  @Override
  public String getNullableResult(ResultSet rs, int columnIndex) throws
SQLException {
    return rs.getString(columnIndex);
  }

  @Override
  public String getNullableResult(CallableStatement cs, int columnIn-
dex) throws SQLException {
    return cs.getString(columnIndex);
  }
}
```

2. 将处理器映射到一个JDBC类型

第二步是将自定义的处理器映射到一个 JDBC 类型上,比如:

```
<!-- 定义在mybatis-config.xml文件中 -->
<typeHandlers>
  <typeHandler handler="org.mybatis.example.ExampleTypeHandler"/>
</typeHandlers>
```

这个自定义的类型处理器将会覆盖已经存在的处理 Java 的 String 类型属性和 VARCHAR 参数及结果的类型处理器。需要注意的是,MyBatis 不会窥探数据库元信息来决定使用哪种类型,所以开发者必须在参数和结果映射中指明哪个是 VARCHAR 类型的字段,以使其能够绑定到正确的类型处理器上。这是因为 MyBatis 直到语句被执行才清楚数据类型。

通过类型处理器的泛型,MyBatis 可以得知该类型处理器所要处理的 Java 类型。不过这种行为可以通过以下两种方式改变。

(1)在类型处理器的配置元素(typeHandler element)上增加一个 javaType 属性(如 javaType = "String")。

(2)在类型处理器的类(TypeHandler class)上增加一个 @MappedTypes 注解来指定与其关联的 Java 类型列表。如果在 javaType 属性中也同时指定,那么注解方式将被忽略。

可以通过以下两种方式指定被关联的 JDBC 类型。

(1)在类型处理器的配置元素上增加一个 jdbcType 属性(如 jdbcType="VARCHAR")。

(2)在类型处理器的类(TypeHandler class)上增加一个 @MappedJdbcTypes 注解来指定与其关联的 JDBC 类型列表。如果在 jdbcType 属性中也同时指定,那么注解方式将被忽略。

当决定在 ResultMap 中使用某一 TypeHandler 时,Java 类型是已知的(从结果类型中获得),但是 JDBC 类型是未知的。因此,Mybatis 使用 javaType=[TheJavaType],jdbcType=null 的组合来选择一个 TypeHandler。这意味着使用 @MappedJdbcTypes 注解可以限制 TypeHandler 的范围,同时必

须为显式的设置，否则 TypeHandler 在 ResultMap 中将是无效的。如果希望在 ResultMap 中使用 TypeHandler，那么设置 @MappedJdbcTypes 注解的 includeNullJdbcType=true 即可。然而从 Mybatis 3.4.0 版本开始，如果只有一个注册的 TypeHandler 来处理 Java 类型，那么它将是 ResultMap 使用 Java 类型时的默认值（即使没有设置 includeNullJdbcType=true）。

最后，可以使用 MyBatis 来查找类型处理器。

```xml
<!-- 定义在mybatis-config.xml文件中 -->
<typeHandlers>
  <package name="org.mybatis.example"/>
</typeHandlers>
```

10.1.5　objectFactory

MyBatis 每次创建结果对象的新实例时，都会使用一个对象工厂（ObjectFactory）实例来完成。默认的对象工厂需要做的仅仅是实例化目标类，要么通过默认构造方法，要么在参数映射存在时通过参数构造方法来实例化。如果想覆盖对象工厂的默认行为，那么可以通过创建自己的对象工厂来实现。

以下示例自定义了一个对象工厂。

```java
public class ExampleObjectFactory extends DefaultObjectFactory {
  public Object create(Class type) {
    return super.create(type);
  }
  public Object create(Class type, List<Class> constructorArgTypes, List<Object> constructorArgs) {
    return super.create(type, constructorArgTypes, constructorArgs);
  }
  public void setProperties(Properties properties) {
    super.setProperties(properties);
  }
  public <T> boolean isCollection(Class<T> type) {
    return Collection.class.isAssignableFrom(type);
  }
}
```

ObjectFactory 接口很简单，它包含两个创建方法：一个是处理默认构造方法的，另一个是处理带参数的构造方法的。最后，setProperties 方法可以被用来配置 ObjectFactory，在初始化 ObjectFactory 实例后，objectFactory 元素体中定义的属性会被传递给 setProperties 方法。

在 mybatis-config.xml 文件中配置使用该对象工厂的示例如下。

```xml
<!-- 定义在mybatis-config.xml文件中 -->
<objectFactory type="org.mybatis.example.ExampleObjectFactory">
  <property name="someProperty" value="100"/>
```

```
</objectFactory>
```

10.1.6 plugins

MyBatis 允许在已映射语句执行过程中的某一点进行拦截调用。默认情况下，MyBatis 允许使用插件的方法来拦截调用，包括以下几种插件。

- Executor (update, query, flushStatements, commit, rollback, getTransaction, close, isClosed)
- ParameterHandler (getParameterObject, setParameters)
- ResultSetHandler (handleResultSets, handleOutputParameters)
- StatementHandler (prepare, parameterize, batch, update, query)

通过 MyBatis 提供的强大机制，使用插件是非常简单的，只需实现 Interceptor 接口，并指定想要拦截的方法签名即可。

以下是一个自定义插件的示例。

```
@Intercepts({@Signature(
  type= Executor.class,
  method = "update",
  args = {MappedStatement.class,Object.class})})
public class ExamplePlugin implements Interceptor {
  public Object intercept(Invocation invocation) throws Throwable {
    return invocation.proceed();
  }
  public Object plugin(Object target) {
    return Plugin.wrap(target, this);
  }
  public void setProperties(Properties properties) {
  }
}
```

在 mybatis-config.xml 文件中，配置使用该插件，示例如下。

```
<!-- 定义在mybatis-config.xml文件中 -->
<plugins>
  <plugin interceptor="org.mybatis.example.ExamplePlugin">
    <property name="someProperty" value="100"/>
  </plugin>
</plugins>
```

上面的插件将会拦截在 Executor 实例中所有的 update 方法调用，这里的 Executor 是负责执行低层映射语句的内部对象。

10.1.7　environments

MyBatis 可以配置成适应多种环境，这种机制有助于将 SQL 映射应用于多种数据库中。例如，开发、测试和生产环境需要有不同的配置；或者共享相同 Schema 的多个生产数据库需要使用相同的 SQL 映射等。

注意：尽管可以配置多个环境，但每个 SqlSessionFactory 实例只能选择其一。

所以，如果要连接两个数据库，就需要创建两个 SqlSessionFactory 实例，每个数据库对应一个。如果是 3 个数据库，就需要 3 个实例，以此类推。

每个数据库对应一个 SqlSessionFactory 实例，为了指定创建哪种环境，只要将其作为可选的参数传递给 SqlSessionFactoryBuilder 即可。可以接受环境配置的两个方法签名示例如下。

```
SqlSessionFactory factory = new SqlSessionFactoryBuilder().build(reader, environment);
SqlSessionFactory factory = new SqlSessionFactoryBuilder().build(reader, environment, properties);
```

如果忽略了环境参数，那么默认环境将会被加载，示例如下。

```
SqlSessionFactory factory = new SqlSessionFactoryBuilder().build(reader);
SqlSessionFactory factory = new SqlSessionFactoryBuilder().build(reader, properties);
```

environment 元素定义了如何配置环境。观察以下示例。

```xml
<environments default="development">
  <environment id="development">
    <transactionManager type="JDBC">
      <property name="…" value="…"/>
    </transactionManager>
    <dataSource type="POOLED">
      <property name="driver" value="${driver}"/>
      <property name="url" value="${url}"/>
      <property name="username" value="${username}"/>
      <property name="password" value="${password}"/>
    </dataSource>
  </environment>
</environments>
```

在该示例中，需要注意以下几个关键点。

（1）默认的环境 ID（如 default="development"）。

（2）每个 environment 元素定义的环境 ID（如 id="development"）。

（3）事务管理器的配置（如 type="JDBC"）。

（4）数据源的配置（如 type="POOLED"）。

（5）默认的环境和环境 ID 是自解释的，因此一目了然。可以对环境随意命名，但一定要保证默认的环境 ID 匹配其中一个环境 ID。

10.1.8　transactionManager

MyBatis 中的事务管理器有以下两种类型。

（1）JDBC：这个配置就是直接使用 JDBC 的提交和回滚设置，它依赖于从数据源得到的连接来管理事务作用域。

（2）MANAGED：这个配置几乎没做什么，它从来不提交或回滚一个连接，而是让容器来管理事务的整个生命周期（如 JEE 应用服务器的上下文）。默认情况下它会关闭连接，然而一些容器并不希望这样，因此需要将 closeConnection 属性设置为 false 来阻止其默认的关闭行为，例如：

```xml
<transactionManager type="MANAGED">
  <property name="closeConnection" value="false"/>
</transactionManager>
```

提示：如果是在 Spring 中使用 MyBatis，就没有必要配置事务管理器，因为 Spring 会使用自带的管理器来覆盖前面的配置。

这两种事务管理器类型都不需要任何属性，它们不过是类型别名，也就是说，可以使用 TransactionFactory 接口的实现类的完全限定名或类型别名代替它们。TransactionFactory 接口定义如下。

```java
public interface TransactionFactory {
  void setProperties(Properties props);
  Transaction newTransaction(Connection conn);
  Transaction newTransaction(DataSource dataSource, TransactionIsola-
tionLevel level, boolean autoCommit);
}
```

任何在 XML 中配置的属性在实例化之后都将被传递给 setProperties() 方法。

也可以创建一个 Transaction 接口的实现类，这个接口也很简单，代码如下。

```java
public interface Transaction {
  Connection getConnection() throws SQLException;
  void commit() throws SQLException;
  void rollback() throws SQLException;
  void close() throws SQLException;
  Integer getTimeout() throws SQLException;
}
```

使用这两个接口，开发者可以完全自定义 MyBatis 对事务的处理。

10.1.9 dataSource

dataSource 元素使用标准的 JDBC 数据源接口来配置 JDBC 连接对象的资源。

许多 MyBatis 的应用程序会按示例来配置数据源。虽然这是可选的，但为了使用延迟加载，数据源是必须配置的。有 3 种内建的数据源类型：UNPOOLED、POOLED 和 JNDI。

1. UNPOOLED

这个数据源的实现只是每次被请求时打开和关闭连接。虽然有点慢，但对于在数据库连接可用性方面没有太高要求的简单应用程序来说，是一个很好的选择。不同的数据库在性能方面的表现也是不一样的，对于某些数据库来说，使用连接池并不重要，这个配置就很适合这种情形。

UNPOOLED 类型的数据源仅需要配置以下 5 种属性。

（1）driver：JDBC 驱动的 Java 类的完全限定名（并不是 JDBC 驱动中可能包含的数据源类）。

（2）url：数据库的 JDBC URL 地址。

（3）username：登录数据库的用户名。

（4）password：登录数据库的密码。

（5）defaultTransactionIsolationLevel：默认的连接事务隔离级别。

作为可选项，也可以传递属性给数据库驱动。若这样做，则属性的前缀为"driver."，例如：

```
driver.encoding=UTF8
```

这将通过 DriverManager.getConnection(url,driverProperties) 方法传递值为 UTF8 的 encoding 属性给数据库驱动。

2. POOLED

这个数据源的实现利用"池"的概念将 JDBC 连接对象组织起来，避免了创建新的连接实例时所必需的初始化和认证。这是一种使并发 Web 应用快速响应请求的流行处理方式。

除了上面提到的 UNPOOLED 的属性外，还有更多属性用来配置 POOLED 的数据源。

（1）poolMaximumActiveConnections：任意时间可以存在的活动（也就是正在使用的）连接数。默认值为 10。

（2）poolMaximumIdleConnections：任意时间可能存在的空闲连接数。

（3）poolMaximumCheckoutTime：在被强制返回之前，池中连接被检出（checked out）时间。默认值为 20 000ms。

（4）poolTimeToWait：一个底层设置，如果获取连接花费了相当长的时间，则连接池会打印状态日志并重新尝试获取一个连接（避免在误配置的情况下一直获取失败）。默认值为 20 000ms。

（5）poolMaximumLocalBadConnectionTolerance：指一个关于坏连接容忍度的底层设置，作用于每一个尝试从缓存池获取连接的线程。如果这个线程获取到的是一个坏连接，那么这个数据源允许此线程尝试重新获取一个新连接，但重新尝试的次数不应该超过 poolMaximumIdleConnections 与

poolMaximumLocalBadConnectionTolerance 之和。默认值为 3。

（6）poolPingQuery：发送到数据库的侦测查询，用来检验连接是否正常工作并准备接受请求。默认为 NO PING QUERY SET，这会导致多数数据库驱动失败时带有一个恰当的错误消息。

（7）poolPingEnabled：是否启用侦测查询。若开启，则需要设置 poolPingQuery 属性为一个可执行的 SQL 语句（最好是一个速度非常快的 SQL 语句）。默认值为 false。

（8）poolPingConnectionsNotUsedFor：配置 poolPingQuery 的频率。可以被设置为和数据库连接超时时间一样，以避免不必要的侦测。默认值为 0，即所有连接每一时刻都被侦测，当然仅当 poolPingEnabled 为 true 时适用。

3. JNDI

这个数据源的实现是为了能在 EJB 或应用服务器这类容器中使用，容器可以集中或在外部配置数据源，然后放置一个 JNDI 上下文的引用。这种数据源配置只需以下两个属性。

（1）initial_context：这个属性用来在 InitialContext 中寻找上下文，即 initialContext.lookup(initial_context)。这是个可选属性，如果忽略，那么 data_source 属性将会直接从 InitialContext 中寻找。

（2）data_source：这是引用数据源实例位置上下文的路径。提供了 initial_context 配置时，会在其返回的上下文中进行查找，如果没有提供，则直接在 InitialContext 中查找。

与其他数据源配置类似，可以通过添加 "env." 前缀直接把属性传递给初始上下文，比如：

```
env.encoding=UTF8
```

这就会在初始上下文（InitialContext）实例化时，向其构造方法传递值为 UTF8 的 encoding 属性。

用户可以通过实现 org.apache.ibatis.datasource.DataSourceFactory 接口来使用第三方数据源，代码如下。

```
public interface DataSourceFactory {
  void setProperties(Properties props);
  DataSource getDataSource();
}
```

org.apache.ibatis.datasource.unpooled.UnpooledDataSourceFactory 可被用作父类来构建新的数据源适配器，下面这段代码是插入 C3P0 数据源所必需的。

```
import org.apache.ibatis.datasource.unpooled.UnpooledDataSourceFactory;
import com.mchange.v2.c3p0.ComboPooledDataSource;

public class C3P0DataSourceFactory extends UnpooledDataSourceFactory {

  public C3P0DataSourceFactory() {
    this.dataSource = new ComboPooledDataSource();
  }
}
```

为了使其工作，应为每个希望 MyBatis 调用的 setter 方法在配置文件中增加对应的属性。以下是一个可以连接至 PostgreSQL 数据库的示例。

```xml
<dataSource type="org.myproject.C3P0DataSourceFactory">
  <property name="driver" value="org.postgresql.Driver"/>
  <property name="url" value="jdbc:postgresql:mydb"/>
  <property name="username" value="postgres"/>
  <property name="password" value="root"/>
</dataSource>
```

10.1.10　databaseIdProvider

MyBatis 可以根据不同的数据库厂商执行不同的语句，这种多厂商的支持是基于映射语句中的 databaseId 属性。MyBatis 会加载不带 databaseId 属性和带有匹配当前数据库 databaseId 属性的所有语句。如果同时找到带有 databaseId 和不带 databaseId 的相同语句，那么后者会被舍弃。为支持多厂商特性，只要向 mybatis-config.xml 文件中加入 databaseIdProvider 即可。

```xml
<databaseIdProvider type="DB_VENDOR" />
```

这里的 DB_VENDOR 会通过 DatabaseMetaData#getDatabaseProductName() 返回的字符串进行设置。通常情况下，由于这个字符串非常长，而且相同产品的不同版本会返回不同的值，因此最好通过设置属性别名来使其变短，示例如下。

```xml
<databaseIdProvider type="DB_VENDOR">
  <property name="SQL Server" value="sqlserver"/>
  <property name="DB2" value="db2"/>
  <property name="Oracle" value="oracle" />
</databaseIdProvider>
```

在提供了属性别名时，DB_VENDOR databaseIdProvider 将被设置为第一个能匹配数据库产品名称的属性键对应的值，如果没有匹配的属性，就会设置为"null"。在这个例子中，如果 getDatabaseProductName() 返回"Oracle (DataDirect)"，databaseId 将被设置为"oracle"。

可以通过实现 org.apache.ibatis.mapping.DatabaseIdProvider 接口，并在 mybatis-config.xml 中注册来构建自己的 DatabaseIdProvider。

```java
public interface DatabaseIdProvider {
  void setProperties(Properties p);
  String getDatabaseId(DataSource dataSource) throws SQLException;
}
```

10.1.11 mappers

现在就要定义 SQL 映射语句了。首先需要告诉 MyBatis 到哪里去找到这些语句。Java 在自动查找方面没有提供一个很好的方法，所以最好使用相对于类路径的资源引用，或者完全限定资源定位符（包括 file:/// 的 URL）、类名和包名等。

以下是查找映射文件的各种方式。

```xml
<!-- 使用相对于类路径的资源引用 -->
<mappers>
  <mapper resource="org/mybatis/builder/AuthorMapper.xml"/>
  <mapper resource="org/mybatis/builder/BlogMapper.xml"/>
  <mapper resource="org/mybatis/builder/PostMapper.xml"/>
</mappers>
<!-- 使用完全限定资源定位符(URL) -->
<mappers>
  <mapper url="file:///var/mappers/AuthorMapper.xml"/>
  <mapper url="file:///var/mappers/BlogMapper.xml"/>
  <mapper url="file:///var/mappers/PostMapper.xml"/>
</mappers>
<!-- 使用映射器接口实现类的完全限定类名 -->
<mappers>
  <mapper class="org.mybatis.builder.AuthorMapper"/>
  <mapper class="org.mybatis.builder.BlogMapper"/>
  <mapper class="org.mybatis.builder.PostMapper"/>
</mappers>
<!-- 将包内的映射器接口实现全部注册为映射器 -->
<mappers>
  <package name="org.mybatis.builder"/>
</mappers>
```

这些配置告诉了 MyBatis 去哪里找映射文件，剩下的细节就应该是每个 SQL 映射文件了，也就是接下来要讨论的内容。

10.2 Mapper映射文件

MyBatis 真正的强大之处在于它的映射语句。由于它的映射语句异常强大，因此 Mapper 映射 XML 文件就显得相对简单。如果将它与具有相同功能的 JDBC 代码进行对比，就会发现它省掉了将近 95% 的代码。MyBatis 就是针对 SQL 构建的，并且比普通的方法做得更好。

Mapper 映射文件主要由以下 8 个元素组成。

（1）cache：给定命名空间的缓存配置。

（2）cache-ref：其他命名空间缓存配置的引用。

（3）resultMap：最复杂也是最强大的元素，用来描述如何从数据库结果集中加载对象。

（4）sql：可被其他语句引用的可重用语句块。

（5）insert：映射插入语句。

（6）update：映射更新语句。

（7）delete：映射删除语句。

（8）select：映射查询语句。

接下来将详细介绍这些元素。

10.2.1 select

查询语句是MyBatis中最常用的元素之一，大多数应用使用查询的频率要远远高于修改。每一个插入、更新或删除操作，通常都对应多个查询操作。这是MyBatis的基本原则之一，也是将焦点放到查询和结果映射的原因。简单查询的select元素是非常简单的，其示例代码如下。

```
<select id="selectPerson" parameterType="int" resultType="hashmap">
  SELECT * FROM PERSON WHERE ID = #{id}
</select>
```

这个语句被称为selectPerson，接受一个int（或Integer）类型的参数，并返回一个HashMap类型的对象，其中的键是列名，值便是结果行中的对应值。其中参数符号"#{id}"是为了告诉MyBatis创建一个预处理语句参数，这样的一个参数在SQL中会由一个"?"来标识，并被传递到一个新的预处理语句中，类似于下面JDBC的代码：

```
// 类似JDBC的代码
String selectPerson = "SELECT * FROM PERSON WHERE ID=?";
PreparedStatement ps = conn.prepareStatement(selectPerson);
ps.setInt(1,id);
```

当然，如果是用原生的JDBC来提取结果并将其映射到对象实例中，需要很多烦琐的代码，而这正是MyBatis节省时间的地方。接下来将深入讲解MyBatis是如何进行参数和结果映射的。

select元素中有很多属性允许用户通过配置来决定每条语句的作用，其代码如下。

```
<select
  id="selectPerson"
  parameterType="int"
  parameterMap="deprecated"
  resultType="hashmap"
  resultMap="personResultMap"
  flushCache="false"
  useCache="true"
  timeout="10000"
  fetchSize="256"
  statementType="PREPARED"
```

```
resultSetType="FORWARD_ONLY">
```

这些属性的含义如表 10-4 所示。

表10-4　Select元素中属性的含义

属　性	描　述
id	命名空间中唯一的标识符，可以被用来引用这条语句
parameterType	将会传入这条语句的参数类的完全限定名或别名。这个属性是可选的，因为 MyBatis 可以通过类型处理器（TypeHandler）推断出具体传入语句的参数，默认值为未设置（unset）
resultType	从这条语句中返回的期望类型的类的完全限定名或别名。注意，若返回的是集合，则应该设置为集合包含的类型，而不是集合本身。可以使用 resultType 或 resultMap，但二者不能同时使用
resultMap	外部 resultMap 的命名引用。结果集的映射是 MyBatis 最强大的特性，如果能将其理解透彻，则许多复杂映射的问题都能迎刃而解。可以使用 resultMap 或 resultType，但二者不能同时使用
flushCache	将其设置为 true 后，只要语句被调用，就会导致本地缓存和二级缓存被清空，默认值为false
useCache	将其设置为 true 后，将会导致本条语句的结果被作为二级缓存，默认值为 true（对于select 元素）
timeout	这个设置是在抛出异常之前，驱动程序等待数据库返回请求结果的秒数。默认值为未设置（unset）（依赖驱动）
fetchSize	这是一个给驱动的提示，尝试让驱动程序每次批量返回的结果行数和这个值相等。默认值为未设置（unset）（依赖驱动）
statementType	STATEMENT、PREPARED 或 CALLABLE 中的一个。这会让 MyBatis 分别使用 Statement、PreparedStatement 或 CallableStatement，默认值为PREPARED
resultSetType	FORWARD_ONLY、SCROLL_SENSITIVE、SCROLL_INSENSITIVE 或 DEFAULT（等价于 unset）中的一个，默认值为 unset（依赖驱动）
databaseId	如果配置了数据库厂商标识（databaseIdProvider），MyBatis 会加载所有不带 databaseId 或匹配当前 databaseId 的语句；若带或不带databaseId的语句都有，则不带的会被忽略
resultOrdered	这个设置仅针对嵌套结果 select 语句适用：如果为 true，就是假设包含了嵌套结果集或分组，这样当返回一个主结果行时，就不会发生对前面结果集的引用情况。这就使得在获取嵌套的结果集时不至于导致内存不够用。默认值为false
resultSets	这个设置仅对多结果集的情况适用。它将列出语句执行后返回的结果集，并给每个结果集一个名称，名称之间用逗号分隔

10.2.2 insert、update和delete

insert、update 和 delete 的实现非常接近，都是用于数据变更，以下是 3 种语句的示例。

```
<insert
  id="insertAuthor"
  parameterType="domain.blog.Author"
  flushCache="true"
  statementType="PREPARED"
  keyProperty=""
  keyColumn=""
  useGeneratedKeys=""
  timeout="20">

<update
  id="updateAuthor"
  parameterType="domain.blog.Author"
  flushCache="true"
  statementType="PREPARED"
  timeout="20">

<delete
  id="deleteAuthor"
  parameterType="domain.blog.Author"
  flushCache="true"
  statementType="PREPARED"
  timeout="20">
```

这些元素属性的含义如表 10-5 所示。

表10-5　insert、update和delte属性的含义

属性	描述
id	命名空间中的唯一标识符，可被用来代表这条语句
parameterType	将要传入语句的参数的完全限定类名或别名。这个属性是可选的，因为 MyBatis 可以通过类型处理器推断出具体传入语句的参数，默认值为未设置（unset）
flushCache	将其设置为 true 后，只要语句被调用，就会导致本地缓存和二级缓存被清空，默认值为true（对于 insert、update 和 delete 语句）
timeout	这个设置是在抛出异常之前，驱动程序等待数据库返回请求结果的秒数，默认值为未设置（unset）（依赖驱动）
statementType	STATEMENT、PREPARED 或 CALLABLE 中的一个。这会让 MyBatis 分别使用 Statement、PreparedStatement 或 CallableStatement，默认值为 PREPARED

续表

属性	描述
useGeneratedKeys	（仅对 insert 和 update 有用）这会令 MyBatis 使用 JDBC 的 getGeneratedKeys 方法来取出由数据库内部生成的主键（类似于 MySQL 和 SQL Server 的关系数据库管理系统的自动递增字段），默认值为false
keyProperty	（仅对 insert 和 update 有用）唯一标记一个属性，MyBatis 会通过 getGeneratedKeys 的返回值，或者通过 insert 语句的 selectKey 子元素设置它的键值，默认值为未设置（unset）。如果希望得到多个生成的列，也可以设置为逗号分隔的属性名称列表
keyColumn	（仅对 insert 和 update 有用）通过生成的键值设置表中的列名，这个设置仅在某些数据库（如 PostgreSQL）中是必需的，当主键列不是表中的第一列时需要设置。如果希望使用多个生成的列，也可以设置为逗号分隔的属性名称列表
databaseId	如果配置了数据库厂商标识（databaseIdProvider），MyBatis 会加载所有不带 databaseId 或匹配当前 databaseId 的语句；若带或不带 databaseId 的语句都有，则不带的会被忽略

以下就是 insert、update 和 delete 语句的使用示例。

```xml
<insert id="insertAuthor">
  insert into Author (id,username,password,email,bio)
  values (#{id},#{username},#{password},#{email},#{bio})
</insert>

<update id="updateAuthor">
  update Author set
    username = #{username},
    password = #{password},
    email = #{email},
    bio = #{bio}
  where id = #{id}
</update>

<delete id="deleteAuthor">
  delete from Author where id = #{id}
</delete>
```

10.2.3 处理主键

MyBatis 在插入语句中有一些额外的属性和子元素用来处理主键的生成，其生成方式有以下两种。

1. 数据库支持自动生成主键

如果数据库本身支持自动生成主键的字段（如 MySQL 和 SQL Server），那么可以设置 useGeneratedKeys="true"，然后再把 keyProperty 设置到目标属性上即可。例如，如果上面的 Author

表已经对 id 使用了自动生成的列类型，那么语句可以修改为：

```
<insert id="insertAuthor" useGeneratedKeys="true"
    keyProperty="id">
  insert into Author (username,password,email,bio)
  values (#{username},#{password},#{email},#{bio})
</insert>
```

如果数据库还支持多行插入，也可以传入一个 Authors 数组或集合，并返回自动生成的主键。

```
<insert id="insertAuthor" useGeneratedKeys="true"
    keyProperty="id">
  insert into Author (username, password, email, bio) values
  <foreach item="item" collection="list" separator=",">
    (#{item.username}, #{item.password}, #{item.email}, #{item.bio})
  </foreach>
</insert>
```

2. 数据库不支持自动生成主键

有些数据库或 JDBC 驱动并不支持自动生成主键，MyBatis 会用另一种方法来生成主键。

下面这个示例会生成一个随机 ID。

```
<insert id="insertAuthor">
  <selectKey keyProperty="id" resultType="int" order="BEFORE">
    select CAST(RANDOM()*1000000 as INTEGER) a from SYSIBM.SYSDUMMY1
  </selectKey>
  insert into Author
    (id, username, password, email,bio, favourite_section)
  values
    (#{id}, #{username}, #{password}, #{email}, #{bio}, #{favouriteSection,jdbcType=VARCHAR})
</insert>
```

其中，selectKey 元素将首先运行，Author 的 id 被设置，然后插入语句会被调用。

在 Oracle 数据库中，还可以使用序列来生成 ID，上面的例子可以改为如下代码。

```
<insert id="insertAuthor">
  <selectKey keyProperty="id" resultType="int" order="BEFORE">
    select Author_seq.NEXTVAL from dual
  </selectKey>
  insert into Author
    (id, username, password, email,bio, favourite_section)
  values
    (#{id}, #{username}, #{password}, #{email}, #{bio}, #{favouriteSection,jdbcType=VARCHAR})
</insert>
```

其中，Author_seq 是表 Author 的序列。

selectKey 元素属性的含义如表 10-6 所示。

```xml
<selectKey
  keyProperty="id"
  resultType="int"
  order="BEFORE"
  statementType="PREPARED">
```

表10-6 selectKey 元素属性的含义

属性	描述
keyProperty	selectKey 语句结果应该被设置的目标属性。如果希望得到多个生成的列，也可以设置为逗号分隔的属性名称列表
keyColumn	匹配属性的返回结果集中的列名称。如果希望得到多个生成的列，也可以设置为逗号分隔的属性名称列表
resultType	结果的类型。MyBatis 通常可以推断出来，但是为了更加精确，写上也不会有什么问题。MyBatis 允许将任何简单类型用作主键的类型，包括字符串。如果希望作用于多个生成的列，那么可以使用一个包含期望属性的 Object 或 Map
order	可以被设置为 BEFORE 或 AFTER。如果设置为 BEFORE，那么它会首先生成主键，设置 keyProperty 后执行插入语句。如果设置为 AFTER，那么先执行插入语句，然后是 selectKey 中的语句，这与 Oracle 数据库的行为相似，在插入语句内部可能有嵌入索引调用
statementType	MyBatis 支持 STATEMENT、PREPARED 和 CALLABLE 语句的映射类型，分别代表 PreparedStatement 和 CallableStatement 类型

10.2.4 sql

sql 元素可以被用来定义可重用的 SQL 代码段，这些代码段可以包含在其他语句中。其不同的属性值可以跟随包含的实例而变化，下面的例子定义了一个 SQL 代码段：

```xml
<sql id="userColumns"> ${alias}.id,${alias}.username,${alias}.password </sql>
```

这个 SQL 代码段可以被包含在其他语句中，例如：

```xml
<select id="selectUsers" resultType="map">
  select
    <include refid="userColumns"><property name="alias" value="t1"/></include>,
    <include refid="userColumns"><property name="alias" value="t2"/></include>
  from some_table t1
    cross join some_table t2
</select>
```

当然，属性值也可以被用在 include 元素的 refid 属性中，如下面的代码：

```xml
<include refid="${include_target}"/>
```

属性值还可以被用在 include 内部语句中,如下面的代码:

```xml
<sql id="sometable">
  ${prefix}Table
</sql>
<sql id="someinclude">
  from
    <include refid="${include_target}"/>
</sql>
<select id="select" resultType="map">
  select
    field1, field2, field3
  <include refid="someinclude">
    <property name="prefix" value="Some"/>
    <property name="include_target" value="sometable"/>
  </include>
</select>
```

上述代码中的 ${prefix}Table 就用在了 include 内部语句中。

10.2.5 参数

像 MyBatis 的其他部分一样,参数也可以指定一个特殊的数据类型。

```
#{property,javaType=int,jdbcType=NUMERIC}
```

javaType 通常可以由参数对象确定,除非该对象是 HashMap。这时所使用的 TypeHandler 应该明确指定 javaType。

也可以指定一个特殊的类型处理器类,比如:

```
#{age,javaType=int,jdbcType=NUMERIC,typeHandler=MyTypeHandler}
```

当然,正常情况下很少需要设置它们。

对于数值类型,还有一个小数保留位数的设置,示例如下:

```
#{height,javaType=double,jdbcType=NUMERIC,numericScale=2}
```

最后,还有一个 mode 属性允许指定 IN、OUT 或 INOUT 参数。如果参数为 OUT 或 INOUT,参数对象属性的真实值将会被改变,就像在获取输出参数时所期望的那样。如果 mode 为 OUT(或 INOUT),而且 jdbcType 为 CURSOR(也就是 Oracle 的 REFCURSOR),就必须指定一个 resultMap 来映射结果集 ResultMap 到参数类型。注意,这里的 javaType 属性是可选的,如果留空并且 jdbcType 是 CURSOR,那么它会自动被设置为 ResultMap。

```
#{department, mode=OUT, jdbcType=CURSOR, javaType=ResultSet, result-
Map=departmentResultMap}
```

MyBatis 也支持很多高级的数据类型，如结构体，但是注册 OUT 参数时必须指定语句类型名称，比如：

```
#{middleInitial, mode=OUT, jdbcType=STRUCT, jdbcTypeName=MY_TYPE, re-
sultMap=departmentResultMap}
```

尽管这些选项都很强大，但大多数时候只需简单地指定属性名，其他的事情 MyBatis 会自己去推断，最多再为可能为空的列指定 jdbcType，例如：

```
#{firstName}
#{middleInitial,jdbcType=VARCHAR}
#{lastName}
```

10.2.6 结果映射

resultMap 元素是 MyBatis 中最重要、最强大的元素。它可以让用户从 90% 的 JDBC ResultSets 数据提取代码中解放出来，并在一些情形下允许做一些 JDBC 不支持的事情。实际上，在对复杂语句进行联合映射时，它很可能可以代替数千行同等功能的代码。

resultMap 的设计思想：简单的语句不需要明确的结果映射，而复杂一点的语句只需描述它们的关系即可。

下面是一个简单映射语句的示例，在这个示例中并没有明确 resultMap。

```
<select id="selectUsers" resultType="map">
  select id, username, hashedPassword
  from some_table
  where id = #{id}
</select>
```

上述语句只是简单地将所有的列映射到 HashMap 的键上，这由 resultType 属性指定。虽然在大部分情况下都够用，但是 HashMap 不是一个很好的领域模型。程序最好使用 JavaBean 或 POJO 作为领域模型。MyBatis 对两者都支持。

下面的示例使用了 JavaBean 作为领域模型。

```
package com.waylau.mybatis.model;
public class User {
  private int id;
  private String username;
  private String hashedPassword;

  public int getId() {
    return id;
```

```
    }
    public void setId(int id) {
      this.id = id;
    }
    public String getUsername() {
      return username;
    }
    public void setUsername(String username) {
      this.username = username;
    }
    public String getHashedPassword() {
      return hashedPassword;
    }
    public void setHashedPassword(String hashedPassword) {
      this.hashedPassword = hashedPassword;
    }
}
```

　　基于 JavaBean 的规范，上面这个类有 3 个属性：id、username 和 hashedPassword。这些属性会对应到 select 语句中的列名。

　　这样的一个 JavaBean 可以被映射到 ResultSet 上，就像映射到 HashMap 上一样简单。

```
select id, username, hashedPassword from some_table where id = #{id}
```

　　更进一步，如果使用类型别名，就可以不用输入类的完全限定名称了，例如：

```
<!-- 配置在mybatis-config.xml文件中 -->
<typeAlias type="com.waylau.mybatis.model.User" alias="User"/>

<!-- 配置在SQL映射文件中 -->
<select id="selectUsers" resultType="User">
  select id, username, hashedPassword
  from some_table
  where id = #{id}
</select>
```

　　这些情况下，MyBatis 会在幕后自动创建一个 ResultMap，再基于属性名来映射列到 JavaBean 的属性上。如果列名和属性名没有精确匹配，就可以在 SELECT 语句中对列使用别名来匹配标签，例如：

```
<select id="selectUsers" resultType="User">
  select
    user_id             as "id",
    user_name           as "userName",
    hashed_password     as "hashedPassword"
  from some_table
  where id = #{id}
</select>
```

resultMap 最优秀的地方在于，虽然用户已经对它相当了解，但是根本不需要显式地用到它。从下面的示例中可以看到，使用外部的 resultMap，也是解决列名不匹配的另一种方式。

```xml
<resultMap id="userResultMap" type="User">
  <id property="id" column="user_id" />
  <result property="username" column="user_name"/>
  <result property="password" column="hashed_password"/>
</resultMap>
```

引用它的语句使用 resultMap 属性即可（注意，这里去掉了 resultType 属性），例如：

```xml
<select id="selectUsers" resultMap="userResultMap">
  select user_id, user_name, hashed_password
  from some_table
  where id = #{id}
</select>
```

10.2.7 自动映射

当使用自动映射查询结果时，MyBatis 会获取 sql 返回的列名，并在 Java 类中查找相同名称的属性（忽略大小写）。这意味着如果 Mybatis 发现了 ID 列和 id 属性，就会将 ID 的值赋给 id。

通常数据库列使用大写单词命名，单词间用下画线分隔；而 Java 属性一般遵循驼峰命名法。为了在这两种命名方式之间启用自动映射，需要将 mapUnderscoreToCamelCase 设置为 true。

自动映射甚至在特定的 resultMap 下也能工作。在这种情况下，对于每一个 resultMap，所有的 ResultSet 提供的列若没有被手工映射，则被自动映射。自动映射处理完毕后，手工映射才会被处理。在接下来的示例中，id 和 userName 列将被自动映射，hashed_password 列将根据配置映射。

```xml
<select id="selectUsers" resultMap="userResultMap">
  select
    user_id             as "id",
    user_name           as "userName",
    hashed_password
  from some_table
  where id = #{id}
</select>
<resultMap id="userResultMap" type="User">
  <result property="password" column="hashed_password"/>
</resultMap>
```

自动映射等级有以下 3 种。

（1）NONE：禁用自动映射，仅设置手动映射属性。

（2）PARTIAL：自动映射结果，除了那些在 join 中定义嵌套结果映射的结果。

（3）FULL：自动映射所有结果。

默认值为 PARTIAL 是有原因的。当使用 FULL 时，自动映射会在处理 join 结果时执行，并且由于 join 会取得若干相同行的不同实体数据，因此可能导致非预期的映射。下面的示例将展示这种风险。

```xml
<select id="selectBlog" resultMap="blogResult">
  select
    B.id,
    B.title,
    A.username,
  from Blog B left outer join Author A on B.author_id = A.id
  where B.id = #{id}
</select>
<resultMap id="blogResult" type="Blog">
  <association property="author" resultMap="authorResult"/>
</resultMap>

<resultMap id="authorResult" type="Author">
  <result property="username" column="author_username"/>
</resultMap>
```

在结果中，Blog 和 Author 均将自动映射。值得注意的是，Author 有一个 id 属性，在 ResultSet 中有一个列名为 id，Author 的 id 将被填充为 Blog 的 id，所以需要谨慎使用 FULL。

通过添加 autoMapping 属性可以忽略自动映射等级配置，也可以启用或禁用自动映射指定的 ResultMap。

```xml
<resultMap id="userResultMap" type="User" autoMapping="false">
  <result property="password" column="hashed_password"/>
</resultMap>
```

10.2.8 缓存

MyBatis 包含一个非常强大的查询缓存特性，可以非常方便地配置和定制。

1. 开启缓存

默认情况下是没有开启缓存的，要开启二级缓存，需要在 SQL 映射文件中添加一行代码：

```xml
<cache/>
```

看上去用法非常简单。该语句实现了如下效果。

（1）映射语句文件中的所有 select 语句将被缓存。

（2）映射语句文件中的所有 insert、update 和 delete 语句会刷新缓存。

（3）缓存会使用 LRU（Least Recently Used，最近最少使用的）算法收回。

（4）缓存不会在任何基于时间的时间表上刷新（即没有刷新间隔）。

（5）缓存将存储 1024 个对列表或对象的引用（无论查询方法返回什么）。

（6）缓存将被视为读/写缓存，这意味着检索的对象不会被共享，并且可以被调用者安全地修改，而不会干扰其他调用者或线程的其他可能的修改。

2. 配置缓存

所有这些属性都可以通过缓存元素的属性来修改，例如：

```
<cache
  eviction="FIFO"
  flushInterval="60000"
  size="512"
  readOnly="true"/>
```

这个配置创建了一个 FIFO 缓存，并每隔 60 秒刷新，将存储 512 个对列表或对象的引用，而且返回的对象被认为是只读的，因此在不同线程中的调用者之间修改它们会导致冲突。

可用的收回策略如下。

（1）LRU（最近最少使用的）：移除最长时间不被使用的对象。默认使用该策略。

（2）FIFO（First Input First Output，先进先出）：按对象进入缓存的顺序来移除。

（3）SOFT（软引用）：移除基于垃圾回收器状态和软引用规则的对象。

（4）WEAK（弱引用）：更积极地移除基于垃圾回收器状态和弱引用规则的对象。

flushInterval（刷新间隔）可以被设置为任意的正整数，而且这些正整数代表一个合理的毫秒形式的时间段。默认情况是不设置的，也就是没有刷新间隔，缓存仅在调用语句时刷新。

size（引用数目）可以被设置为任意正整数，要记住所要缓存的对象数目和运行环境的可用内存资源数目。默认值是 1024。

readOnly（只读）属性可以被设置为 true 或 false。只读的缓存会给所有调用者返回缓存对象的相同实例，因此这些对象不能被修改。这提供了很重要的性能优势。可读写的缓存会返回缓存对象的拷贝（通过序列化）。这会慢一些，但很安全，因此默认为 false。

3. 使用自定义缓存

可以通过实现自己的缓存或为第三方缓存方案创建适配器来完全覆盖缓存行为，下面是使用自定义缓存的示例。

```
<cache type="com.domain.something.MyCustomCache"/>
```

以上示例展示了如何使用一个自定义的缓存实现。type 属性指定的类必须实现 org.mybatis.cache.Cache 接口。Cache 接口定义如下。

```
public interface Cache {
  String getId();
  int getSize();
  void putObject(Object key, Object value);
  Object getObject(Object key);
  boolean hasKey(Object key);
```

```
  Object removeObject(Object key);
  void clear();
}
```

下面的代码会在缓存实现中调用一个名称为 "setCacheFile(String file)" 的方法。

```
<cache type="com.domain.something.MyCustomCache">
  <property name="cacheFile" value="/tmp/my-custom-cache.tmp"/>
</cache>
```

可以使用所有简单类型作为 JavaBeans 的属性，MyBatis 会进行转换。

MyBatis 从 3.4.2 版本开始已经支持在所有属性设置完毕后调用一个初始化方法。如果想要使用这个特性，那么可以在自定义缓存类中实现 org.apache.ibatis.builder.InitializingObject 接口。

```
public interface InitializingObject {
  void initialize() throws Exception;
}
```

语句可以按下面的方式配置缓存。

```
<select … flushCache="false" useCache="true"/>
<insert … flushCache="true"/>
<update … flushCache="true"/>
<delete … flushCache="true"/>
```

4. 引用缓存

如果想在命名空间中共享相同的缓存配置和实例，可以使用 cache-ref 元素来引用另一个缓存。

```
<cache-ref namespace="com.someone.application.data.SomeMapper"/>
```

10.3 动态SQL

如果用户有使用 JDBC 或其他类似框架的经验，就能体会到根据不同条件拼接 SQL 语句的痛苦。拼接 SQL 是危险和烦琐的，拼接时不仅要确保添加必要的空格，还要注意去掉列表中最后一个列名的逗号，同时还要避免 SQL 注入风险。而利用 MyBatis 动态 SQL 这一特性，可以彻底摆脱这种痛苦。

MyBatis 动态 SQL 基于功能强大的 OGNL 表达式。开发者使用 MyBatis 动态 SQL，只需记住以下几种表达式元素即可。

10.3.1 if

动态 SQL 通常要做的事情是根据条件包含 where 子句的一部分，比如：

```
<select id="findActiveBlogWithTitleLike"
    resultType="Blog">
  SELECT * FROM BLOG
  WHERE state = 'ACTIVE'
  <if test="title != null">
    AND title like #{title}
  </if>
</select>
```

这条语句提供了一种可选的查找文本功能。如果没有传入"title"，那么所有处于"ACTIVE"状态的 BLOG 都会返回；反之，如果传入了"title"，就会对"title"列进行模糊查找并返回 BLOG 结果。

10.3.2 choose（when，otherwise）

有时并不想应用所有的条件语句，而只想从中选择一项。针对这种情况，MyBatis 提供了 choose 元素，它有点类似于 Java 中的 switch 语句。

以下就是一个使用 choose 的示例。

```
<select id="findActiveBlogLike"
    resultType="Blog">
  SELECT * FROM BLOG WHERE state = 'ACTIVE'
  <choose>
    <when test="title != null">
      AND title like #{title}
    </when>
    <when test="author != null and author.name != null">
      AND author_name like #{author.name}
    </when>
    <otherwise>
      AND featured = 1
    </otherwise>
  </choose>
</select>
```

在上面的示例中，如果提供了"title"，就按"title"查找；如果提供了"author"，就按"author"查找；如果两者都没有提供，就返回所有符合条件的 BLOG。

10.3.3 trim（where，set）

观察下面的示例。

```
<select id="findActiveBlogLike"
```

```
    resultType="Blog">
  SELECT * FROM BLOG
  WHERE
  <if test="state != null">
    state = #{state}
  </if>
  <if test="title != null">
    AND title like #{title}
  </if>
  <if test="author != null and author.name != null">
    AND author_name like #{author.name}
  </if>
</select>
```

如果这些条件没有一个能匹配上,那么这条 SQL 的最终代码为:

```
SELECT * FROM BLOG
WHERE
```

这会导致查询失败。如果仅第二个条件匹配,那么这条 SQL 的最终代码为:

```
SELECT * FROM BLOG
WHERE
AND title like 'someTitle'
```

这个查询也会失败。因为它不能简单地用条件句式来解决。

而 MyBatis 通过简单的修改就能达到目的,其代码如下。

```
<select id="findActiveBlogLike"
     resultType="Blog">
  SELECT * FROM BLOG
  <where>
    <if test="state != null">
        state = #{state}
    </if>
    <if test="title != null">
        AND title like #{title}
    </if>
    <if test="author != null and author.name != null">
        AND author_name like #{author.name}
    </if>
  </where>
</select>
```

where 元素只会在至少有一个子元素返回 SQL 子句的情况下插入"WHERE"子句。而且,若语句的开头为"AND"或"OR",where 元素也会将它们去除。

如果 where 元素没有按正常套路"出牌",就可以通过自定义 trim 元素来定制 where 元素的功能。例如,与 where 元素等价的自定义 trim 元素为:

续表

```
<trim prefix="WHERE" prefixOverrides="AND |OR ">
    ...
</trim>
```

prefixOverrides 属性会忽略通过管道分隔的文本序列（注意，此例中的空格也是必要的）。它的作用是移除所有 prefixOverrides 属性中指定的内容，并且插入 prefix 属性中指定的内容。

类似的用于动态更新语句的解决方案称为 set。set 元素可用于动态包含需要更新的列，而舍去其他的列，例如：

```
<update id="updateAuthorIfNecessary">
  update Author
    <set>
      <if test="username != null">username=#{username},</if>
      <if test="password != null">password=#{password},</if>
      <if test="email != null">email=#{email},</if>
      <if test="bio != null">bio=#{bio}</if>
    </set>
  where id=#{id}
</update>
```

这里，set 元素会动态前置 SET 关键字，同时也会删掉无关的逗号。因为用了条件语句后，很可能会在生成的 SQL 语句后留下这些逗号。

10.3.4　foreach

动态 SQL 的另一个常用的操作是对一个集合进行遍历，通常是在构建 IN 条件语句时，例如：

```
<select id="selectPostIn" resultType="domain.blog.Post">
  SELECT *
  FROM POST P
  WHERE ID in
  <foreach item="item" index="index" collection="list"
      open="(" separator="," close=")">
        #{item}
  </foreach>
</select>
```

以将任何可迭代对象（如 List、Set 等）、Map 对象或数组对象传递给 foreach 作为集合参数。当使用可迭代对象或数组时，index 是当前迭代的次数，item 的值是本次迭代获取的元素。当使用 Map 对象（或者 Map.Entry 对象的集合）时，index 是键，item 是值。

10.3.5　bind

bind 元素可以从 OGNL 表达式中创建一个变量并将其绑定到上下文，例如：

```xml
<select id="selectBlogsLike" resultType="Blog">
  <bind name="pattern" value="'%' + _parameter.getTitle() + '%'" />
  SELECT * FROM BLOG
  WHERE title LIKE #{pattern}
</select>
```

10.3.6 多数据库支持

一个配置了"_databaseId"变量的databaseIdProvider可用于动态代码中,这样就可以根据不同的数据库厂商构建特定的语句。比如下面的示例:

```xml
<insert id="insert">
  <selectKey keyProperty="id" resultType="int" order="BEFORE">
    <if test="_databaseId == 'oracle'">
      select seq_users.nextval from dual
    </if>
    <if test="_databaseId == 'db2'">
      select nextval for seq_users from sysibm.sysdummy1
    </if>
  </selectKey>
  insert into users values (#{id}, #{name})
</insert>
```

10.4 常用API

下面介绍 MyBatis 常用的 API。

10.4.1 SqlSessionFactoryBuilder

SqlSessionFactoryBuilder 有 5 种 build() 方法,每一种都允许从不同的资源中创建一个 SqlSession 实例。

```
SqlSessionFactory build(InputStream inputStream)
SqlSessionFactory build(InputStream inputStream, String environment)
SqlSessionFactory build(InputStream inputStream, Properties properties)
SqlSessionFactory build(InputStream inputStream, String env, Properties props)
SqlSessionFactory build(Configuration config)
```

第一种方法是最常用的,它使用一个参照 XML 文档或特定的 mybatis-config.xml 文件的 Reader 实例。可选的参数是 environment 和 properties。由 environment 决定加载哪种环境,包括数据源和事务管理器。比如：

```xml
<environments default="development">
  <environment id="development">
    <transactionManager type="JDBC">
      …
    <dataSource type="POOLED">
      …
  </environment>
  <environment id="production">
    <transactionManager type="MANAGED">
      …
    <dataSource type="JNDI">
      …
  </environment>
</environments>
```

如果调用了带 environment 参数的 build 方法，那么 MyBatis 将会使用 configuration 对象来配置这个 environment。当然，如果指定了一个不合法的 environment，就会得到错误提示。如果调用了不带 environment 参数的 build 方法，就会使用默认的 environment（在上面的示例中指定为 default="development" 的代码）。

如果调用了参数为 properties 实例的方法，那么 MyBatis 就会加载那些 properties（属性配置文件）。那些属性可以用 ${propName} 语法形式多次在配置文件中使用。

属性可以从 mybatis-config.xml 中被引用，或者直接指定它。MyBatis 将会按照下面的优先级加载它们。

（1）读取在 properties 元素体中指定的属性。

（2）读取在 properties 元素的类路径 resource 或 url 中指定的属性，且会覆盖已经指定了的重复属性。

（3）读取作为方法参数传递的属性，且会覆盖已经从 properties 元素体和 resource 或 url 属性中加载了的重复属性。

因此，作为方法参数传递的属性优先级最高，在 resource 或 url 中指定的属性优先级中等，在 properties 元素体中指定的属性优先级最低。

下面是一个从 mybatis-config.xml 文件创建 SqlSessionFactory 的示例。

```
String resource = "org/mybatis/builder/mybatis-config.xml";
InputStream inputStream = Resources.getResourceAsStream(resource);
SqlSessionFactoryBuilder builder = new SqlSessionFactoryBuilder();
SqlSessionFactory factory = builder.build(inputStream);
```

注意：这里使用了 Resources 工具类，这个类在 org.apache.ibatis.io 包中。Resources 类能从类路径、文件系统或一个 Web URL 中加载资源文件。Resources 工具类提供了如下接口。

```
URL getResourceURL(String resource)
URL getResourceURL(ClassLoader loader, String resource)
InputStream getResourceAsStream(String resource)
InputStream getResourceAsStream(ClassLoader loader, String resource)
Properties getResourceAsProperties(String resource)
Properties getResourceAsProperties(ClassLoader loader, String resource)
Reader getResourceAsReader(String resource)
Reader getResourceAsReader(ClassLoader loader, String resource)
File getResourceAsFile(String resource)
File getResourceAsFile(ClassLoader loader, String resource)
InputStream getUrlAsStream(String urlString)
Reader getUrlAsReader(String urlString)
Properties getUrlAsProperties(String urlString)
Class classForName(String className)
```

最后一个 build 方法的参数为 Configuration 实例。Configuration 类包含用户可能需要了解的 SqlSessionFactory 实例的所有内容。下面是一个简单的示例，演示了如何手动配置 Configuration 实例，然后将其传递给 build() 方法来创建 SqlSessionFactory。

```
DataSource dataSource = BaseDataTest.createBlogDataSource();
TransactionFactory transactionFactory = new JdbcTransactionFactory();

Environment environment = new Environment("development", transaction-
Factory, dataSource);

Configuration configuration = new Configuration(environment);
configuration.setLazyLoadingEnabled(true);
configuration.setEnhancementEnabled(true);
configuration.getTypeAliasRegistry().registerAlias(Blog.class);
configuration.getTypeAliasRegistry().registerAlias(Post.class);
configuration.getTypeAliasRegistry().registerAlias(Author.class);
configuration.addMapper(BoundBlogMapper.class);
configuration.addMapper(BoundAuthorMapper.class);

SqlSessionFactoryBuilder builder = new SqlSessionFactoryBuilder();
SqlSessionFactory factory = builder.build(configuration);
```

这样，就能获得用来创建 SqlSession 实例的 SqlSessionFactory 了。

10.4.2　SqlSessionFactory

SqlSessionFactory 有 6 个方法可以创建 SqlSession 实例。通常情况下，在选择这些方法时，用户需要考虑以下几点。

（1）事务处理：需要在 session 中使用事务或使用自动提交功能（auto-commit）吗？（通常意味着很多数据库和 / 或 JDBC 驱动没有事务）。

（2）连接：需要依赖 MyBatis 获得来自数据源的配置吗？还是使用自己提供的配置？

（3）执行语句：需要 MyBatis 复用预处理语句和（或）批量更新语句（包括插入和删除）吗？

基于以上需求，有下列已重载的多个 openSession() 方法可供使用。

```
SqlSession openSession()
SqlSession openSession(boolean autoCommit)
SqlSession openSession(Connection connection)
SqlSession openSession(TransactionIsolationLevel level)
SqlSession openSession(ExecutorType execType,TransactionIsolationLevel
level)
SqlSession openSession(ExecutorType execType)
SqlSession openSession(ExecutorType execType, boolean autoCommit)
SqlSession openSession(ExecutorType execType, Connection connection)
Configuration getConfiguration();
```

默认的 openSession() 方法没有参数，它会创建有以下特性的 SqlSession。

（1）会开启一个事务（也就是不自动提交）。

（2）将从当前环境配置的 DataSource 实例中获取 Connection 对象。

（3）事务隔离级别将会使用驱动或数据源的默认设置。

（4）预处理语句不会被复用，也不会批量处理更新。

还有一个参数需要特别关注，就是 ExecutorType。这个枚举类型定义了以下 3 个值。

（1）ExecutorType.SIMPLE：为每个语句的执行创建一个新的预处理语句。

（2）ExecutorType.REUSE：复用预处理语句。

（3）ExecutorType.BATCH：批量执行所有更新语句，如果 SELECT 在它们中间执行，那么必要时请把它们区分开来，以保证行为的易读性。

10.4.3　SqlSession

在 SqlSession 类中有超过 20 种方法，可将它们分为以下几个易于理解的组。

1. 执行语句方法

这些方法被用来执行定义在 SQL 映射的 XML 文件中的 SELECT、INSERT、UPDATE 和 DELETE 语句。它们都会自行解释，每一句都使用语句的 ID 属性和参数对象，参数可以是原生类型（自动装箱或包装类）、JavaBean、POJO 或 Map。

```
<T> T selectOne(String statement, Object parameter)
<E> List<E> selectList(String statement, Object parameter)
<K,V> Map<K,V> selectMap(String statement, Object parameter, String map-
Key)
int insert(String statement, Object parameter)
int update(String statement, Object parameter)
int delete(String statement, Object parameter)
```

selectOne 和 selectList 的不同仅仅是 selectOne 必须返回一个对象或 null 值，如果返回值多于一个，就会抛出异常。如果不知道返回对象的数量，就可以使用 selectList。如果需要查看返回对象是否存在，那么可行的方案是返回一个值（0 或 1）。

selectMap 稍微特殊一点，因为它会将返回对象的其中一个属性作为 key 值，将对象作为 value 值，从而将多结果集转为 Map 类型值。因为并不是所有语句都需要参数，所以这些方法都重载成不需要参数的形式，其代码如下。

```
<T> T selectOne(String statement)
<E> List<E> selectList(String statement)
<K,V> Map<K,V> selectMap(String statement, String mapKey)
int insert(String statement)
int update(String statement)
int delete(String statement)
```

最后，还有 select 方法的 3 个高级版本，它们允许用户限制返回行数的范围，或者提供自定义结果控制逻辑，这通常在数据集合庞大的情形下使用，其代码如下。

```
<E> List<E> selectList (String statement, Object parameter, RowBounds rowBounds)
<K,V> Map<K,V> selectMap(String statement, Object parameter, String mapKey, RowBounds rowbounds)
void select (String statement, Object parameter, ResultHandler<T> handler)
void select (String statement, Object parameter, RowBounds rowBounds, ResultHandler<T> handler)
```

RowBounds 参数会告诉 MyBatis 略过指定数量的记录，还有限制返回结果的数量。RowBounds 类有一个构造方法来接收 offset 和 limit。另外，它们是不可以二次赋值的。其示例代码如下。

```
int offset = 100;
int limit = 25;
RowBounds rowBounds = new RowBounds(offset, limit);
```

所以在这方面，不同的驱动能够取得不同级别的高效率。为了取得最佳的表现，请使用结果集的 SCROLL_SENSITIVE 或 SCROLL_INSENSITIVE 类型。

ResultHandler 参数允许用户按喜欢的方式处理每一行。例如，可以将它添加到 List 中、创建 Map 和 Set，或者丢弃每个返回值，它取代了仅保留执行语句后总结果列表的"死板"结果。也可以使用 ResultHandler 做很多事，并且这是 MyBatis 自身内部会使用的方法。

ResultHandler 的接口很简单，代码如下。

```
package org.apache.ibatis.session;
public interface ResultHandler<T> {
  void handleResult(ResultContext<? extends T> context);
}
```

ResultContext 参数允许用户访问结果对象本身、被创建的对象数目，以及返回值为 Boolean 的 stop 方法，也可以使用此 stop 方法来使 MyBatis 停止加载更多的结果。

使用 ResultHandler 时需要注意以下两种限制。

（1）从被 ResultHandler 调用的方法返回的数据不会被缓存。

（2）当使用结果映射集（resultMap）时，MyBatis 大多数情况下需要数行结果来构造外键对象。如果用户正在使用 ResultHandler，就可以给出外键（association）或集合（collection）尚未赋值的对象。

2. 批量立即更新方法

有一个方法可以刷新（执行）存储在 JDBC 驱动类中的批量更新语句。当用户将 ExecutorType.BATCH 作为 ExecutorType 使用时，可以采用此方法。

```
List<BatchResult> flushStatements()
```

3. 事务控制方法

控制事务作用域有 4 种方法。当然，如果用户已经设置了自动提交或正在使用外部事务管理器，就没有任何效果了。然而，如果用户正在使用 JDBC 事务管理器，由 Connection 实例来控制，那么这 4 种方法就能派上用场。

```
void commit()
void commit(boolean force)
void rollback()
void rollback(boolean force)
```

默认情况下，MyBatis 不会自动提交事务，除非它侦测到有插入、更新或删除操作改变了数据库。如果用户已经做出了一些改变但没有使用这些方法，那么可以传递 true 值到 commit 和 rollback 方法来保证事务被正常处理。很多时候并不用调用 rollback()，因为 MyBatis 会在没有调用 commit 时就完成回滚操作。然而，如果用户需要在支持多提交和回滚的 session 中获得更多细粒度控制，就可以使用回滚操作来达到目的。

4. 本地缓存

Mybatis 用到了两种缓存：本地缓存（Local Cache）和二级缓存（Second Level Cache）。

每当一个新 session 被创建，MyBatis 就会创建一个与之相关联的本地缓存。任何在 session 中执行过的查询语句本身都会被保存在本地缓存中，那么，相同的查询语句和相同的参数所产生的更改就不会二次影响数据库了。本地缓存会被增删改、提交事务、关闭事务，以及关闭 session 所清空。

默认情况下，本地缓存数据可在整个 session 周期内使用，这一缓存需要被用来解决循环引用错误和加快重复嵌套查询的速度，所以它可以不被禁用。但可以设置 localCacheScope=STATEMENT，表示缓存仅在语句执行时有效。

可以随时调用以下方法来清空本地缓存。

```
void clearCache()
```

5. 确保SqlSession被关闭

```
void close()
```

必须保证最重要的事情是关闭所打开的任何session，其最佳工作模式如下。

```
SqlSession session = sqlSessionFactory.openSession();
try {
    // …
    session.insert(…);
    session.update(…);
    session.delete(…);
    session.commit();
} finally {
    session.close();
}
```

还有，如果用户正在使用jdk 1.7以上及MyBatis 3.2以上的版本，就可以使用try-with-resources语句。

```
try (SqlSession session = sqlSessionFactory.openSession()) {
    // …
    session.insert(…);
    session.update(…);
    session.delete(…);
    session.commit();
}
```

10.4.4 注解

最初设计时，MyBatis是一个XML驱动的框架。配置信息是基于XML的，而且映射语句也是定义在XML中的。而到了MyBatis 3，就有新选择了。MyBatis 3构建在全面且强大的基于Java语言的配置API之上。这个配置API是基于XML的MyBatis配置，也是新的基于注解的配置。注解提供了一种简单的方式来实现简单映射语句，而不会引入大量的消耗。

注意：Java注解的表达力和灵活性是十分有限的，因此，在MyBatis中，Java注解并不能适用所有的场景。在复杂的场景下，建议使用XML来进行映射。

下面的示例展示了如何使用@SelectKey注解在插入前读取数据库序列的值。

```
@Insert("insert into table3 (id, name) values(#{nameId}, #{name})")
@SelectKey(statement="call next value for TestSequence", keyProperty="n-
ameId", before=true, resultType=int.class)
int insertTable3(Name name);
```

下面的示例展示了如何使用 @SelectKey 注解在插入后读取数据库识别列的值。

```
@Insert("insert into table2 (name) values(#{name})")
@SelectKey(statement="call identity()", keyProperty="nameId", before=-
false, resultType=int.class)
int insertTable2(Name name);
```

下面的示例展示了如何使用 @Flush 注解调用 SqlSession#flushStatements() 方法。

```
@Flush
List<BatchResult> flush();
```

下面的示例展示了如何通过指定 @Result 的 id 属性来命名结果集。

```
@Results(id = "userResult", value = {
  @Result(property = "id", column = "uid", id = true),
  @Result(property = "firstName", column = "first_name"),
  @Result(property = "lastName", column = "last_name")
})
@Select("select * from users where id = #{id}")
User getUserById(Integer id);

@Results(id = "companyResults")
@ConstructorArgs({
  @Arg(property = "id", column = "cid", id = true),
  @Arg(property = "name", column = "name")
})
@Select("select * from company where id = #{id}")
Company getCompanyById(Integer id);
```

下面的示例展示了单一参数使用 @SqlProvider 注解。

```
@SelectProvider(type = UserSqlBuilder.class, method = "buildGetUsersBy-
Name")
List<User> getUsersByName(String name);

class UserSqlBuilder {
  public static String buildGetUsersByName(final String name) {
    return new SQL(){{
      SELECT("*");
      FROM("users");
      if (name != null) {
        WHERE("name like #{value} || '%'");
      }
      ORDER_BY("id");
    }}.toString();
  }
}
```

下面的示例展示了多参数使用 @SqlProvider 注解。

```java
@SelectProvider(type = UserSqlBuilder.class, method = "buildGetUsersBy-Name")
List<User> getUsersByName(
    @Param("name") String name, @Param("orderByColumn") String orderByColumn);

class UserSqlBuilder {
  // 不使用@Param
  public static String buildGetUsersByName(
      final String name, final String orderByColumn) {
    return new SQL(){{
      SELECT("*");
      FROM("users");
      WHERE("name like #{name} || '%'");
      ORDER_BY(orderByColumn);
    }}.toString();
  }

  // 使用了@Param
  public static String buildGetUsersByName(@Param("orderByColumn") final String orderByColumn) {
    return new SQL(){{
      SELECT("*");
      FROM("users");
      WHERE("name like #{name} || '%'");
      ORDER_BY(orderByColumn);
    }}.toString();
  }
}
```

其中，如果不使用@Param，就应该定义与接口相同的参数；如果使用@Param，就只能定义要使用的参数。

第11章
Lite技术集成

学习技术并不是为了学而学，而应学以致用。接下来将会基于前面所介绍的技术，从零开始搭建一个互联网应用框架——Lite。顾名思义，Lite意味着轻量、快速、高效。

11.1 技术集成概述

当前的互联网应用都会使用外部成熟的框架，毕竟"站在巨人的肩膀上"可以缩短研发周期、降低开发成本、减少资源的重复投入等。但技术的选型需要考虑诸多因素，如学习成本、易用性、稳定性、社区的完善性等。Lite 在做技术选型时也是如此。以下是 Lite 所要集成的技术。

（1）Java：毫无疑问，Java 是企业级应用的首选，是开发大型互联网应用的利器。

（2）Servlet：最为成功的 Java EE 规范之一，在 Java 应用中使用非常广泛。

（3）Jetty：高性能 Servlet 容器，采用内嵌模式，可以轻松构建微服务。

（4）Spring：事实上的 Java EE 标准，每个 Java 开发者的首选框架。

（5）Spring Web MVC：应用最为广泛的 MVC 框架。

（6）Jackson JSON：应用广泛的 JSON 序列化工具。

（7）Spring Security：企业级安全管理框架。

（8）MyBatis：轻量级 ORM 解决方案。

（9）MySQL：最为流行的开源数据库。

（10）Apache Commons DBCP：高性能数据库连接池方案。

（11）JUnit Jupiter Engine：新一代 JUnit 单元测试框架。

（12）logback：高性能日志框架。

（13）SLF4J：日志门面抽象框架。

（14）NGINX：高性能的 HTTP 和反向代理服务器。

（15）Redis：高性能缓存服务器。

（16）Angular：现代 Web 应用程序前端开发框架，核心功能包括 MVC、模块化、自动化双向数据绑定、语义化标签、服务、依赖注入等。

（17）Spring Boot：用于简化 Spring 应用的配置，达到开箱即用的效果。

有关上述技术的具体版本，可以参阅"附录：本书所采用的技术及相关版本"部分。

接下来将这些"零散"的技术"拼装"起来。

11.2 MySQL 的安装及基本操作

本节将简单介绍 MySQL 在 Windows 系统下的安装及基本使用方法。其他环境的安装，如 Linux、Mac 等系统都与之类似，可以参照本节的步骤进行安装。

11.2.1 下载安装包

可以从 https://dev.mysql.com/downloads/mysql/8.0.html 免费下载最新的 MySQL 8 版本的安装包。MySQL 8 带来了全新的体验，如支持 NoSQL、JSON 等，其性能比 MySQL 5.7 版本提升了 2 倍以上。

本例下载的安装包为 mysql-8.0.12-winx64.zip。

11.2.2 解压安装包

解压安装包至安装目录，如 D 盘根目录下。

本例为 D:\mysql-8.0.12-winx64。

11.2.3 创建my.ini

my.ini 是 MySQL 安装的配置文件，配置内容如下。

```
[mysqld]
# 安装目录
basedir=D:\\mysql-8.0.12-winx64
# 数据存放目录
datadir=D:\\mysqlData\\data
```

其中，basedir 指定了 MySQL 的安装目录；datadir 指定了数据目录。

将 my.ini 放置在 MySQL 安装目录的根目录下。需要注意的是，要先创建 D:\mysqlData 目录，data 目录是由 MySQL 创建的。

11.2.4 初始化安装

初始化安装需要执行以下命令行。

```
mysqld --defaults-file=D:\mysql-8.0.12-winx64\my.ini --initialize --console
```

如果控制台输出如下内容，就说明已经安装成功。

```
>mysqld --defaults-file=D:\mysql-8.0.12-winx64\my.ini --initialize
--console
2018-08-20T16:14:45.287448Z 0 [System] [MY-013169] [Server] D:\mysql-
8.0.12-winx64\bin\mysqld.exe (mysqld 8.0.12) initializing of server in
progress as process 5012
2018-08-20T16:14:45.289628Z 0 [ERROR] [MY-010457] [Server] --initialize
specified but the data directory has files in it. Aborting.
2018-08-20T16:14:45.299329Z 0 [ERROR] [MY-010119] [Server] Aborting
2018-08-20T16:14:45.301316Z 0 [System] [MY-010910] [Server] D:\mysql-
8.0.12-winx64\bin\mysqld.exe: Shutdown complete (mysqld 8.0.12)  MySQL
```

```
Community Server - GPL.

D:\mysql-8.0.12-winx64\bin>mysqld --defaults-file=D:\mysql-8.0.12-
winx64\my.ini --initialize --console
2018-08-20T16:15:25.729771Z 0 [System] [MY-013169] [Server] D:\mysql-
8.0.12-winx64\bin\mysqld.exe (mysqld 8.0.12) initializing of server in
progress as process 18148
2018-08-20T16:15:43.569562Z 5 [Note] [MY-010454] [Server] A temporary
password is generated for root@localhost: L-hk!rBuk9-.
2018-08-20T16:15:55.811470Z 0 [System] [MY-013170] [Server] D:\mysql-
8.0.12-winx64\bin\mysqld.exe (mysqld 8.0.12) initializing of server has
completed
```

其中,"L-hk!rBuk9-."就是 root 用户的初始化密码。稍后可以对该密码进行更改。

11.2.5 启动、关闭MySQL server

执行 mysqld 命令就能启动 MySQL server,或者执行 mysqld --console 命令来查看完整的启动信息。

```
>mysqld --console
2018-08-20T16:18:23.698153Z 0 [Warning] [MY-010915] [Server] 'NO_ZERO_
DATE', 'NO_ZERO_IN_DATE' and 'ERROR_FOR_DIVISION_BY_ZERO' sql modes
should be used with strict mode. They will be merged with strict mode
in a future release.
2018-08-20T16:18:23.698248Z 0 [System] [MY-010116] [Server] D:\mysql-
8.0.12-winx64\bin\mysqld.exe (mysqld 8.0.12) starting as process 16304
2018-08-20T16:18:27.624422Z 0 [Warning] [MY-010068] [Server] CA certifi-
cate ca.pem is self signed.
2018-08-20T16:18:27.793310Z 0 [System] [MY-010931] [Server] D:\mysql-
8.0.12-winx64\bin\mysqld.exe: ready for connections. Version: '8.0.12'
socket: ''  port: 3306  MySQL Community Server - GPL.
```

可以通过执行 mysqladmin -u root shutdown 命令来关闭 MySQL server。

11.2.6 使用MySQL客户端

使用 mysql 命令登录,账号为 root,密码为 "L-hk!rBuk9-."。

```
>mysql -u root -p
Enter password: ************
Welcome to the MySQL monitor.  Commands end with ; or \g.
Your MySQL connection id is 11
Server version: 8.0.12

Copyright (c) 2000, 2018, Oracle and/or its affiliates. All rights re-
served.
```

```
Oracle is a registered trademark of Oracle Corporation and/or its
affiliates. Other names may be trademarks of their respective
owners.

Type 'help;' or '\h' for help. Type '\c' to clear the current input
statement.
```

执行下面的语句修改密码，其中"123456"为新密码。

```
mysql> ALTER USER 'root'@'localhost' IDENTIFIED BY '123456';
Query OK, 0 rows affected (0.13 sec)
```

11.2.7　MySQL常用指令

以下总结了 MySQL 常用的指令。

1. 显示已有的数据库

要显示已有的数据库，执行以下指令。

```
mysql> show databases;
+--------------------+
| Database           |
+--------------------+
| information_schema |
| mysql              |
| performance_schema |
| sys                |
+--------------------+
4 rows in set (0.08 sec)
```

2. 创建新的数据库

要创建新的数据库，执行以下指令。

```
mysql> CREATE DATABASE lite;
Query OK, 1 row affected (0.19 sec)
```

3. 使用数据库

要使用数据库，执行以下指令。

```
mysql> USE lite;
Database changed
```

4. 建表

要建表，执行以下指令。

```
mysql> CREATE TABLE t_user (user_id BIGINT NOT NULL, username
VARCHAR(20));
```

```
Query OK, 0 rows affected (0.82 sec)
```

5. 查看表

要查看数据库中的所有表,执行以下指令。

```
mysql> SHOW TABLES;
+----------------+
| Tables_in_lite |
+----------------+
| t_user         |
+----------------+
1 row in set (0.00 sec)
```

要查看表的详情,执行以下指令。

```
mysql> DESCRIBE t_user;
+----------+-------------+------+-----+---------+-------+
| Field    | Type        | Null | Key | Default | Extra |
+----------+-------------+------+-----+---------+-------+
| user_id  | bigint(20)  | NO   |     | NULL    |       |
| username | varchar(20) | YES  |     | NULL    |       |
+----------+-------------+------+-----+---------+-------+
2 rows in set (0.00 sec)
```

6. 插入数据

要插入数据,则执行以下指令。

```
mysql> INSERT INTO t_user(user_id, username) VALUES(1, '老卫');
Query OK, 1 row affected (0.08 sec)
```

11.3 Spring与MyBatis集成

在使用 MyBatis 框架时,经常会与 Spring 搭配使用。

利用 Spring 的依赖注入,可以创建线程安全的、基于事务的 SqlSession 和映射器(mapper),并将它们直接注入 Spring 应用的 bean 中,因此可以省去开发者手动管理 MyBatis 生命周期的烦琐过程。

MyBatis 社区提供了 MyBatis-Spring 开源项目(http://www.mybatis.org/spring/),用于方便地实现 Spring 与 MyBatis 的适配与集成。

接下来将演示如何在 Spring 应用中集成 MyBatis 框架,并配置数据库连接池。项目源码可以在 spring-mybatis 目录下找到。

11.3.1 添加依赖

创建一个名为 spring-mybatis 的 Maven 项目，并在应用中添加如下依赖。

```xml
<dependency>
    <groupId>org.springframework</groupId>
    <artifactId>spring-context</artifactId>
    <version>${spring.version}</version>
</dependency>

<!-- 数据库相关的依赖 -->
<dependency>
    <groupId>org.springframework</groupId>
    <artifactId>spring-jdbc</artifactId>
    <version>${spring.version}</version>
</dependency>
<dependency>
    <groupId>org.mybatis</groupId>
    <artifactId>mybatis</artifactId>
    <version>${mybatis.version}</version>
</dependency>
<dependency>
    <groupId>org.mybatis</groupId>
    <artifactId>mybatis-spring</artifactId>
    <version>${mybatis-spring.version}</version>
</dependency>
<dependency>
    <groupId>mysql</groupId>
    <artifactId>mysql-connector-java</artifactId>
    <version>${mysql-connector-java.version}</version>
</dependency>
<dependency>
    <groupId>org.apache.commons</groupId>
    <artifactId>commons-dbcp2</artifactId>
    <version>${dbcp2.version}</version>
</dependency>

<!-- 测试相关的依赖 -->
<dependency>
    <groupId>org.junit.jupiter</groupId>
    <artifactId>junit-jupiter-engine</artifactId>
    <version>${junit.version}</version>
</dependency>
<dependency>
    <groupId>org.springframework</groupId>
    <artifactId>spring-test</artifactId>
    <version>${spring.version}</version>
</dependency>
```

其中，mybatis-spring 依赖用于添加 MyBatis-Spring 适配器；mysql-connector-java 依赖是 MySQL 数据库的 JDBC 驱动；commons-dbcp2 依赖用于创建数据库连接池。同时，添加的 junit-jupiter-engine 和 spring-test 用于单元测试。

11.3.2　初始化数据库信息

执行 mysqld --console 命令来启动 MySQL server，并按下面的步骤创建数据库信息。

1. 创建新的数据库

要创建新的数据库 lite，执行以下指令。

```
mysql> CREATE DATABASE lite;
Query OK, 1 row affected (0.19 sec)
```

2. 使用数据库

要使用数据库 lite，执行以下指令。

```
mysql> USE lite;
Database changed
```

3. 建表

要建表，执行以下指令。

```
mysql> CREATE TABLE t_user (user_id BIGINT NOT NULL, username
VARCHAR(20));
Query OK, 0 rows affected (0.82 sec)
```

11.3.3　数据库连接池配置

mybatis-spring 应用采用基于 XML 的配置方式，数据库连接池配置如下。

```xml
<?xml version="1.0" encoding="UTF-8"?>
<beans xmlns="http://www.springframework.org/schema/beans"
    xmlns:xsi="http://www.w3.org/2001/XMLSchema-instance"
    xmlns:context="http://www.springframework.org/schema/context"
    xsi:schemaLocation="
        http://www.springframework.org/schema/beans
        http://www.springframework.org/schema/beans/spring-beans.xsd
        http://www.springframework.org/schema/context
        http://www.springframework.org/schema/context/spring-context.xsd
        http://www.springframework.org/schema/aop
        http://www.springframework.org/schema/aop/spring-aop.xsd">


    <bean class="org.springframework.beans.factory.config.PropertyPlace-
```

```xml
holderConfigurer">
        <property name="locations" value="classpath:jdbc.properties"/>
    </bean>
    
    <!-- 数据源 -->
    <bean id="dataSource" class="org.apache.commons.dbcp2.BasicDataSource"
        destroy-method="close">
        <property name="driverClassName" value="${jdbc.driverClassName}" />
        <property name="url" value="${jdbc.url}" />
        <property name="username" value="${jdbc.username}" />
        <property name="password" value="${jdbc.password}" />
    </bean>
    
    <!-- MyBatis 工厂 -->
    <bean id="sqlSessionFactory" class="org.mybatis.spring.SqlSessionFactoryBean">
        <property name="dataSource" ref="dataSource" />
    </bean>
    
    <!-- 扫描 MyBatis Mapper 接口所在的包 -->
    <bean class="org.mybatis.spring.mapper.MapperScannerConfigurer">
        <property name="basePackage" value="com.waylau.mybatis.mapper" />
    </bean>
</beans>
```

其中，数据库连接池采用了 Apache Commons DBCP 框架。

数据源信息放置在独立的 .properties 文件中，内容如下。

```
jdbc.driverClassName=com.mysql.cj.jdbc.Driver
jdbc.url=jdbc:mysql://localhost:3306/lite?useSSL=false&serverTimezone=UTC
jdbc.username=root
jdbc.password=123456
```

11.3.4 编写业务代码

应用的业务代码由以下几部分组成。

1. 领域对象

创建 User 类，代表领域对象。

```java
package com.waylau.mybatis.domain;

public class User {

    private Long userId;
```

```java
    private String username;

    public User(Long userId, String username) {
        this.userId = userId;
        this.username = username;
    }

    // …省略getter/setter方法

    @Override
    public String toString() {
        return "User [userId=" + userId + ", username=" + username + "]";
    }
}
```

2. Mapper

UserMapper 接口定义如下。

```java
package com.waylau.mybatis.mapper;

import com.waylau.mybatis.domain.User;

public interface UserMapper {

    int createUser(User user);

    User getUser(Long userId);

    int updateUser(User user);

}
```

相应的 Mapper XML 文件定义如下。

```xml
<?xml version="1.0" encoding="UTF-8"?>
<!DOCTYPE mapper
    PUBLIC "-//mybatis.org//DTD Mapper 3.0//EN"
    "http://mybatis.org/dtd/mybatis-3-mapper.dtd">

<mapper namespace="com.waylau.mybatis.mapper.UserMapper">

    <select id="getUser" resultType="com.waylau.mybatis.domain.User">
        select user_id as userId, username
        from t_user where user_id = #{userId}
    </select>

    <update id="updateUser"
```

```xml
        parameterType="com.waylau.mybatis.domain.User">
        update t_user set username = #{username}
        where user_id = #{userId ,jdbcType=NUMERIC}
    </update>

    <insert id="createUser"
        parameterType="com.waylau.mybatis.domain.User">
        insert into t_user(user_id, username)
        values(
        #{userId ,jdbcType=NUMERIC},
        #{username}
        )
    </insert>
</mapper>
```

该 UserMapper 实现了对用户的创建、查询、修改。

11.3.5　编写测试用例

测试用例代码如下。

```java
package com.waylau.mybatis.mapper;

import static org.junit.jupiter.api.Assertions.assertEquals;
import org.junit.jupiter.api.Test;
import org.springframework.beans.factory.annotation.Autowired;
import org.springframework.test.context.junit.jupiter.SpringJUnitConfig;
import com.waylau.mybatis.domain.User;

@SpringJUnitConfig(locations = "/spring.xml")
class UserMapperTests {

    @Autowired
    private UserMapper userMapper;

    @Test
    void testCreateUser() {
        User user = new User(1L, "老卫");
        int i = userMapper.createUser(user);
        assertEquals(1, i);
    }

    @Test
    void testGetUser() {
        User user = userMapper.getUser(1L);
        User user2 = new User(1L, "老卫");
        assertEquals(user.toString(), user2.toString());
```

```
    }

    @Test
    void testUpdateUser() {
        User user = userMapper.getUser(1L);
        user.setUsername("waylau");
        int i = userMapper.updateUser(user);
        assertEquals(1, i);

        User user2 = userMapper.getUser(1L);
        assertEquals(user.toString(), user2.toString());
    }
}
```

其中，@SpringJUnitConfig 注解用于简化在 JUnit 中初始化 Spring ApplicationContext 的过程。
上述 3 个测试用例，分别用于测试 UserMapper 的创建、查询、修改。
右击该类运行，如果 JUnit 显示绿色，代表测试通过。

11.4　集成Spring Web MVC

本节将演示如何集成 Spring Web MVC。在 11.3 节 spring-mybatis 应用代码的基础上，创建一个新的 "spring-mvc" 应用，以演示整个集成过程。

本节项目源码可以在 spring-mvc 目录下找到。

11.4.1　添加依赖

在 spring-mvc 应用的原有 pom.xml 文件中添加如下依赖。

```
<!-- Web相关的依赖 -->
<dependency>
    <groupId>org.springframework</groupId>
    <artifactId>spring-webmvc</artifactId>
    <version>${spring.version}</version>
</dependency>
<dependency>
    <groupId>org.eclipse.jetty</groupId>
    <artifactId>jetty-servlet</artifactId>
    <version>${jetty.version}</version>
    <scope>provided</scope>
</dependency>
<dependency>
```

```xml
        <groupId>com.fasterxml.jackson.core</groupId>
        <artifactId>jackson-core</artifactId>
        <version>${jackson.version}</version>
</dependency>
<dependency>
        <groupId>com.fasterxml.jackson.core</groupId>
        <artifactId>jackson-databind</artifactId>
        <version>${jackson.version}</version>
</dependency>
...
```

其中，spring-webmvc 依赖用于在应用中启用 Spring Web MVC 功能；jetty-servlet 依赖提供 Servlet 容器支持；jackson-core 和 jackson-databind 用于 JSON 序列化。

11.4.2 添加控制器

添加控制器，代码如下。

```java
package com.waylau.spring.controller;

import org.springframework.beans.factory.annotation.Autowired;
import org.springframework.web.bind.annotation.DeleteMapping;
import org.springframework.web.bind.annotation.GetMapping;
import org.springframework.web.bind.annotation.PathVariable;
import org.springframework.web.bind.annotation.PostMapping;
import org.springframework.web.bind.annotation.PutMapping;
import org.springframework.web.bind.annotation.RequestBody;
import org.springframework.web.bind.annotation.RequestMapping;
import org.springframework.web.bind.annotation.RestController;

import com.waylau.mybatis.domain.User;
import com.waylau.mybatis.mapper.UserMapper;

@RestController
@RequestMapping("/users")
public class UserController {
    @Autowired
    private UserMapper userMapper;

    /**
     * 根据id查询用户
     * @param id
     * @return
     */
    @GetMapping("/{id}")
    public User getUserById(@PathVariable Long id) {
        return userMapper.getUser(id);
    }
```

```java
/**
 * 创建用户
 * @param user
 * @return
 */
@PostMapping
public int createUser(@RequestBody User user) {
    return userMapper.createUser(user);
}

/**
 * 删除用户
 * @param id
 */
@DeleteMapping("/{id}")
public int deleteUser(@PathVariable Long id) {
    return userMapper.deleteUser(id);
}

/**
 * 修改用户
 * @param user
 * @return
 */
@PutMapping
public int  updateUser(@RequestBody User user) {
    return userMapper.updateUser(user);
}
}
```

为了保持代码简单，UserController 直接依赖了 UserMapper。

11.4.3 修改UserMapper

在原有的 UserMapper 代码的基础上，增加删除用户的接口。完整的 UserMapper 定义如下。

```java
package com.waylau.mybatis.mapper;

import com.waylau.mybatis.domain.User;

public interface UserMapper {

    int createUser(User user);

    User getUser(Long userId);

    int updateUser(User user);
```

```
    int deleteUser(Long userId);
}
```

UserMapper.xml 定义如下。

```xml
<?xml version="1.0" encoding="UTF-8"?>
<!DOCTYPE mapper
    PUBLIC "-//mybatis.org//DTD Mapper 3.0//EN"
    "http://mybatis.org/dtd/mybatis-3-mapper.dtd">

<mapper namespace="com.waylau.mybatis.mapper.UserMapper">

    <select id="getUser" resultType="com.waylau.mybatis.domain.User">
        select user_id as userId, username
        from t_user where user_id = #{userId}
    </select>

    <update id="updateUser"
        parameterType="com.waylau.mybatis.domain.User">
        update t_user set username = #{username}
        where user_id = #{userId ,jdbcType=NUMERIC}
    </update>

    <insert id="createUser"
        parameterType="com.waylau.mybatis.domain.User">
        insert into t_user(user_id, username)
        values(
        #{userId ,jdbcType=NUMERIC},
        #{username}
        )
    </insert>

    <delete id="deleteUser">
        delete from t_user where
        user_id = #{userId ,jdbcType=NUMERIC}
    </delete>

</mapper>
```

11.4.4　引入Servlet容器

本应用采用内嵌 Servlet 容器的方式来运行。

嵌入服务器的代码在之前的章节中也做过介绍，完整代码如下。

```java
package com.waylau.spring;

import org.eclipse.jetty.server.Server;
```

```java
import org.eclipse.jetty.servlet.ServletContextHandler;
import org.eclipse.jetty.servlet.ServletHolder;
import org.springframework.web.context.ContextLoaderListener;
import org.springframework.web.context.WebApplicationContext;
import org.springframework.web.context.support.AnnotationConfigWebApplicationContext;
import org.springframework.web.servlet.DispatcherServlet;

import com.waylau.spring.config.AppConfiguration;

public class JettyServer {
    private static final int DEFAULT_PORT = 8080;
    private static final String CONTEXT_PATH = "/";
    private static final String MAPPING_URL = "/*";

    public void run() throws Exception {
        Server server = new Server(DEFAULT_PORT);
        server.setHandler(servletContextHandler(webApplicationContext()));
        server.start();
        server.join();
    }

    private ServletContextHandler servletContextHandler(WebApplicationContext ct) {
        // 启用Session管理器
        ServletContextHandler handler =
                new ServletContextHandler(ServletContextHandler.SESSIONS);

        handler.setContextPath(CONTEXT_PATH);
        handler.addServlet(new ServletHolder(new DispatcherServlet(ct)),
                MAPPING_URL);
        handler.addEventListener(new ContextLoaderListener(ct));

        return handler;
    }

    private WebApplicationContext webApplicationContext() {
        AnnotationConfigWebApplicationContext context =
                new AnnotationConfigWebApplicationContext();
        context.register(AppConfiguration.class);
        return context;
    }
}
```

其中，AppConfiguration 为应用配置类。

11.4.5 创建应用配置

1. AppConfiguration

AppConfiguration 为主应用配置,其代码如下。

```
package com.waylau.spring.config;

import org.springframework.context.annotation.ComponentScan;
import org.springframework.context.annotation.Configuration;
import org.springframework.context.annotation.Import;
import org.springframework.context.annotation.ImportResource;

@Configuration
@ComponentScan(basePackages = { "com.waylau" })
@Import({ MvcConfiguration.class})
@ImportResource("classpath*:*spring.xml")
public class AppConfiguration {

}
```

AppConfiguration 同时导入了两个配置:基于 Java 的 MvcConfiguration 配置类和基于 XML 的配置。

2. MvcConfiguration

MvcConfiguration 定义如下。

```
package com.waylau.spring.config;

import java.util.List;

import org.springframework.context.annotation.Configuration;
import org.springframework.http.converter.HttpMessageConverter;
import org.springframework.http.converter.json.MappingJackson2HttpMessageConverter;
import org.springframework.web.servlet.config.annotation.EnableWebMvc;
import org.springframework.web.servlet.config.annotation.WebMvcConfigurer;

@EnableWebMvc // 启用MVC
@Configuration
public class MvcConfiguration implements WebMvcConfigurer {

    public void extendMessageConverters(List<HttpMessageConverter<?>> cs) {
        // 使用Jackson JSON来进行消息转换
        cs.add(new MappingJackson2HttpMessageConverter());
    }
}
```

以上代码主要用于启用 MVC 功能，以及添加对 Jackson JSON 序列化的支持。

3. spring.xml

spring.xml 沿用了 spring-mybatis 定义的配置，其代码如下。

```
<?xml version="1.0" encoding="UTF-8"?>
<beans xmlns="http://www.springframework.org/schema/beans"
    xmlns:xsi="http://www.w3.org/2001/XMLSchema-instance"
    xmlns:context="http://www.springframework.org/schema/context"
    xsi:schemaLocation="
        http://www.springframework.org/schema/beans
        http://www.springframework.org/schema/beans/spring-beans.xsd
        http://www.springframework.org/schema/context
        http://www.springframework.org/schema/context/spring-context.xsd
        http://www.springframework.org/schema/aop
        http://www.springframework.org/schema/aop/spring-aop.xsd">


    <bean class="org.springframework.beans.factory.config.PropertyPlaceholderConfigurer">
        <property name="locations" value="classpath:jdbc.properties"/>
    </bean>

    <!-- 数据源 -->
    <bean id="dataSource" class="org.apache.commons.dbcp2.BasicDataSource" destroy-method="close">
        <property name="driverClassName" value="${jdbc.driverClassName}" />
        <property name="url" value="${jdbc.url}" />
        <property name="username" value="${jdbc.username}" />
        <property name="password" value="${jdbc.password}" />
    </bean>

    <!-- MyBatis 工厂 -->
    <bean id="sqlSessionFactory" class="org.mybatis.spring.SqlSessionFactoryBean">
        <property name="dataSource" ref="dataSource" />
    </bean>

    <!-- 扫描 MyBatis Mapper 接口所在的包 -->
    <bean class="org.mybatis.spring.mapper.MapperScannerConfigurer">
      <property name="basePackage" value="com.waylau.mybatis.mapper" />
    </bean>
</beans>
```

11.4.6 运行应用

主应用程序代码保持不变,其代码如下。

```
package com.waylau.spring;

public class Application {

    public static void main(String[] args) throws Exception {
        new JettyServer().run();
    }
}
```

右击该类运行后,便可以在浏览器中访问应用的接口。

1. 创建用户

使用 POST 请求访问 http://localhost:8080/users,请求参数如下。

```
{"userId":1,"username":"waylau"}
```

成功创建用户之后,能看到返回数字 1。

2. 查询用户

使用 GET 请求访问 http://localhost:8080/users/1,响应内容如下。

```
{
    "userId": 1,
    "username": "waylau"
}
```

3. 修改用户

使用 PUT 请求访问 http://localhost:8080/users,请求参数如下。

```
{"userId":1,"username":"老卫"}
```

再次使用 GET 请求访问 http://localhost:8080/users/1,响应内容如下。

```
{
    "userId": 1,
    "username": "老卫"
}
```

从以上代码可以看到,已经成功修改用户。

4. 删除用户

使用 DELETE 请求访问 http://localhost:8080/users/1,成功删除后,响应数字 1。

11.5 集成Spring Security

本节将演示如何集成 Spring Security。在 11.4 节 spring-mvc 应用代码的基础上，创建一个新的 "spring-security" 应用，以演示整个集成过程。

本节的项目源码可以在 spring-security 目录下找到。

11.5.1 添加依赖

在 spring-security 应用的原有 pom.xml 文件中添加如下依赖。

```xml
<repositories>
    <repository>
        <id>spring-snapshots</id>
        <name>Spring Snapshots</name>
        <url>https://repo.spring.io/snapshot</url>
    </repository>
</repositories>

<dependencies>
    <!-- 安全相关的依赖 -->
    <dependency>
        <groupId>org.springframework.security</groupId>
        <artifactId>spring-security-web</artifactId>
        <version>${spring-security.version}</version>
    </dependency>
    <dependency>
        <groupId>org.springframework.security</groupId>
        <artifactId>spring-security-config</artifactId>
        <version>${spring-security.version}</version>
    </dependency>
</dependencies>
...
```

其中，spring-security-web 和 spring-security-config 用于在应用中启用 Spring Security 功能。由于 Spring Serutiy 的依赖属于 snapshot，因此还需要额外引入 Spring 的仓库。

11.5.2 添加安全配置类

添加安全配置类 WebSecurityConfig，其代码如下。

```
package com.waylau.spring.config;
```

```java
import org.springframework.context.annotation.Bean;
import org.springframework.security.config.annotation.web.builders.HttpSecurity;
import org.springframework.security.config.annotation.web.configuration.EnableWebSecurity;
import org.springframework.security.config.annotation.web.configuration.WebSecurityConfigurerAdapter;
import org.springframework.security.core.userdetails.User;
import org.springframework.security.core.userdetails.UserDetailsService;
import org.springframework.security.provisioning.InMemoryUserDetailsManager;

@EnableWebSecurity // 启用Spring Security功能
public class WebSecurityConfig
    extends WebSecurityConfigurerAdapter {

    /**
     * 自定义配置
     */
    @Override
    protected void configure(HttpSecurity http) throws Exception {
        http.authorizeRequests().anyRequest().authenticated() //所有请求都需认证
        .and()
        .formLogin() // 使用form表单登录
        .and()
        .httpBasic(); // HTTP基本认证
    }

    @SuppressWarnings("deprecation")
    @Bean
    public UserDetailsService userDetailsService() {
        InMemoryUserDetailsManager manager =
                new InMemoryUserDetailsManager();

        manager.createUser(
                User.withDefaultPasswordEncoder()   // 密码编码器
                .username("waylau")    // 用户名
                .password("123")       // 密码
                .roles("USER")         // 角色
                .build()
                );
        return manager;
    }
}
```

WebSecurityConfig 配置的含义已经在第 8 章进行了详细介绍,此处不再赘述。

11.5.3 修改AppConfiguration

修改主配置类 AppConfiguration，将 WebSecurityConfig 配置导入该类中，代码如下。

```
package com.waylau.spring.config;

import org.springframework.context.annotation.ComponentScan;
import org.springframework.context.annotation.Configuration;
import org.springframework.context.annotation.Import;
import org.springframework.context.annotation.ImportResource;

@Configuration
@ComponentScan(basePackages = { "com.waylau" })
@Import({ WebSecurityConfig.class, MvcConfiguration.class})
@ImportResource("classpath*:*spring.xml")
public class AppConfiguration {

}
```

11.5.4 修改嵌入服务器

嵌入服务器的代码需要做以下调整，才能引入 Spring Security 特有的过滤器，其完整代码如下。

```
package com.waylau.spring;

import java.util.EnumSet;
import javax.servlet.DispatcherType;
import org.eclipse.jetty.server.Server;
import org.eclipse.jetty.servlet.FilterHolder;
import org.eclipse.jetty.servlet.ServletContextHandler;
import org.eclipse.jetty.servlet.ServletHolder;
import org.springframework.web.context.ContextLoaderListener;
import org.springframework.web.context.WebApplicationContext;
import org.springframework.web.context.support.AnnotationConfigWebApplicationContext;
import org.springframework.web.filter.DelegatingFilterProxy;
import org.springframework.web.servlet.DispatcherServlet;
import com.waylau.spring.config.AppConfiguration;

public class JettyServer {
    private static final int DEFAULT_PORT = 8080;
    private static final String CONTEXT_PATH = "/";
    private static final String MAPPING_URL = "/*";

    public void run() throws Exception {
        Server server = new Server(DEFAULT_PORT);
        server.setHandler(servletContextHandler(webApplicationContext()));
```

```
        server.start();
        server.join();
    }

    private ServletContextHandler servletContextHandler(WebApplication-
Context ct) {
        // 启用Session管理器
        ServletContextHandler handler =
                new ServletContextHandler(ServletContextHandler.SES-
SIONS);

        handler.setContextPath(CONTEXT_PATH);
        handler.addServlet(new ServletHolder(new DispatcherServlet(ct)),
                MAPPING_URL);
        handler.addEventListener(new ContextLoaderListener(ct));

        // 添加Spring Security过滤器
        FilterHolder filterHolder=new FilterHolder(DelegatingFilterProxy.
class);
        filterHolder.setName("springSecurityFilterChain");
        handler.addFilter(filterHolder, MAPPING_URL,
                EnumSet.of(DispatcherType.REQUEST));

        return handler;
    }

    private WebApplicationContext webApplicationContext() {
        AnnotationConfigWebApplicationContext context =
                new AnnotationConfigWebApplicationContext();
        context.register(AppConfiguration.class);
        return context;
    }
}
```

11.5.5 运行应用

主应用程序代码保持不变，其代码如下。

```
package com.waylau.spring;

public class Application {

    public static void main(String[] args) throws Exception {
        new JettyServer().run();
    }

}
```

右击 Application 类运行后，便可以在浏览器中访问应用接口进行测试。当使用 GET 请求访问 http://localhost:8080/users/1 时，可以看到请求被安全机制拦截了，重定向到了图 11-1 所示的登录界面。

图11-1　登录界面

当输入账号密码并成功登录后，就能正常访问接口了，如图 11-2 所示。

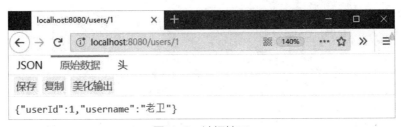

图11-2　访问接口

11.6　集成日志框架

不管是单机应用，还是 Cloud Natvie 分布式应用，日志都不可或缺。对于开发阶段而言，日志有助于跟踪程序执行的过程；对于运维而言，日志是排查问题的关键。

本节将介绍常用的日志框架，以及演示如何集成这些日志框架。在 11.5 节 spring-security 应用代码的基础上，创建一个新的"spring-log"应用，以演示整个集成过程。

本节的项目源码可以在 spring-log 目录下找到。

11.6.1　日志框架概述

日志是来自正在运行的进程的事件流。对于传统的 Java 应用程序而言，有许多框架和库可用于日志记录。Java Util Logging 是 Java 自身所提供的现成选项。除此之外，Log4j、logbac 和 SLF4J 是一些流行的日志框架。

（1）Log4j 是 Apache 旗下的 Java 日志记录工具，它是由 Ceki Gülcü 首创的。

（2）Log4j 2 是 Log4j 的升级产品。

（3）Commons Logging 是 Apache 基金会所属的项目，是一套 Java 日志接口，原来名为 Jakarta Commons Logging，后更名为 Commons Logging。

（4）SLF4J（Simple Logging Facade for Java）类似于 Commons Logging，是一套简易的 Java 日志门面，本身并无日志的实现。同样也是 Ceki Gülcü 首创的。

（5）logback 是 SLF4J 的实现，与 SLF4J 是同一个作者。

（6）JUL（Java Util Logging）是自 Java 1.4 以来的官方日志实现。

这些框架都能很好地支持 UDP 及 TCP 协议。应用程序将日志条目发送到控制台或文件系统，通常使用文件回收技术来避免日志填满所有磁盘空间。

日志处理的最佳实践之一是关闭生产中的大部分日志条目，因为磁盘 IO 的成本很高。磁盘 IO 不但会减慢应用程序的运行速度，还会严重影响其可伸缩性。将日志写入磁盘也需要较高的磁盘容量，当磁盘空间用完之后，就有可能降低应用程序的性能。日志框架提供了在运行时控制日志记录的选项，以限制必须打印的内容和不打印的内容。这些框架中的大部分都对日志记录控件提供了细粒度的控制。此外，它还提供了在运行时更改这些配置的选项。

然而，日志中可能包含重要的信息，如果分析得当，可能具有很高的价值。因此，限制日志条目本质上也限制了人们理解应用程序行为的能力。

本书主要介绍 logback 和 SLF4J 的使用。

11.6.2 添加依赖

在 spring-log 应用的原有 pom.xml 文件中添加如下依赖。

```
<dependency>
    <groupId>ch.qos.logback</groupId>
    <artifactId>logback-classic</artifactId>
    <version>${logback-classic.version}</version>
</dependency>
...
```

其中，logback-classic 是需要引入的 logbac 依赖，该依赖自身就包括了 SLF4J，因此无须显式引入 SLF4J 的依赖。

11.6.3 添加日志配置文件

在应用的 resource 目录下添加 logback.xml 文件，用于配置 logback。配置内容如下。

```xml
<?xml version="1.0" encoding="UTF-8"?>
<configuration>

  <appender name="STDOUT" class="ch.qos.logback.core.ConsoleAppender">
    <layout class="ch.qos.logback.classic.PatternLayout">
      <Pattern>%d{HH:mm:ss.SSS} [%thread] %-5level %logger{36} - %msg%n</Pattern>
    </layout>
  </appender>

  <root level="info">
    <appender-ref ref="STDOUT" />
  </root>
</configuration>
```

11.6.4 使用日志框架

使用日志框架非常简单，先在使用的类中使用工厂方式创建日志对象。

```
static final Logger logger =
    LoggerFactory.getLogger(JettyServer.class);
```

然后就可以使用 logger 对象进行日志的记录了。比如：

```
logger.info("Lite start at {}", port);
```

```
logger.error("Lite start exception!", e);
```

info 和 error 代表了不同的日志级别。一般而言，平常的日志就用 info，如果是一些错误信息，则需要用 error 级别的日志。

11.6.5 修改嵌入服务器

在嵌入服务器的代码中使用日志，完整代码如下。

```
import org.slf4j.Logger;
import org.slf4j.LoggerFactory;

public class JettyServer {
    static final Logger logger = LoggerFactory.getLogger(JettyServer.class);

    private static final int DEFAULT_PORT = 8080;
    private static final String CONTEXT_PATH = "/";
    private static final String MAPPING_URL = "/*";
```

```
    public void run() throws Exception {
        Server server = new Server(DEFAULT_PORT);
        server.setHandler(servletContextHandler(webApplicationContext()));
        server.start();

        logger.info("Server started!"); // 记录服务器启动信息

        server.join();
    }

    // …省略非核心内容
}
```

11.6.6 运行应用

主应用程序代码保持不变,代码如下。

```
package com.waylau.spring;

public class Application {

    public static void main(String[] args) throws Exception {
        new JettyServer().run();
    }
}
```

右击 Application 类运行后,可以看到如下日志输出内容。

```
00:16:22.145 [main] INFO  com.waylau.spring.JettyServer - Server started!
```

第12章 Lite架构分层

分层结构是目前最为流行、应用最广泛的应用软件的设计方式。在应用了分层结构的系统中，各个子系统按照层次的形式组织起来，上层使用下层的各种服务，而下层支撑上层。每一层都对自己的上层隐藏本层的细节。

从本章开始，将进入Lite框架的开发，本章主要介绍如何对Lite框架进行分层设计。

12.1 分层架构概述

为了便于对程序进行管理，人们倾向于在开发时对应用程序进行分割、分层。本节将讨论以下话题。

（1）为什么要对应用程序进行分层？
（2）如果不分层，系统将会出现哪些问题？
（3）常见的分层方式有哪些？

12.1.1 应用的分层

随着面向对象程序设计和设计模式的出现，人们发现，很多建筑学理论都可以用来指导软件工程（即程序）的开发。例如，在开发楼房时，人们会先对要盖的楼房进行评估和核算（软件项目管理）；根据需求设计楼房的图纸（软件设计），再根据设计把楼房的地基、骨架搭建出来（搭建框架）；根据不同工种将人员进行分工，有的去砌墙，有的去贴砖（前端编码、后台编码）；最后进行验收测试并交付。

软件应用开发与建筑学的分层目的是一致的，旨在根据不同的业务、不同的技术、不同的组织，结合灵活性、可维护性、可扩展性等多种因素，将应用系统划分为不同的部分，并使它们相互协作，从而体现出应用的最大化价值。对于一个有良好分层的应用来说，其一般具备如下特点。

1. 按业务功能进行分层

分层就是将业务功能相关的类或组件放置在一起，而将业务功能不相关的类或组件隔离开。

例如，将与用户直接交互的部分分为"表示层"，将实现逻辑计算或业务处理的部分分为"业务层"，将与数据库打交道的部分分为"数据访问层"。

2. 良好的层次关系

设计良好的架构分层是上层依赖于下层，而下层支撑起上层，但却不能直接访问上层，层与层之间通过协作来共同完成特定的功能。

3. 每一层都能保持独立

层能够被单独构造，也能被单独替换，最终不会影响整体功能。例如，将整个数据持久层的技术从 Hibernate 转成 EclipseLink，但不会对上层业务逻辑功能造成影响。

12.1.2 不分层的应用架构

为了更好地理解分层的好处，先来看一下不分层的应用架构是如何运作的。

以 Web 应用程序为例。在 Web 应用程序开发的早期，所有的逻辑代码并没有明显的层次区分，因此代码之间的调用是相互交错的，整体看上去错综复杂。例如，在早期使用诸如 ASP、JSP 及

PHP等动态网页技术时，常常会将所有的页面逻辑、业务逻辑及数据库访问逻辑放在一起，很多时候就在JSP页面中写SQL语句了，编码风格完全是过程化的。

以下代码就是一个JSP访问SQL Server数据库的示例。

```jsp
<%@ page language="java" contentType="text/html;
    charset=UTF-8" pageEncoding="UTF-8"%>
<%@ page import = "java.sql.*"%>
<!DOCTYPE html PUBLIC "-//W3C//DTD XHTML 1.0 Transitional//EN"
"http://www.w3.org/TR/xhtml1/DTD/xhtml1-transitional.dtd">
<html xmlns="http://www.w3.org/1999/xhtml">
<head>
</head>
<body>
<%

// 创建数据库连接
Class.forName("com.microsoft.sqlserver.jdbc.SQLServerDriver");
String url="jdbc:sqlserver://localhost:1433;databaseName=Book;user=sa;-password=";
PreparedStatement pstmt;
String sql = "insert into students (UserName,WebSite) values(?,?)";
int returnValue = 0;
try{
    pstmt = (PreparedStatement) conn.prepareStatement(sql);
    pstmt.setString(1, student.getName());
    pstmt.setString(2, student.getSex());
    returnValue = pstmt.executeUpdate();

    // 判断是添加成功还是失败
    if(returnValue == 1){
        out.print("<li>添加成功  ");
        out.print("<li>returnValue = " + returnValue);
    } else {
        out.print("<li>添加失败  ");
    }
}catch(Exception ex){
    out.print(ex.getLocalizedMessage());
}finally{
    try{
        if(pstmt != null){
            pstmt.close();
            pstmt = null;
        }
        if(cn != null){
            conn.close();
            conn = null;
        }
    }catch(Exception e){
        e.printStackTrace();
```

```
    }
}
%>
</body>
</html>
```

先不讨论这段代码是否正确，从功能实现上来讲，这段代码既处理了数据库的访问操作，又做了页面的展示，还夹杂了业务逻辑判断。还好这段代码不长，读下来还能够理解，但如果是更加复杂的功能，这段代码肯定是非常不清晰的，维护起来也相当麻烦。

早期不分层的架构主要存在如下弊端。

（1）代码不够清晰，难以阅读。

（2）代码职责不明，难以扩展。

（3）代码错综复杂，难以维护。

（4）代码没做分工，难以组织。

12.1.3　应用的三层架构

目前比较常用的、典型的应用软件倾向于使用三层架构（Three-Tier Architecture），即表示层、业务层和数据访问层。

（1）表示层（Presentation Layer）：提供与用户交互的界面。GUI（图形用户界面）和 Web 页面是表示层的两个典型例子。

（2）业务层（Business Layer）：也称为业务逻辑层，用于实现各种业务逻辑。例如，处理数据验证时，根据特定的业务规则和任务来响应特定的行为。

（3）数据访问层（Data Access Layer）：也称为数据持久层，负责存放和管理应用的持久性业务数据。

图 12-1 所示为三层架构，图中的每一层都需要不同的技能。

图12-1　三层架构

（1）表示层需要 HTML、CSS、JavaScript 之类的前端技能，以及具备 UI 设计能力。

（2）业务层需要编程语言技能，以便使计算机可以处理业务规则。

（3）数据访问层需要具有数据定义语言（DDL）和数据操作语言（DML），以及数据库设计形式的 SQL 技能。

虽然一个人有可能拥有上述所有技能，但这样的人是相当罕见的。在具有大型软件应用程序的大型组织中，应该将应用程序分割为单独的层，使得每个层都可以由具有相关专业技能的不同团队来开发和维护。

一个良好分层的架构系统应遵循如下分层原则。

（1）每个层的代码必须包含可以单独维护的单独文件。

（2）每个层只能包含属于该层的代码。因此，业务逻辑只能驻留在业务层，表示逻辑只能驻留在表示层，而数据访问逻辑只能驻留在数据访问层。

（3）表示层只能接收来自外部代理的请求，并向外部代理返回响应。这通常由一个人来完成，但也可能由一个软件来完成。

（4）表示层只能向业务层发送请求，并从业务层接收响应。它不能直接访问数据库或数据访问层。

（5）业务层只能接收来自表示层的请求，并返回对表示层的响应。

（6）业务层只能向数据访问层发送请求，并从数据访问层接收响应。它不能直接访问数据库。

（7）数据访问层只能从业务层接收请求并返回响应。它不能发出请求到除了它支持的数据库管理系统（DBMS）以外的任何地方。

（8）每层应完全不知道其他层的内部工作原理。例如，业务层对数据访问层一无所知，并且不知道也不关心数据访问对象的内部工作原理，它必须与表示层无关，不知道也不关心表示层是如何处理数据的。表示层获取数据并构造 HTML 文档、PDF 文档、CSV 文件或以某种其他方式处理它，但是这应该与业务层完全无关。

（9）每层应当可以用具有类似特征的替代组件来交换这个层，使得整体可以继续工作。

简言之，应用在一开始设计时，就要考虑系统的架构设计及如何将系统进行有效分层。由于系统架构设计属于比较高级别的话题，有这方面兴趣的读者，可以参阅笔者所著的《分布式系统常用技术及案例分析》。

12.1.4　MVC与三层架构的差异

用户可能认为 MVC 的任何实现都可以被自动视为三层架构的实现，但并非如此。图 12-2 所示为 MVC 各个组成部分所处的位置，以及其与三层架构之间的差异。

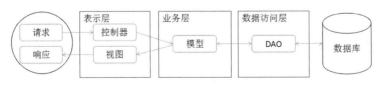

图12-2　三层架构与MVC的差异

在每个 MVC 的实现中，都有一个经常会犯的基本错误，就是数据库连接的地方不对。在三层架构中,当接收到来自业务层的请求时,与数据库的所有通信(包括打开连接)都在数据访问层内完成。表示层没有与数据库的任何通信,它只能通过业务层与其通信。对于初学者来说,在使用 MVC 框架时,经常会将数据库连接放在控制器内,将连接对象传递给模型,然后在必要时使用它。这是一个错误!在控制器中会打开连接到数据库的服务器实例,以便与数据库通信。这些连接被传递到不实际使用该连接的不同 Model 组件,这会导致连接资源的浪费。

这种实现还有一个经常犯的错误:认为数据验证(有时称为数据过滤)应该在从控制器中传递到模型之前执行。这不符合三层架构的定义,定义中规定这样的数据验证逻辑,连同业务逻辑和任务特定的行为,应仅存在于业务层或模型组件内。将数据验证从模型中取出并放入控制器中,会导致控制器和模型之间紧耦合,因为控制器不能与不同的模型一起使用。而紧耦合的缺点在于,更新一个模块的结果会导致其他模块的结果变化,难以重用特定的关联模块。松耦合将使控制器可以被多个模型类共享,而不是固定为一个。

正确的实现方式是,使用三层架构开发的应用程序应该将其所有逻辑(数据验证、业务规则和任务特定行为)局限于业务层。在表示层和数据访问层中不应该有应用逻辑,改变这两个层中的任意一个,都不会影响任何应用逻辑。

12.1.5 Lite的分层架构

Lite 框架同样也是采用三层架构,分别为表示层、业务层和数据访问层。每一层的职责和功能划分如图 12-3 所示。

图12-3 Lite的分层架构

1. 表示层

表示层提供与用户交互的界面,分为 Controller(控制器)和 View(视图),它们是 MVC 中重要的两部分,用于处理请求和响应界面。

在 RESTful 系统架构中,表示层通常可以承担 RESTful API,将数据以 JSON 格式进行响应。

2. 业务层

业务层也称为业务逻辑层,用于实现各种业务逻辑。例如,处理数据验证,根据特定的业务规则和任务来响应特定的行为。

MVC 中的 Model（模型）涵盖的范围比较广，如 Entity（实体）、VO（值对象）、Domain（领域）、Service（服务）都可以称为领域模型的一种。

3. 数据访问层

数据访问层也称为数据持久层，负责存放和管理应用的持久性业务数据。数据访问层可以用 DAO（数据访问对象）或 Repository（仓储）来表示。在 MyBatis 框架中，该层的业务功能一般由 Mapper（映射）来承担。

12.2 数据访问层

数据访问层（Data Access Layer，DAL）也称为持久层，其功能主要是负责数据库的访问。简单来说，就是实现对数据表的 Select（查询）、Insert（插入）、Update（更新）、Delete（删除）等操作。更复杂的操作还包含事务管理、数据并发处理、查询解析机制等。

为了实现关系型数据与面向对象语言之间的数据的转换，往往需要在数据访问层加入 ORM 框架，以实现对象和数据表之间的映射，以及对象实体的持久化。Java EE 针对数据持久化制定了 JPA 规范，该规范的相关实现有 Hibernate、EclipseLink 等。未实现 JPA 规范的 ORM 框架有 MyBatis 等。

数据访问层设计模式有以下两种。

12.2.1 仓储模式

仓储（Repsoitory）模式用于协调领域和数据映射层，起到了类似于内存中业务对象集合的作用。

这种模式将业务层与底层数据存储机制进行了分离，业务层只知道调用相应的 Respositoty 就可以实现数据库操作，它并不关心采用的是什么数据访问技术和存储介质，也不关心底层数据模型的结构差异如何。常见的 JPA、Spring Data JPA 等框架均实现了该模式。

12.2.2 数据访问对象模式

数据访问对象（Data Access Object，DAO）模式是一种简单地将数据访问层与系统其他部分相隔离的设计方式。DAO 和 Respository 类似，但是 DAO 没有隐藏数据底层的相关信息，而且数据库中的每一个表中都存在一个 DAO。

这种模式的实现方式有 MyBatis 及 Spring 的 JdbcTemplate 等。

本书主要介绍以 MyBatis 为技术核心的 DAO 模式。读者如果想了解 Spring Data JPA 或 Hibernate 方面的内容，可以参阅笔者所著的《Spring Boot 企业级应用开发实战》。

12.3　事务处理

数据库事务（Database Transaction）是指作为单个逻辑工作单元执行的一系列操作，要么完全执行，要么完全不执行。事务处理可以确保除非事务性单元内的所有操作都成功完成，否则不会永久更新面向数据的资源。通过将一组相关操作组合为一个全部成功或全部失败的单元，可以简化错误恢复，并使应用程序更加可靠。

一个逻辑工作单元要成为事务，必须满足 ACID 属性（原子性、一致性、隔离性和持久性）。事务是数据库运行中的逻辑工作单位，由关系型数据库中的事务管理子系统负责事务的处理。

在大型应用中，事务的处理不可或缺。

MyBatis 提供了 JDBC 和 MANAGED 两类事务管理器，并实现了编程式的事务管理，主要接口如下。

```
public interface Transaction {
  Connection getConnection() throws SQLException;
  void commit() throws SQLException;
  void rollback() throws SQLException;
  void close() throws SQLException;
  Integer getTimeout() throws SQLException;
}
```

在与 Spring 集成的应用中，MyBatis 无须配置任何事务管理器，因为 Spring 会将所有的事务纳入自己的事务管理中。

12.4　权限验证

Lite 框架主要以 Spring Security 技术为核心，来实现用户的权限验证。权限验证层会在用户发起请求时对请求进行拦截，以验证用户的身份。如果身份非法，则会拒绝用户的请求。

互联网应用一般是多用户的，主要分为管理员和普通用户两类角色。管理员一般对系统做维护，以新闻类应用为例，管理员的主要操作是发布新闻、编辑新闻、对新闻进行分类等。而普通用户角色可以访问整个应用的新闻内容，但不能对新闻内容做变更。

12.5 接口访问层

在互联网应用盛行的今天，客户端的形式也多种多样，有手机 App、小程序，也有传统的 PC 端应用。那么，如何才能用统一的方式来减少开发工作量呢？

RESTful 架构是比较好的用于统一接口的方式。通过后台提供一致的接口，让不同的客户端应用来访问统一的后台接口，从而减少了适配不同客户端的工作量。

那么，到底什么是 RESTful 架构呢？

12.5.1 REST概述

表述性状态转移（REpresentation State Transfer, REST）描述了一个架构样式的网络系统，如 Web 应用程序。有关 RESTful 架构系统（或称为 REST 式的架构系统）的描述，首次出现在 2000 年 Roy Fielding 的博士论文 *Architectural Styles and the Design of Network-based Software Architectures*[1] 中。Roy Fielding 同时还是 HTTP 规范的主要编写者之一，也是 Apache HTTP 服务器项目的共同创立者。所以这篇论文一经发表，就引起了极大的反响。

很多公司或组织开始宣称自己的应用或服务实现了 REST API，但该论文实际上只是描述了一种架构风格，并未对具体的实现做出规范。所以，各大厂商不免会误用或滥用 REST。在这种背景下，Roy Fielding 不得不再次发文进行澄清。同时他还指出，应用状态引擎只能是超文本驱动的，否则就不是 REST 或 REST API。据此，他给出了 REST API 应该具备的条件。

（1）REST API 不应该依赖于任何通信协议，尽管要成功映射到某个协议可能会依赖于元数据的可用性、所选的方法等。

（2）REST API 不应该包含对通信协议的任何改动，除非是补充或确定标准协议中未规定的部分。

（3）REST API 应该将大部分的描述工作放在定义用于表示资源和驱动应用状态的媒体类型上，或者定义现有标准媒体类型的扩展名和（或）支持超文本的标记。

（4）REST API 绝不应该定义一个固定的资源名或层次结构（客户端和服务器之间的明显耦合）。

（5）REST API 永远也不应该有影响客户端的"类型化"资源。

（6）REST API 不应该要求有先验知识（Prior Knowledge），除了初始 URI 和适合目标用户的一组标准化的媒体类型外（即它能被任何潜在使用该 API 的客户端理解）。

需要强调的是，REST 并非标准，而是一种开发 Web 应用的架构风格，可以将其理解为一种设

---

[1] 论文地址 http://www.ics.uci.edu/~fielding/pubs/dissertation/top.htm。

计模式。REST 基于 HTTP、URI 及 XML 这些现有的广泛流行的协议和标准，伴随着 REST 的应用，HTTP 协议得到了更加正确的使用。

12.5.2 REST的特征

REST 是指一组架构约束条件和原则，满足这些约束条件和原则的应用程序或设计就是 REST。

相较于基于 SOAP 和 WSDL 的 Web 服务，REST 模式提供了更为简洁的实现方案。REST Web 服务（RESTful Web Services）是松耦合的，特别适用于为客户创建在互联网传播的轻量级的 Web 服务 API。REST 应用是以资源表述的转移（The Transfer of Representations of Resources）为中心来做请求和响应的。数据和功能均被视为资源，并使用统一资源标识符（URI）来访问。网页里面的链接就是典型的 URI。该资源由文档表述，并通过使用一组简单的、定义明确的操作来执行。

例如，一个 REST 资源可能是一个城市的天气情况。该资源的表述可能是一个 XML 文档、图像文件或 HTML 页面。客户端可以检索资源的特定表述，也可以通过更新其数据来修改对应的资源，或者完全删除该资源。

目前，越来越多的 Web 服务开始采用 REST 风格设计和实现，比较著名的 REST 服务包括 Google AJAX 搜索 API、Amazon Simple Storage Service（Amazon S3）等。

基于 REST 的 Web 服务遵循一些基本的设计原则，使 RESTful 应用更加简单、轻量，开发速度也更快。

（1）通过 URI 标识资源：系统中的每一个对象或资源都可以通过一个唯一的 URI 来进行寻址，URI 的结构应该简单、可预测且易于理解，如定义目录结构式的 URI。

（2）统一接口：以遵循 RFC-2616 所定义的协议的方式显式地使用 HTTP 方法，建立创建、检索、更新和删除操作与 HTTP 方法之间的一对一映射。

（3）若在服务器上创建资源，应该使用 POST 方法。

（4）若检索某个资源，应该使用 GET 方法。

（5）若更新或添加资源，应该使用 PUT 方法。

（6）若删除某个资源，应该使用 DELETE 方法。

（7）资源多重表述：URI 所访问的每个资源都可以使用不同的形式加以表示（如 XML 或 JSON），具体的表现形式取决于访问资源的客户端，客户端与服务提供者使用一种内容协商的机制（请求头与 MIME 类型）来选择合适的数据格式，最小化彼此之间的数据耦合。

（8）无状态：对服务器端的请求应该是无状态的，完整、独立的请求不要求服务器在处理请求时检索任何类型的应用程序上下文或状态。无状态约束使服务器的变化对客户端是不可见的，因为在两次连续的请求中，客户端并不依赖于同一台服务器。一个客户端从某台服务器上收到一份包含链接的文档，当它要做一些处理时，这台服务器宕机了，可能是硬盘损坏，也可能是软件需要升级重启——如果这个客户端访问了从这台服务器接收的链接，它不会察觉到后台的服务器已经改变了。通过超链接实现有状态交互，即请求消息是自包含的（每次交互都包含完整的信息），有多种技术实现了不同请求间状态信息的传输，如 URI 重定向、Cookies 和隐藏表单字段等，状态可以嵌入应答消息中，这样一来，状态在接下来的交互中仍然有效。

12.5.3 Java REST规范

REST 在 Java 中的规范主要是 JAX-RS（Java API for RESTful Web Services），该规范使 Java 程序员可以用一套固定的接口来开发 REST 应用，避免了依赖于第三方框架。同时，JAX-RS 使用 POJO 编程模型和基于标注的配置，并集成了 JAXB，从而可以有效地缩短 REST 应用的开发周期。Java EE 6引入了对JSR-311(https://jcp.org/en/jsr/detail?id=311)的支持，Java EE 7 支持JSR-339(http://jcp.org/en/jsr/detail?id=339 ）规范。而在最新发布的 Java EE 8 中，则支持了JSR-370（http://jcp.org/en/jsr/detail?id=370）。

JAX-RS 定义的 API 位于 javax.ws.rs 包中。该规范的具体实现有 Jersey（https://jersey.java.net）、Apache CXF（http://cxf.apache.org）及 JBoss RESTEasy（http://resteasy.jboss.org）等。未实现该规范的其他 REST 框架包括 Spring Web MVC 等。

本书主要以 Spring Web MVC 技术为核心来构建 RESTful 服务。有关 Jersey 方面的内容，可以参阅笔者所著的电子开源书[1]。

12.6 实战：Lite框架的搭建

本节详细介绍如何从 0 开始搭建属于自己的互联网应用框架。该框架名为 Lite，意思是开源、简单、轻量。同时，在本书的后续章节，笔者也会展示如何基于 Lite 框架来开发一个真实的互联网应用。

12.6.1 内嵌Sevlet容器

Lite 主要是基于 Jetty 来实现内嵌 Sevlet 容器的目的，以下是 Jetty 的依赖。

```
<dependency>
    <groupId>org.eclipse.jetty</groupId>
    <artifactId>jetty-servlet</artifactId>
    <version>${jetty.version}</version>
</dependency>
```

---

[1] 电子开源书地址 https://waylau.com/books/。

接下来介绍如何编写启动服务器。

1. 编写启动服务器

定义如下 Lite 启动服务器的接口。

```java
public interface LiteServer {

    /**
     * Default port is 8080
     */
    void run();

    void run(int port);

    void run(String[] args);

}
```

该接口有一个默认的实现 LiteJettyServer,其代码如下。

```java
package com.waylau.lite.jetty;

import java.util.EnumSet;

import javax.servlet.DispatcherType;

import org.eclipse.jetty.server.Server;
import org.eclipse.jetty.servlet.FilterHolder;
import org.eclipse.jetty.servlet.ServletContextHandler;
import org.eclipse.jetty.servlet.ServletHolder;
import org.slf4j.Logger;
import org.slf4j.LoggerFactory;
import org.springframework.web.context.ContextLoaderListener;
import org.springframework.web.context.WebApplicationContext;
import org.springframework.web.context.support.AnnotationConfigWebApplicationContext;
import org.springframework.web.filter.DelegatingFilterProxy;
import org.springframework.web.servlet.DispatcherServlet;

import com.waylau.lite.LiteConfig;
import com.waylau.lite.LiteServer;
import com.waylau.lite.exception.LiteRuntimeException;
import com.waylau.lite.util.CommandLineArgs;
import com.waylau.lite.util.CommandLineArgsParser;


public class LiteJettyServer implements LiteServer {

    static final Logger logger = LoggerFactory.getLogger(LiteJettyServer.class);
```

```java
    private static final String CONTEXT_PATH = "/";
    private static final String MAPPING_URL = "/*";
    private static final String PORT_NAME = "port";
    private static final int PORT = 8080;
    private Class<?> annotatedClass;

    public LiteJettyServer() {
    }

    @Override
    public void run() {
        this.run(PORT);
    }

    @Override
    public void run(String[] args) {
        CommandLineArgs commandLineArgs = CommandLineArgsParser.parse(args);
        Integer port = commandLineArgs.getIntArg(PORT_NAME);

        // 判断是否通过命令行传port
        if (port == null) {
            this.run();
        } else {
            this.run(port);
        }

    }

    @Override
    public void run(int port) {
        Server server = new Server(port);
        server.setHandler(this.servletContextHandler(this.webApplicationContext()));

        try {
            server.start();
        } catch (Exception e) {
            logger.error("Lite start exception!", e);
            throw new LiteRuntimeException("Lite started exception!", e);
        }

        logger.info("Lite start at {}", port);

        try {
            server.join();
        } catch (InterruptedException e) {
```

```
            logger.error("Lite join exception!", e);
            throw new LiteRuntimeException("Lite join exception!", e);
        }

    }

    private ServletContextHandler servletContextHandler(WebApplication-
Context ct) {
        // 启用Session管理器
        ServletContextHandler handler =
                new ServletContextHandler(ServletContextHandler.SES-
SIONS);

        handler.setContextPath(CONTEXT_PATH);
        handler.addServlet(new ServletHolder(new DispatcherServlet(ct)),
                MAPPING_URL);
        handler.addEventListener(new ContextLoaderListener(ct));

        // 添加Spring Security过滤器
        FilterHolder filterHolder=new FilterHolder(DelegatingFilterProxy.
class);
        filterHolder.setName("springSecurityFilterChain");
        handler.addFilter(filterHolder, MAPPING_URL,
                EnumSet.of(DispatcherType.REQUEST));

        return handler;
    }

    private WebApplicationContext webApplicationContext() {
        AnnotationConfigWebApplicationContext context = new Annotation-
ConfigWebApplicationContext();
        context.register(LiteConfig.class);    // 注入Lite应用配置
        return context;
    }
}
```

LiteJettyServer 主要通过 context.register(LiteConfig.class) 方法来注入 Spring 的上下文配置信息。

LiteJettyServer 默认启动在 8080 端口，当然也可以通过命令行来传参，实现自定义端口号。

2. 通过命令行传参

CommandLineArgsParser 及 CommandLineArgs 工具类实现了通过命令行传参。CommandLineArgsParser 核心代码如下。

```
package com.waylau.lite.util;

public class CommandLineArgsParser {

    public static CommandLineArgs parse(String… args) {
        CommandLineArgs commandLineArgs = new CommandLineArgs();
```

```java
        for (String arg : args) {
            if (arg.startsWith("--")) {
                String argText = arg.substring(2, arg.length());
                String argName;
                String argValue = null;
                if (argText.contains("=")) {
                    argName = argText.substring(0, argText.indexOf("="));
                    argValue = argText.substring(argText.indexOf("=") + 1,
                        argText.length());
                } else {
                    argName = argText;
                }
                if (argName.isEmpty()
                    || (argValue != null && argValue.isEmpty())) {
                    throw new IllegalArgumentException("Invalid argument syntax: " + arg);
                }
                commandLineArgs.addArg(argName, argValue);
            }

        }
        return commandLineArgs;
    }
}
```

CommandLineArgs 代码如下。

```java
package com.waylau.lite.util;

import java.util.HashMap;
import java.util.Map;
import org.springframework.lang.Nullable;

public class CommandLineArgs {

    private final Map<String, String> args = new HashMap<>();

    public void addArg(String argName, @Nullable String argValue) {
        this.args.put(argName, argValue);
    }

    public boolean contains(String argName) {
        return this.args.containsKey(argName);
    }

    @Nullable
    public String getArg(String argName) {
```

```java
        return this.args.get(argName);
    }

    public Integer getIntArg(String argName) {
        String argValue = this.getArg(argName);
        return argValue == null ? null : Integer.valueOf(argValue);
    }
}
```

可以接受如下格式的命令行参数。

```
--port=8081
```

12.6.2　3种启动方式

到目前为止，LiteJettyServer 可以通过以下 3 种方式来启动。

1. 不指定端口

不指定端口，则默认启动在 8080 端口，示例代码如下。

```
new LiteJettyServer().run(); // 默认端口8080
```

2. 以编程方式指定端口

以编程方式指定端口，示例代码如下。

```
new LiteJettyServer().run(8081);
```

3. 在命令行指定端口号

在命令行指定端口号，示例代码如下。

```
new LiteJettyServer().run(args);
```

在命令行指定端口号，命令参数如下。

```
--port=8081
```

12.6.3　支持打包为可执行的jar

将应用打包为可执行的 jar，可以通过 maven-shade-plugin 插件来实现。应用的 pom.xml 文件配置如下。

```xml
...
<!-- 可执行的jar插件 -->
<plugin>
    <groupId>org.apache.maven.plugins</groupId>
```

```xml
        <artifactId>maven-shade-plugin</artifactId>
        <version>${maven-shade-plugin.version}</version>
        <configuration>
            <createDependencyReducedPom>false</createDependencyReducedPom>
        </configuration>
        <executions>
            <execution>
                <phase>package</phase>
                <goals>
                    <goal>shade</goal>
                </goals>
                <configuration>
                    <transformers>
                        <!-- 不覆盖Spring的同名文件  而是追加合并同名文件 -->
                        <transformer implementation="org.apache.maven.plugins.shade.resource.AppendingTransformer">
                            <resource>META-INF/spring.handlers</resource>
                        </transformer>
                        <transformer implementation="org.apache.maven.plugins.shade.resource.AppendingTransformer">
                            <resource>META-INF/spring.schemas</resource>
                        </transformer>
                        <transformer implementation="org.apache.maven.plugins.shade.resource.AppendingTransformer">
                            <resource>META-INF/spring.tooling</resource>
                        </transformer>
                        <transformer
                                implementation="org.apache.maven.plugins.shade.resource.ManifestResourceTransformer">
                            <mainClass>com.waylau.lite.App</mainClass>
                        </transformer>
                    </transformers>
                </configuration>
            </execution>
        </executions>
</plugin>
```

通过下面的命令来编译。

```
mvn clean install
```

通过下面的命令来运行。

```
java -jar target/lite-1.0.0.jar --port=8081
```

12.6.4 支持REST API

下面通过 Spring Web MVC 来实现 REST API，所需依赖如下。

```xml
<!-- Web相关的依赖 -->
<dependency>
    <groupId>org.springframework</groupId>
    <artifactId>spring-webmvc</artifactId>
    <version>${spring.version}</version>
</dependency>
<dependency>
    <groupId>org.eclipse.jetty</groupId>
    <artifactId>jetty-servlet</artifactId>
    <version>${jetty.version}</version>
</dependency>
<dependency>
    <groupId>com.fasterxml.jackson.core</groupId>
    <artifactId>jackson-core</artifactId>
    <version>${jackson.version}</version>
</dependency>
<dependency>
    <groupId>com.fasterxml.jackson.core</groupId>
    <artifactId>jackson-databind</artifactId>
    <version>${jackson.version}</version>
</dependency>
<dependency>
    <groupId>org.springframework</groupId>
    <artifactId>spring-aspects</artifactId>
    <version>${spring.version}</version>
</dependency>
...
```

其中，Jackson JSON 实现数据的 JSON 序列化。

接下来演示如何编写控制器。

1. 编写控制器

编写一个非常简单的控制器，其代码如下。

```java
package com.waylau.lite.mvc.controller;

import org.springframework.web.bind.annotation.GetMapping;
import org.springframework.web.bind.annotation.RequestMapping;
import org.springframework.web.bind.annotation.RestController;
import com.waylau.lite.Lite;


@RestController
@RequestMapping("/lite")
public class LiteController {

    @GetMapping
    public Lite sayHi() {
        return new Lite("waylau.com", "1.0.0");
```

        }
}
```

在访问 "/lite" 接口时，响应 Lite("waylau.com", "1.0.0") 对象的 JSON 格式数据。

### 2. 启动MVC配置

MVC 配置类 LiteMvcConfig 代码如下。

```java
package com.waylau.lite.mvc;

import java.util.List;

import org.springframework.context.annotation.Configuration;
import org.springframework.http.converter.HttpMessageConverter;
import org.springframework.http.converter.json.MappingJackson2HttpMessageConverter;
import org.springframework.web.servlet.config.annotation.EnableWebMvc;
import org.springframework.web.servlet.config.annotation.WebLiteMvcConfigurer;

@EnableWebMvc
@Configuration
public class LiteMvcConfig implements WebLiteMvcConfigurer {

    public void extendMessageConverters(List<HttpMessageConverter<?>> converters) {

        // 添加 Jackson JSON的支持
        converters.add(new MappingJackson2HttpMessageConverter());
    }
}
```

把 LiteMvcConfig 类添加到主应用中，代码如下。

```java
package com.waylau.lite;

import org.springframework.context.annotation.ComponentScan;
import org.springframework.context.annotation.Configuration;
import org.springframework.context.annotation.Import;
import org.springframework.context.annotation.ImportResource;

import com.waylau.lite.mvc.LiteMvcConfig;

@Configuration
@ComponentScan(basePackages = { "com.waylau.lite" })
@Import({LiteMvcConfig.class})
public class LiteConfig {
```
```

}

3. 启动应用

启动应用后，访问 http://localhost:8080/lite，可以看到 JSON 的响应内容如下。

```
{"author":"waylau.com","version":"1.0.0"}
```

12.6.5 支持JUnit5测试

Lite 集成了 JUnit5 等最新的测试框架。

在 pom.xml 中添加如下依赖。

```xml
<!-- 测试相关的依赖 -->
<dependency>
    <groupId>org.apache.commons</groupId>
    <artifactId>commons-dbcp2</artifactId>
    <version>${dbcp2.version}</version>
</dependency>
<dependency>
    <groupId>org.junit.jupiter</groupId>
    <artifactId>junit-jupiter-engine</artifactId>
    <version>${junit.version}</version>
</dependency>
<dependency>
    <groupId>org.springframework</groupId>
    <artifactId>spring-test</artifactId>
    <version>${spring.version}</version>
</dependency>
<dependency>
    <groupId>com.jayway.jsonpath</groupId>
    <artifactId>json-path</artifactId>
    <version>${json-path.version}</version>
</dependency>
<dependency>
    <groupId>org.hamcrest</groupId>
    <artifactId>hamcrest-core</artifactId>
    <version>${hamcrest-core.version}</version>
</dependency>
...
```

其中，json-path 依赖用于测试 JSON 格式的数据。

下面演示如何编写测试用例。

1. 编写测试用例

编写如下测试用例来测试 LiteController 类。

```
package com.waylau.lite.mvc.controller;
```

```java
import static org.springframework.test.web.servlet.request.MockMvcRequestBuilders.get;
import static org.springframework.test.web.servlet.result.MockMvcResultMatchers.content;
import static org.springframework.test.web.servlet.result.MockMvcResultMatchers.jsonPath;
import static org.springframework.test.web.servlet.result.MockMvcResultMatchers.status;
import org.junit.jupiter.api.BeforeEach;
import org.junit.jupiter.api.Test;
import org.springframework.http.MediaType;
import org.springframework.test.context.junit.jupiter.web.SpringJUnitWebConfig;
import org.springframework.test.web.servlet.MockMvc;
import org.springframework.test.web.servlet.setup.MockMvcBuilders;
import org.springframework.web.context.WebApplicationContext;
import com.waylau.lite.LiteConfig;

@SpringJUnitWebConfig(classes = LiteConfig.class)
class LiteControllerTests {

    private MockMvc mockMvc;

    @BeforeEach
    void setup(WebApplicationContext wac) {
        this.mockMvc = MockMvcBuilders.webAppContextSetup(wac).build();
    }

    @Test
    void testSayHi() throws Exception {
        this.mockMvc.perform(get("/lite")
                .accept(MediaType.parseMediaType("application/json;charset=UTF-8")))
            .andExpect(status().isOk())
            .andExpect(content().contentType("application/json;charset=UTF-8"))
            .andExpect(jsonPath("$.version").isNotEmpty())
            .andExpect(jsonPath("$.author").value("waylau.com"));
    }
}
```

其中：

（1）jsonPath("$.version").isNotEmpty() 用于验证响应的 JSON 数据中的 version 属性值是否为空。

（2）jsonPath("$.author").value("waylau.com") 用于验证响应的 JSON 数据中的 author 属性值是否为 "waylau.com"。

2. 运行测试用例

在测试用例上右击，通过 JUnit Test 来运行该类，绿色代表测试通过。

12.6.6　支持传入外部上下文

LiteJettyServer 主要通过 context.register(LiteConfig.class) 方法来注入 Spring 的上下文配置信息。但该 LiteConfig 配置只属于 Lite 框架，那么其他基于 Lite 框架的应用，又该如何自定义 Spring 的上下文呢？有以下几种方案。

1. 导入外部配置文件

使用 ImportResource 导入外部的配置文件，其代码如下。

```
package com.waylau.lite;

import org.springframework.context.annotation.ComponentScan;
import org.springframework.context.annotation.Configuration;
import org.springframework.context.annotation.Import;
import org.springframework.context.annotation.ImportResource;

import com.waylau.lite.mvc.LiteMvcConfig;


@Configuration
@ComponentScan(basePackages = { "com.waylau.lite" })
@Import({LiteMvcConfig.class})
@ImportResource("classpath*:*spring.xml")
public class LiteConfig {

}
```

通过 @ImportResource("classpath*:*spring.xml") 就能将所有以"spring.xml"结尾的 Spring 配置文件加载进来。这种方式适用于基于 XML 配置的 Spring 应用，且 Spring 配置文件命名规范统一以"spring.xml"结尾。

2. 支持将外部配置类注册进容器

通过将外部配置类注册进容器来实现基于 Java 的配置文件。观察 LiteJettyServer 代码，添加如下方法。

```
private Class<?> annotatedClass;

public LiteJettyServer(Class<?> annotatedClass) {
    this.annotatedClass = annotatedClass;
}

private WebApplicationContext webApplicationContext() {
    AnnotationConfigWebApplicationContext context = new AnnotationConfig-
```

```
WebApplicationContext();
    context.register(LiteConfig.class);

    if (this.annotatedClass != null) {
        context.register(this.annotatedClass); // 支持外部上下文
    }

    return context;
}
```

这样，在实例化 LiteJettyServer 时，就可以将外部的配置类注册进上下文中，具体用法如下。

```
public class App {
    public static void main(String[] args) {
        new LiteJettyServer(AppConfig.class).run(args);
    }
}
```

其中，AppConfig 就是一个外部的上下文配置类。

12.6.7 实现数据访问层

数据访问层的设置主要集中在应用的 lite-spring.xml 文件中，其具体用法如下。

1. 启用AOP

为了启用 AOP，在应用中添加如下配置类。

```
package com.waylau.lite.aop;

import org.springframework.context.annotation.Configuration;
import org.springframework.context.annotation.EnableAspectJAutoProxy;

@Configuration
@EnableAspectJAutoProxy
public class LiteAopConfig {

}
```

并将 LiteAopConfig 配置类添加到主应用配置中。

```
package com.waylau.lite;

import org.springframework.context.annotation.ComponentScan;
import org.springframework.context.annotation.Configuration;
import org.springframework.context.annotation.Import;
import org.springframework.context.annotation.ImportResource;

import com.waylau.lite.aop.LiteAopConfig;
import com.waylau.lite.mvc.LiteMvcConfig;
```

```
@Configuration
@ComponentScan(basePackages = { "com.waylau.lite" })
@Import({LiteMvcConfig.class, LiteAopConfig.class})
@ImportResource("classpath*:*spring.xml")
public class LiteConfig {

}
```

2. 配置数据库连接池

配置数据库连接池，代码如下。

```xml
<bean class="org.springframework.beans.factory.config.PropertyPlaceholderConfigurer">
        <property name="locations" value="classpath:lite.properties"/>
    </bean>

<!-- 数据源 -->
<bean id="dataSource" class="org.apache.commons.dbcp2.BasicDataSource" destroy-method="close">
        <property name="driverClassName" value="${lite.jdbc.driverClassName}" />
        <property name="url" value="${lite.jdbc.url}" />
        <property name="username" value="${lite.jdbc.username}" />
        <property name="password" value="${lite.jdbc.password}" />
</bean>
```

其中，lite.properties 文件内容如下。

```
lite.jdbc.driverClassName=com.mysql.cj.jdbc.Driver
lite.jdbc.url=jdbc:mysql://localhost:3306/test
lite.jdbc.username=root
lite.jdbc.password=123456
```

3. 基于AOP的事务管理

设置基于 AOP 的事务管理，代码如下。

```xml
<!-- 事务管理 -->
<bean id="txManager" class="org.springframework.jdbc.datasource.DataSourceTransactionManager">
    <property name="dataSource" ref="dataSource"/>
</bean>

<!-- 事务Advice -->
<tx:advice id="txAdvice" transaction-manager="txManager">
    <tx:attributes>
        <!--将以get,find,list开头的方法设置为"只读" -->
        <tx:method name="get*" read-only="true" propagation="NOT_SUP-
```

```xml
PORTED" />
        <tx:method name="find*" read-only="true" propagation="NOT_SUP-
PORTED" />
        <tx:method name="list*" read-only="true" propagation="NOT_SUP-
PORTED" />

<!--其余方法使用默认事务 -->
<tx:method name="*" propagation="REQUIRED"/>
    </tx:attributes>
</tx:advice>

<!--配置 事务Advice -->
<aop:config>
    <aop:pointcut id="serviceOperation" expression="execution(* com..
service.*.*(..)) || execution(* org..service.*.*(..))"/>
    <aop:advisor advice-ref="txAdvice" pointcut-ref="serviceOperation"/>
</aop:config>
```

Service 中以 get、find、list 开头的方法都设置为"只读"（没有事务），而其他方法的事务传播机制为"REQUIRED"。

4. 集成MyBatis

在 Spring 中配置 MyBatis 的 SqlSessionFactory。

```xml
<!-- MyBatis 工厂 -->
<bean id="sqlSessionFactory" class="org.mybatis.spring.SqlSessionFacto-
ryBean">
    <property name="dataSource" ref="dataSource" />
</bean>
```

在用户程序的配置中，添加要扫描注入的 Mapper 接口。如扫描 "com.waylau.lite.mall.mapper" 包中的 Mapper 接口。

```xml
<!-- 扫描 MyBatis Mapper 接口所在的包 -->
<bean class="org.mybatis.spring.mapper.MapperScannerConfigurer">
  <property name="basePackage" value="com.waylau.lite.mall.mapper" />
</bean>
```

12.6.8　实现安全认证

安全认证主要通过 Spring Security 来实现，需添加如下依赖。

```xml
<!-- 安全相关的依赖 -->
<dependency>
    <groupId>org.springframework.security</groupId>
    <artifactId>spring-security-web</artifactId>
    <version>${spring-security.version}</version>
</dependency>
```

```xml
<dependency>
    <groupId>org.springframework.security</groupId>
    <artifactId>spring-security-config</artifactId>
    <version>${spring-security.version}</version>
</dependency>
...
```

创建配置类 LiteSecurityConfig。

```java
package com.waylau.lite.security;

import org.springframework.security.config.annotation.web.builders.HttpSecurity;
import org.springframework.security.config.annotation.web.configuration.EnableWebSecurity;
import org.springframework.security.config.annotation.web.configuration.LiteSecurityConfigurerAdapter;


@EnableWebSecurity // 启用Spring Security功能
public class LiteSecurityConfig
    extends LiteSecurityConfigurerAdapter {

    /**
     * 自定义配置
     */
    @Override
    protected void configure(HttpSecurity http) throws Exception {

        // 允许所有人访问
        http.authorizeRequests().anyRequest().anonymous();
    }

}
```

这里默认所有的接口都允许访问，不涉及安全认证，安全认证交给具体的用户程序去设置。

将 LiteSecurityConfig 添加到主应用配置类中。

```java
package com.waylau.lite;

import org.springframework.context.annotation.ComponentScan;
import org.springframework.context.annotation.Configuration;
import org.springframework.context.annotation.Import;
import org.springframework.context.annotation.ImportResource;

import com.waylau.lite.aop.LiteAopConfig;
import com.waylau.lite.mvc.LiteMvcConfig;
import com.waylau.lite.security.LiteSecurityConfig;
```

```
@Configuration
@ComponentScan(basePackages = { "com.waylau.lite" })
@Import({LiteMvcConfig.class, LiteAopConfig.class, LiteSecurityConfig.
class})
@ImportResource("classpath*:*spring.xml")
public class LiteConfig {

}
```

12.7 发布Lite框架到Maven中央仓库

为了能让更多的开发者方便地使用 Lite 框架，可以将 Lite 发布到 Maven 中央仓库，这样开发者就能通过 Maven 轻松集成 Lite 框架。

本节介绍发布 Lite 框架到 Maven 中央仓库的详细过程。

12.7.1 注册Sonatype账号

Maven 中央仓库由 Sonatype 公司管理，因此首先需要注册 Sonatype 账号。访问 https://issues.sonatype.org/secure/Signup!default.jspa 页面进行注册，其界面如图 12-4 所示。

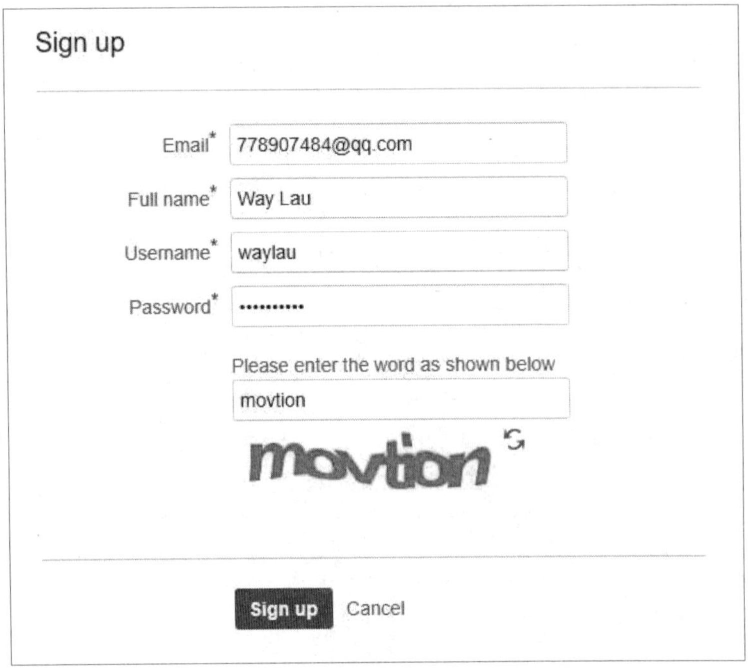

图12-4　注册Sonatype账号

看到图 12-5 所示的提示，说明注册成功，可以用刚刚注册的用户名和密码进行登录。

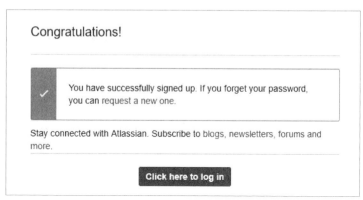

图12-5　注册Sonatype账号成功

12.7.2　创建一个Issue

访问 https://issues.sonatype.org/secure/CreateIssue!default.jspa 页面来创建一个 Issue，界面如图 12-6 所示。其中：

（1）将 Project 设置为"Community Support - Open Source Project Repository Hosting (OSSRH)"。

（2）将 Issue Type 设置为"New Project"。

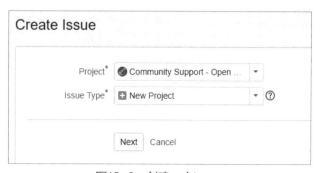

图12-6　创建一个Issue

单击"Next"按钮进行下一步操作。

此时，就需要对 Issue 信息进行详细的描述，如图 12-7 所示。

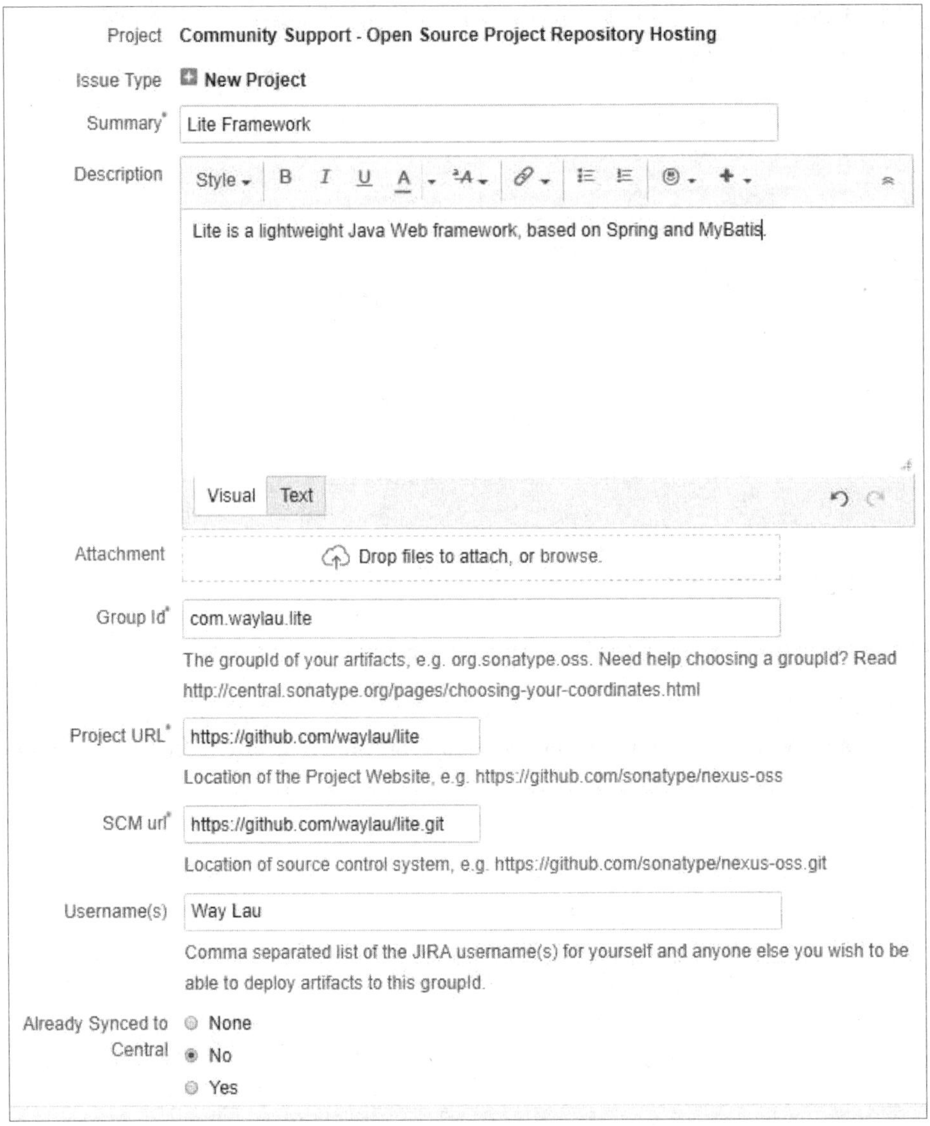

图12-7　详细描述Issue

单击"Create"按钮完成创建。此时就能在Issue面板中看到Issue的状态信息，如图12-8所示。

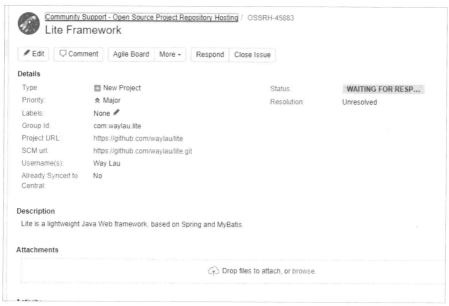

图12-8　Issue面板

12.7.3　等待Issue审核

审核时间无法确定，一般一个工作日内可以收到邮件答复。

当看到类似以下回复信息时，说明 Issue 审核成功。

```
    [ https://issues.sonatype.org/browse/OSSRH-45883?page=com.atlas-
sian.jira.plugin.system.issuetabpanels:all-tabpanel ]

Joel Orlina resolved OSSRH-45883.
---------------------------------
    Resolution: Fixed

Configuration has been prepared, now you can:
* Deploy snapshot artifacts into repository https://oss.sonatype.org/
content/repositories/snapshots
* Deploy release artifacts into the staging repository https://oss.
sonatype.org/service/local/staging/deploy/maven2
* Promote staged artifacts into repository 'Releases'
* Download snapshot and release artifacts from group https://oss.
sonatype.org/content/groups/public
* Download snapshot, release and staged artifacts from staging group
https://oss.sonatype.org/content/groups/staging

please comment on this ticket when you promoted your first release,
thanks
```

在 Issue 面板中，能够看到 Issue 的状态信息为"RESOLVED"（已解决），如图 12-9 所示。

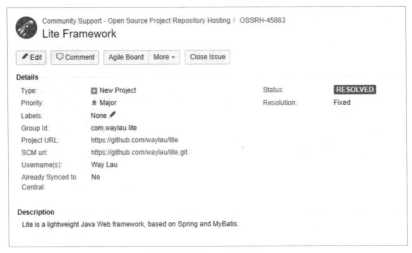

图12-9　Issue已解决

12.7.4　下载安装Gpg4win

Gpg4win 用于提供 GPG/PGP 的签名文件。访问 https://gpg4win.org/download.html 页面，下载 Gpg4win 安装包。

安装包下载完毕之后，在其上双击，运行安装文件执行安装。

12.7.5　生成密钥对

Gpg4win 安装完成后，就能够看到 Kleopatra 界面，单击"新建密钥对"按钮，如图 12-10 所示。

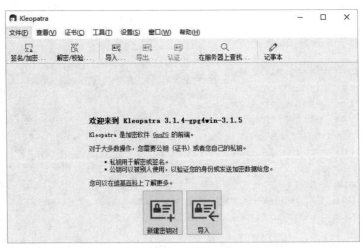

图12-10　Kleopatra（Gpg4win）界面

按照密钥创建向导填写姓名和邮箱，如图 12-11 所示。

图12-11　填写姓名和邮箱

单击"下一步"按钮之后，会要求输入密码，如图 12-12 所示。

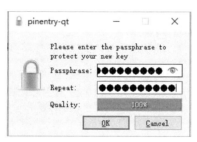

图12-12　填写密码

此时，会看到图 12-13 所示的界面，提示生成密钥对成功。

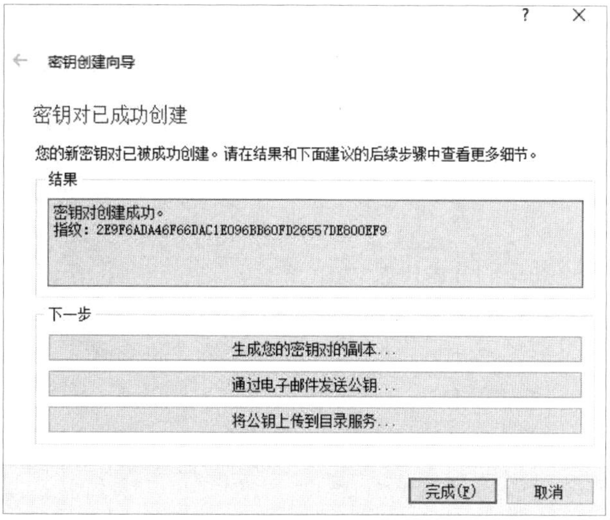

图12-13　生成密钥对

也可以在命令行查看已经生成的密钥对。

```
>gpg --list-keys
C:/Users/Administrator/AppData/Roaming/gnupg/pubring.kbx
---------------------------------------------------------
pub   rsa2048 2019-01-29 [SC] [expires: 2021-01-29]
      2E9F6ADA46F66DAC1E096BB60FD26557DE800EF9
uid           [ultimate] waylau <778907484@qq.com>
sub   rsa2048 2019-01-29 [E] [expires: 2021-01-29]
```

同时，需要将公钥发送到服务器，以完成认证。其中"2E9F6ADA46F66DAC1E096BB60FD26557DE800EF9"为上一步中生成的密钥对中的 key。

```
>gpg --keyserver hkp://keyserver.ubuntu.com:11371 --send-keys
2E9F6ADA46F66DAC1E096BB60FD26557DE800EF9
gpg: sending key 0FD26557DE800EF9 to hkp://keyserver.ubuntu.com:11371
```

12.7.6 修改pom文件

修改项目的 pom 文件，添加如下内容。

1. 配置项目坐标

groupId 与创建 issue 时填写的内容相同。

```xml
<groupId>com.waylau</groupId>
<artifactId>lite</artifactId>
<version>1.0.0</version>
```

2. 项目名称、描述和URL

补充项目的其他信息，如名称、描述、URL。

```xml
<name>${project.groupId}:${project.artifactId}</name>
<description>Lite is a lightweight Java Web framework, based on Spring and MyBatis.
</description>
<url>https://github.com/waylau/lite</url>
```

3. License信息

按照自己的需要，填好相应的开源协议。

```xml
<licenses>
    <license>
        <name>MIT License</name>
        <url>http://www.opensource.org/licenses/mit-license.php</url>
    </license>
</licenses>
```

4. 开发者信息

补充开发者的信息。

```xml
<developers>
    <developer>
        <id>waylau</id>
        <name>Way Lau</name>
        <email>waylau521@gmall.com</email>
    </developer>
</developers>
```

5. SCM信息

补充 SCM 信息。托管在 GitHub 上的可以按照以下代码填写。

```xml
<developers>
    <developer>
        <id>waylau</id>
        <name>Way Lau</name>
        <email>waylau521@gmall.com</email>
    </developer>
</developers>
```

6. 配置分发管理服务器

添加如下配置。

```xml
<distributionManagement>
    <snapshotRepository>
        <id>ossrh</id>
        <url>https://oss.sonatype.org/content/repositories/snapshots</url>
    </snapshotRepository>
    <repository>
        <id>ossrh</id>
        <url>https://oss.sonatype.org/service/local/staging/deploy/maven2</url>
    </repository>
</distributionManagement>
```

7. 添加插件

添加如下插件。

（1）gpg 签名插件 maven-gpg-plugin。

（2）Nexus 发布插件 nexus-staging-maven-plugin。

（3）Sources 插件 maven-source-plugin。

（4）Javadoc 插件 maven-javadoc-plugin。

配置内容如下。

```xml
<plugins>
    <!-- gpg sign Plugin -->
    <plugin>
        <groupId>org.apache.maven.plugins</groupId>
        <artifactId>maven-gpg-plugin</artifactId>
        <version>${maven-gpg-plugin.version}</version>
        <executions>
            <execution>
                <id>sign-artifacts</id>
                <phase>verify</phase>
                <goals>
                    <goal>sign</goal>
                </goals>
            </execution>
        </executions>
    </plugin>
    <!-- Nexus Staging Plugin -->
    <plugin>
        <groupId>org.sonatype.plugins</groupId>
        <artifactId>nexus-staging-maven-plugin</artifactId>
        <version>${nexus-staging-maven-plugin.version}</version>
        <extensions>true</extensions>
        <configuration>
            <serverId>ossrh</serverId>
            <nexusUrl>https://oss.sonatype.org/</nexusUrl>
            <autoReleaseAfterClose>true</autoReleaseAfterClose>
        </configuration>
    </plugin>
    <!-- Sources Plugin -->
    <plugin>
        <groupId>org.apache.maven.plugins</groupId>
        <artifactId>maven-source-plugin</artifactId>
        <version>${maven-source-plugin.version}</version>
        <executions>
            <execution>
                <id>attach-sources</id>
                <goals>
                    <goal>jar-no-fork</goal>
                </goals>
            </execution>
        </executions>
    </plugin>
    <!-- Javadoc Plugin -->
    <plugin>
        <groupId>org.apache.maven.plugins</groupId>
        <artifactId>maven-javadoc-plugin</artifactId>
        <version>${maven-javadoc-plugin.version}</version>
        <executions>
            <execution>
```

```xml
            <id>attach-javadocs</id>
            <goals>
                <goal>jar</goal>
            </goals>
        </execution>
    </executions>
</plugin>
</plugins>
```

12.7.7　修改Maven的settings.xml文件

在 Maven 的 settings.xml 文件中，添加如下配置信息。

```xml
<server>
    <id>ossrh</id>
    <username>waylau</username>
    <password>yourpassword</password>
</server>
```

其中：

（1）id 值"ossrh"要与 nexus-staging-maven-plugin 中的 serverId 值一致。

（2）username 和 password 分别是 Sonatype 的账号和密码。

12.7.8　发布到仓库

执行 mvn clean deploy 命令发布项目到仓库。

```
>mvn clean deploy
[INFO] Scanning for projects…
[INFO] Inspecting build with total of 1 modules…
[INFO] Installing Nexus Staging features:
[INFO]   … total of 1 executions of maven-deploy-plugin replaced with nexus-staging-maven-plugin
[INFO]
[INFO] ------------------------< com.waylau.lite:lite >------------------------
[INFO] Building com.waylau.lite:lite 1.0.0
[INFO] --------------------------------[ jar ]---------------------------------
[INFO]
[INFO] --- maven-clean-plugin:2.5:clean (default-clean) @ lite ---
[INFO] Deleting D:\workspaceGithub\lite\target
[INFO]
[INFO] --- maven-resources-plugin:2.6:resources (default-resources) @ lite ---
[INFO] Using 'UTF-8' encoding to copy filtered resources.
```

```
[INFO] Copying 3 resources
[INFO]
[INFO] --- maven-compiler-plugin:3.8.0:compile (default-compile) @ lite ---
[INFO] Changes detected - recompiling the module!
[INFO] Compiling 12 source files to D:\workspaceGithub\lite\target\classes
[INFO]
[INFO] --- maven-resources-plugin:2.6:testResources (default-testResources) @ lite ---
[INFO] Using 'UTF-8' encoding to copy filtered resources.
[INFO] skip non existing resourceDirectory D:\workspaceGithub\lite\src\test\resources
[INFO]
[INFO] --- maven-compiler-plugin:3.8.0:testCompile (default-testCompile) @ lite ---
[INFO] Changes detected - recompiling the module!
[INFO] Compiling 1 source file to D:\workspaceGithub\lite\target\test-classes
[INFO]
[INFO] --- maven-surefire-plugin:2.12.4:test (default-test) @ lite ---
[INFO]
[INFO] --- maven-jar-plugin:2.4:jar (default-jar) @ lite ---
[INFO] Building jar: D:\workspaceGithub\lite\target\lite-1.0.0.jar
[INFO]
[INFO] --- maven-shade-plugin:3.2.1:shade (default) @ lite ---
…

Uploaded to ossrh: https://oss.sonatype.org:443/service/local/staging/deployByRepositoryId/comwaylau-1002/com/waylau/lite/1.0.0/lite-1.0.0.pom.asc (499 B at 752 B/s)
[INFO]  * Upload of locally staged artifacts finished.
[INFO]  * Closing staging repository with ID "comwaylau-1002".

Waiting for operation to complete…
……

[INFO] Remote staged 1 repositories, finished with success.
[INFO] Remote staging repositories are being released…

Waiting for operation to complete…
……

[INFO] Remote staging repositories released.
[INFO] ------------------------------------------------------------------------
[INFO] BUILD SUCCESS
[INFO] ------------------------------------------------------------------------
```

```
[INFO] Total time: 05:48 min
[INFO] Finished at: 2019-01-29T21:51:25+08:00
[INFO] ------------------------------------------------------------
--------
```

发布成功之后，就可以在 https://oss.sonatype.org/ 查询到自己部署的 Artifact。一般在几十分钟后，Artifact 就会同步到 Maven 中央仓库及其他镜像站。此时，就可以通过 Maven 来添加 Lite 框架的依赖了。

```
<dependency>
  <groupId>com.waylau</groupId>
  <artifactId>lite</artifactId>
  <version>1.0.0</version>
</dependency>
```

图 12-14 所示为在 Maven 中央仓库查询 Lite 的界面。

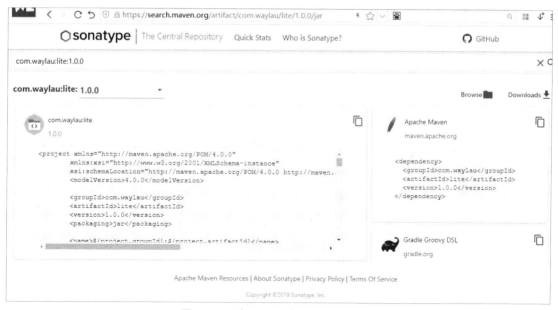

图12-14　在Maven中央仓库查询Lite

12.7.9　查看Lite源码

Lite 现在已经开源，有兴趣的读者可以访问 https://github.com/waylau/lite 查看源代码。

第13章
实战：基于Lite框架的互联网应用

本章将演示如何基于Lite框架来实现一个真实的互联网应用。该应用名为"lite-news"，是一款新闻资讯类应用。整个应用分为客户端（lite-news-ui）和服务端（lite-news-server）两部分。

13.1 lite-news概述

lite-news 是一款新闻资讯类手机应用，所实现的功能与市面上的头条新闻等类似，主要功能是让用户阅读实时的新闻信息。

lite-news 采用当前互联网应用所流行的前后台分离的技术。前台（lite-news-ui）采用以 Angular 为主要技术的前端框架；后台（lite-news-server）则完全基于 Lite 框架。前后台通过 REST API 进行通信，整体架构如图 13-1 所示。

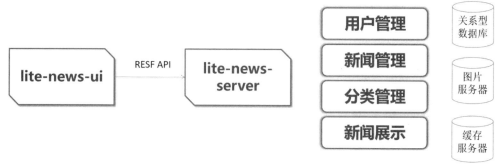

图13-1 lite-news整体架构

从图 13-1 中也可看出，lite-news 主要包含的功能有用户管理、新闻管理、分类管理、新闻展示等。其中，新闻数据存储在关系型数据库中，新闻图片存储于图片服务器中。同时，还设置了缓存服务器，用于存储经常被访问的数据。

13.1.1 初始化数据库

创建名为"lite_news"的数据库。

```
mysql> CREATE DATABASE lite_news;
Query OK, 1 row affected (0.13 sec)
```

并使用该数据库。

```
mysql> USE lite_news; Database changed
```

13.1.2 创建用户表

创建名为"t_user"的用户表，用于存储用户信息。

```
mysql> CREATE TABLE t_user (user_id BIGINT PRIMARY KEY AUTO_INCREMENT,
username VARCHAR(20) COMMENT '用户名称', password VARCHAR(80) COMMENT '密
码', email VARCHAR(20)) COMMENT='用户';
Query OK, 0 rows affected (0.22 sec)
```

其中，PRIMARY KEY 用于标识 user_id 为表的主键；AUTO_INCREMENT 用于设置主键为自增长。

注意：为了简化业务模型，这里的用户只设置了用户名称（username）、密码（password）、email 等字段。在实际项目中，可以按需对字段进行扩展。

13.1.3　创建新闻表

创建名为 "t_news" 的新闻表，该表用于存储新闻信息。

```
mysql> CREATE TABLE t_news (news_id BIGINT PRIMARY KEY AUTO_INCREMENT,
title VARCHAR(100)  COMMENT '标题', author VARCHAR(20) COMMENT '作者',
creation DATETIME COMMENT '创建时间', content TEXT COMMENT '正文') COM-
MENT='新闻';
Query OK, 0 rows affected (0.38 sec)
```

13.1.4　创建后台应用lite-news-server

lite-news-server 作为后台应用（服务器端），是一个典型的 Maven 项目。lite-news-server 使用了 Lite 框架，因此需要在 pom.xml 文件中添加 Lite 框架的依赖坐标。

```xml
<dependency>
    <groupId>com.waylau</groupId>
    <artifactId>lite</artifactId>
    <version>${lite.version}</version>
</dependency>
```

13.1.5　创建Spring配置文件

创建 spring.xml 文件，主要配置 MyBatis 的信息。

```xml
<?xml version="1.0" encoding="UTF-8"?>
<beans xmlns="http://www.springframework.org/schema/beans"
    xmlns:xsi="http://www.w3.org/2001/XMLSchema-instance"
    xmlns:context="http://www.springframework.org/schema/context"
    xsi:schemaLocation="
        http://www.springframework.org/schema/beans
        http://www.springframework.org/schema/beans/spring-beans.xsd
        http://www.springframework.org/schema/context
        http://www.springframework.org/schema/context/spring-context.xsd
        http://www.springframework.org/schema/aop
        http://www.springframework.org/schema/aop/spring-aop.xsd">

    <!-- 扫描 MyBatis Mapper 接口所在的包 -->
```

```xml
    <bean class="org.mybatis.spring.mapper.MapperScannerConfigurer">
        <property name="basePackage"
            value="com.waylau.lite.news.mapper" />
    </bean>

    <!-- MyBatis 工厂 -->
    <bean id="sqlSessionFactory"
        class="org.mybatis.spring.SqlSessionFactoryBean">
        <property name="dataSource" ref="dataSource" />
    </bean>
</beans>
```

在上述配置中，定义了 Mapper 接口所在的包及 MyBatis 工厂（sqlSessionFactory）。其中，MyBatis 工厂所引用的数据源是 Lite 框架自带的。

除了使用 XML 方式配置 Spring 外，还可以通过 JavaConfig 方式进行配置，示例如下。

```java
@ComponentScan(basePackages = { "com.waylau.lite.news" })
@Import({WebSecurityConfig.class})
public class AppConfig extends LiteConfig {

}
```

13.1.6 自定义Lite配置文件

可以通过创建 lite.properties 文件来覆盖 Lite 的默认配置信息。以下是 lite.properties 定义的内容。

```
lite.jdbc.driverClassName=com.mysql.cj.jdbc.Driver
lite.jdbc.url=jdbc:mysql://localhost:3306/lite_news?useSSL=false&-
serverTimezone=UTC
lite.jdbc.username=root
lite.jdbc.password=123456
```

上述配置涵盖了与 JDBC 相关的配置信息。

13.1.7 自定义日志格式

如果想要自定义日志的格式信息，可以新增 logback.xml，其配置示例如下。

```xml
<?xml version="1.0" encoding="UTF-8"?>
<configuration>

  <appender name="STDOUT" class="ch.qos.logback.core.ConsoleAppender">
    <layout class="ch.qos.logback.classic.PatternLayout">
      <Pattern>%d{HH:mm:ss.SSS} [%thread] %-5level %logger{36} - %msg%n</Pattern>
```

```xml
    </layout>
  </appender>

  <root level="debug">
    <appender-ref ref="STDOUT" />
  </root>
</configuration>
```

13.2 模型设计

用户表和新闻表创建完成之后，即可进行模型的设计。本节建议采用 POJO 的编程模式，针对用户表和新闻表分别建立以下用户模型和新闻模型。

13.2.1 用户模型设计

用户模型设计用 User 类表示，其代码如下。

```java
package com.waylau.lite.news.domain;

public class User {

    private Long userId;
    private String username;
    private String password;
    private String email;

    public Long getUserId() {
        return userId;
    }

    public void setUserId(Long userId) {
        this.userId = userId;
    }

    public String getUsername() {
        return username;
    }

    public void setUsername(String username) {
        this.username = username;
    }

    public String getPassword() {
```

```
        return password;
    }

    public void setPassword(String password) {
        this.password = password;
    }

    public String getEmail() {
        return email;
    }

    public void setEmail(String email) {
        this.email = email;
    }

}
```

13.2.2 新闻模型设计

新闻模型设计用 News 类表示,其代码如下。

```
package com.waylau.lite.news.domain;

import java.util.Date;

public class News {

    private Long newsId;
    private String title;
    private String author;
    private Date creation;
    private String content;

    public Long getNewsId() {
        return newsId;
    }

    public void setNewsId(Long newsId) {
        this.newsId = newsId;
    }

    public String getTitle() {
        return title;
    }

    public void setTitle(String title) {
        this.title = title;
    }
```

```java
    public String getAuthor() {
        return author;
    }

    public void setAuthor(String author) {
        this.author = author;
    }

    public Date getCreation() {
        return creation;
    }

    public void setCreation(Date creation) {
        this.creation = creation;
    }

    public String getContent() {
        return content;
    }

    public void setContent(String content) {
        this.content = content;
    }
}
```

13.3 接口设计与实现

接口设计主要涉及两个方面：内部接口设计和外部接口设计。其中，内部接口设计又细分为服务接口和 DAO 接口；外部接口设计主要指提供给外部应用访问的 REST 接口。

13.3.1 DAO接口设计

DAO 层采用的是 MyBatis 的 Mapper 方式。

1. 用户Mapper

UserMapper 定义如下。

```
package com.waylau.lite.news.mapper;

import com.waylau.lite.news.domain.User;
```

```java
public interface UserMapper {

    void createUser(User user);

    void deleteUser(Long userId);

    void updateUser(User user);

    User getUser(Long userId);

}
```

其对应的 Mapper 的 XML 定义如下。

```xml
<?xml version="1.0" encoding="UTF-8"?>
<!DOCTYPE mapper
    PUBLIC "-//mybatis.org//DTD Mapper 3.0//EN"
    "http://mybatis.org/dtd/mybatis-3-mapper.dtd">

<mapper namespace="com.waylau.lite.news.mapper.UserMapper">

    <insert id="createUser"
        parameterType="com.waylau.lite.news.domain.User">
        insert into t_user (username, password, email)
        values( #{username}, #{password}, #{email})
    </insert>

    <delete id="deleteUser" parameterType="long">
        delete from t_user
        where user_id = #{userId}
    </delete>

    <update id="updateUser"
        parameterType="com.waylau.lite.news.domain.User">
        update t_user set
        password = #{password},
        email = #{email}
        where user_id = #{userId}
    </update>

    <select id="getUser" parameterType="long"
        resultType="com.waylau.lite.news.domain.User">
        select user_id as userId, username, password, email
        from t_user where user_id = #{userId}
    </select>

</mapper>
```

2. 新闻Mapper

NewsMapper 定义如下。

```java
package com.waylau.lite.news.mapper;

import com.waylau.lite.news.domain.News;

public interface NewsMapper {

    void createNews(News news);

    void updateNews(News news);

    void deleteNews(Long newsId);

    News getNews(Long newsId);
}
```

其对应的 Mapper 的 XML 定义如下。

```xml
<?xml version="1.0" encoding="UTF-8"?>
<!DOCTYPE mapper
    PUBLIC "-//mybatis.org//DTD Mapper 3.0//EN"
    "http://mybatis.org/dtd/mybatis-3-mapper.dtd">

<mapper namespace="com.waylau.lite.news.mapper.NewsMapper">

    <insert id="createNews"
        parameterType="com.waylau.lite.news.domain.News">
        insert into t_news (title, author, creation, content)
        values( #{title}, #{author}, #{creation}, #{content})
    </insert>

    <delete id="deleteNews" parameterType="long">
        delete from t_news
        where news_id = #{newsId}
    </delete>

    <update id="updateNews"
        parameterType="com.waylau.lite.news.domain.News">
        update t_news set
        title = #{title},
        author = #{author},
        content = #{content}
        where news_id = #{newsId}
    </update>

    <select id="getNews" parameterType="long"
```

```
            resultType="com.waylau.lite.news.domain.News">
        select news_id as newsId, title, author, content, creation
        from t_news where news_id = #{newsId}
    </select>
</mapper>
```

13.3.2 Service接口设计

用户 Service 和新闻 Service 接口定义如下。

1. 用户Service

UserService 定义如下。

```
package com.waylau.lite.news.service;

import com.waylau.lite.news.domain.User;

public interface UserService {

    void createUser(User user);

    void deleteUser(Long userId);

    void updateUser(User user);

    User getUser(Long userId);
}
```

其对应的 UserService 的实现类 UserServiceImpl 定义如下。

```
package com.waylau.lite.news.service.impl;

import org.springframework.beans.factory.annotation.Autowired;
import org.springframework.stereotype.Service;

import com.waylau.lite.news.domain.User;
import com.waylau.lite.news.mapper.UserMapper;
import com.waylau.lite.news.service.UserService;

@Service
public class UserServiceImpl implements UserService {

    @Autowired
    private UserMapper userMapper;

    @Override
    public User getUser(Long userId) {
```

```java
        return userMapper.getUser(userId);
    }

    @Override
    public void updateUser(User user) {
        userMapper.updateUser(user);
    }

    @Override
    public void createUser(User user) {
        userMapper.createUser(user);
    }

    @Override
    public void deleteUser(Long userId) {
        userMapper.deleteUser(userId);
    }
}
```

2. 新闻Service

NewsService 定义如下。

```java
package com.waylau.lite.news.service;

import com.waylau.lite.news.domain.News;

public interface NewsService {

    void createNews(News news);

    void deleteNews(Long newsId);

    void updateNews(News news);

    News getNews(Long newsId);
}
```

其对应的 NewsService 的实现类 NewsServiceImpl 定义如下。

```java
package com.waylau.lite.news.service.impl;

import org.springframework.beans.factory.annotation.Autowired;
import org.springframework.stereotype.Service;

import com.waylau.lite.news.domain.News;
import com.waylau.lite.news.mapper.NewsMapper;
import com.waylau.lite.news.service.NewsService;
```

```java
@Service
public class NewsServiceImpl implements NewsService {

    @Autowired
    private NewsMapper newsMapper;

    @Override
    public News getNews(Long newsId) {
        return newsMapper.getNews(newsId);
    }

    @Override
    public void updateNews(News news) {
        newsMapper.updateNews(news);
    }

    @Override
    public void createNews(News news) {
        newsMapper.createNews(news);
    }

    @Override
    public void deleteNews(Long newsId) {
        newsMapper.deleteNews(newsId);
    }

}
```

13.3.3　REST接口设计

用户 REST 接口和新闻 REST 接口定义如下。

1. 用户REST接口

用户 REST 接口定义如下。

（1）增加用户：POST /users。

（2）删除用户：DELETE /users/{userId}。

（3）编辑用户：PUT /users。

（4）查询用户：GET /users/{userId}。

其具体实现方式如下。

```java
package com.waylau.lite.news.controller;

import org.springframework.beans.factory.annotation.Autowired;
import org.springframework.web.bind.annotation.DeleteMapping;
import org.springframework.web.bind.annotation.GetMapping;
```

```java
import org.springframework.web.bind.annotation.PathVariable;
import org.springframework.web.bind.annotation.PostMapping;
import org.springframework.web.bind.annotation.PutMapping;
import org.springframework.web.bind.annotation.RequestMapping;
import org.springframework.web.bind.annotation.RestController;

import com.waylau.lite.news.domain.User;
import com.waylau.lite.news.service.UserService;

@RestController
@RequestMapping("/users")
public class UserController {

    @Autowired
    private UserService userService;

    @GetMapping("/{userId}")
    public User getUser(@PathVariable("userId") Long userId) {
        return userService.getUser(userId);
    }

    @PutMapping
    public void updateUser(@RequestBody User user) {
        userService.updateUser(user);
    }

    @PostMapping
    public void createUser(@RequestBody User user) {
        userService.createUser(user);
    }

    @DeleteMapping("/{userId}")
    public void deleteUser(@PathVariable("userId") Long userId) {
        userService.deleteUser(userId);
    }
}
```

2. 新闻REST接口

新闻 REST 接口定义如下。

（1）增加新闻：POST /news。

（2）删除新闻：DELETE /news/{newsId}。

（3）编辑新闻：PUT /news。

（4）查询新闻：GET /news/{newsId}。

其具体实现方式如下。

```java
package com.waylau.lite.news.controller;
```

```java
import java.util.Date;

import org.springframework.beans.factory.annotation.Autowired;
import org.springframework.web.bind.annotation.DeleteMapping;
import org.springframework.web.bind.annotation.GetMapping;
import org.springframework.web.bind.annotation.PathVariable;
import org.springframework.web.bind.annotation.PostMapping;
import org.springframework.web.bind.annotation.PutMapping;
import org.springframework.web.bind.annotation.RequestBody;
import org.springframework.web.bind.annotation.RequestMapping;
import org.springframework.web.bind.annotation.RestController;

import com.waylau.lite.news.domain.News;
import com.waylau.lite.news.service.NewsService;

@RestController
@RequestMapping("/news")
public class NewsController {

    @Autowired
    private NewsService newsService;

    @GetMapping("/{newsId}")
    public News getNews(@PathVariable("newsId") Long newsId) {
        return newsService.getNews(newsId);
    }

    @PutMapping
    public void updateNews(@RequestBody News news) {
        newsService.updateNews(news);
    }

    @PostMapping
    public void createNews(@RequestBody News news) {
        // 当前用户
        User currentUser =
            (User)SecurityContextHolder.getContext().getAuthentication().getPrincipal();
        news.setAuthor(currentUser.getUsername());
        // 当前时间
        news.setCreation(new Date());

        newsService.createNews(news);
    }

    @DeleteMapping("/{newsId}")
    public void deleteNews(@PathVariable("newsId") Long newsId) {
        newsService.deleteNews(newsId);
```

```
    }
}
```

13.4 实现权限管理

org.springframework.security.core.userdetails.UserDetailsService 接口可以获取用户认证信息,其定义如下。

```
public interface UserDetailsService {

    UserDetails loadUserByUsername(String username) throws UsernameNot-
FoundException;
}
```

可以自定义 UserDetailsService,并在应用中实现认证。

13.4.1 实现UserDetailsService

用 UserServiceImpl 实现 UserDetailsService 接口,并重写 loadUserByUsername 方法。loadUserByUsername 方法需要根据用户名称来查询用户信息,因此需要在 UserMapper 中增加相应的方法。

新的根据用户名称查询用户信息的接口定义如下。

```
public interface UserMapper {
    // …

    User getUserByUsername(String username);

}
```

对应的 Mapper 的 XML 定义如下。

```xml
<select id="getUserByUsername" parameterType="string"
    resultType="com.waylau.lite.news.domain.User">
    select user_id as userId, username, password, email
    from t_user where username = #{username}
</select>
```

用 UserServiceImpl 重写 loadUserByUsername 方法的代码如下。

```java
private final String ADMIN = "ROLE_ADMIN";
```

```java
@Autowired
private UserMapper userMapper;

// …

@Override
public UserDetails loadUserByUsername(String username)
        throws UsernameNotFoundException {
    User user = userMapper.getUserByUsername(username);
    UserDetails userDetails;

    if (user != null) {
        // 默认都是ADMIN角色
        List<GrantedAuthority> authorities = List.of(new SimpleGranted-
Authority(ADMIN));

        // 构建认证信息
        userDetails = new org.springframework.security.core.userdetails.
            User(user.getUsername(),
                user.getPassword(), authorities);
    } else {
        throw new UsernameNotFoundException(username + " is not found");
    }

    return userDetails;
}
```

在上述代码中，org.springframework.security.core.userdetails.User 是 UserDetails 接口的实现，也是 Spring Security 内置的用户信息类，因此需要将 User 类转换为 org.springframework.security.core.userdetails.User 类。

注意，在 Spring Security 中，角色都是以 "ROLE_" 开头。为求简单，将应用的用户分为两类：普通用户和管理员用户。在本书中，将普通用户定义为无须登录系统即可访问新闻内容的用户，而将管理员用户定义为能够访问后台具备编辑新闻、增加新闻等操作权限的用户。简言之，普通用户在应用中是没有账号的，而管理员用户是有账号的（在 t_user 表中存在记录）。因此，t_user 表中都是 "ROLE_ADMIN" 角色的用户。

13.4.2 定义安全配置类

创建安全配置类 WebSecurityConfig，其代码如下。

```java
package com.waylau.lite.news.security;

import org.springframework.beans.factory.annotation.Autowired;
import org.springframework.context.annotation.Bean;
import org.springframework.core.annotation.Order;
```

```java
import org.springframework.security.authentication.AuthenticationProvider;
import org.springframework.security.authentication.dao.DaoAuthenticationProvider;
import org.springframework.security.config.annotation.authentication.builders.AuthenticationManagerBuilder;
import org.springframework.security.config.annotation.web.builders.HttpSecurity;
import org.springframework.security.config.annotation.web.configuration.EnableWebSecurity;
import org.springframework.security.config.annotation.web.configuration.WebSecurityConfigurerAdapter;
import org.springframework.security.core.userdetails.UserDetailsService;
import org.springframework.security.crypto.bcrypt.BCryptPasswordEncoder;
import org.springframework.security.crypto.password.PasswordEncoder;


@EnableWebSecurity // 启用Spring Security功能
@Order(1)   // 覆盖Lite框架中的配置
public class WebSecurityConfig
    extends WebSecurityConfigurerAdapter {

    @Autowired
    private UserDetailsService userDetailsService;

    @Autowired
    private PasswordEncoder passwordEncoder;

    @Bean
    public PasswordEncoder passwordEncoder() {
        return new BCryptPasswordEncoder();    // 使用 BCrypt 加密
    }

    @Bean
    public AuthenticationProvider authenticationProvider() {
        DaoAuthenticationProvider authenticationProvider = new DaoAuthenticationProvider();
        authenticationProvider.setUserDetailsService(userDetailsService);
        authenticationProvider.setPasswordEncoder(passwordEncoder); // 设置密码加密方式
        return authenticationProvider;
    }

    /**
     * 认证信息管理
     * @param auth
     * @throws Exception
```

```
    */
    @Autowired
    public void configureGlobal(AuthenticationManagerBuilder auth) throws 
Exception {
        auth.userDetailsService(userDetailsService);
        auth.authenticationProvider(authenticationProvider());
    }

    /**
     * 自定义配置
     */
    @Override
    protected void configure(HttpSecurity http) throws Exception {
        http.authorizeRequests().antMatchers("/css/**", "/js/**", "/
fonts/**", "/index").permitAll() // 都可以访问
                .antMatchers("/admins/**").hasRole("ADMIN") // 需要相应的角色才能访问
                .and()
                .formLogin()    //基于 Form 表单登录验证
                .and()
                .httpBasic(); // HTTP基本认证
    }

}
```

其中：

（1）使用 BCryptPasswordEncoder 作为密码编码器。

（2）除了限制 "/admins/**" 及其子路径外，其他页面都无须认证即可访问。

（3）要访问 "/admins/**" 及其子路径，需要拥有 ADMIN 角色的权限。

13.4.3　定义后台管理控制器

定义后台管理控制器 AdminController，其代码如下。

```
@RestController
@RequestMapping("/admins")
public class AdminController {

    @GetMapping("/hi")
    public String hi() {
        return "hello";
    }

}
```

上面的 hi 接口纯粹是为了进行测试，所以实现比较简单。

13.4.4　初始化用户数据

在 t_user 表中初始化用户 waylau 的数据。

```
insert into t_user (username, password, email) values( 'waylau',
'$2a$10$qZLF2y/a1yhfFZ.Lwhq9f.sOADw6QlBRRGL4XLodTHJJupkAEBB8i', 'way-
lau521@gmail.com')
```

其中，password 的值是经过 BCryptPasswordEncoder 加密的，原始值为"123456"，其加密过程的实现可以参考如下代码。

```
String rawPassword = "123456";
BCryptPasswordEncoder passwordEncoder = new BCryptPasswordEncoder();

// 加密
String encodedPassword = passwordEncoder.encode(rawPassword);
System.out.println("明文 " + rawPassword + " -> 密文 " + encodedPass-
word);
```

13.4.5　测试认证

启动应用后，访问 http://localhost:8080/admins/hi 页面，此时会提示输入用户名和密码，认证登录界面如图 13-2 所示。

如果密码输错，就会提示"Bad credentials"（认证失败），界面如图 13-3 所示。

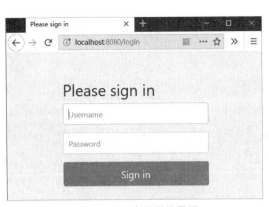

图13-2　认证登录界面

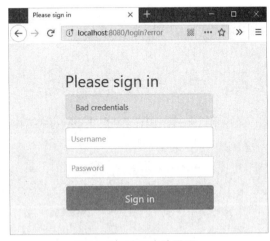

图13-3　认证失败界面

如果用户名和密码都输入正确，就能够正常访问 http://localhost:8080/admins/hi 页面了，响应内容如图 13-4 所示。

第 13 章 实战：基于 Lite 框架的互联网应用

图13-4 成功访问界面

13.5 前端lite-news-ui设计

lite-news-ui 是客户端应用，主要使用 Angular 等技术框架实现。

13.5.1 应用概述

lite-news-ui 是一个汇聚热点新闻的 Web 应用。该应用采用 Angular 作为主要实现技术，通过调用 lite-news-server 所提供的 REST 服务接口，将新闻数据在应用中进行展示。

13.5.2 需求分析

lite-news-ui 应用主要面向手机用户，即屏幕应能在宽屏、窄屏之间实现响应式缩放。lite-news-ui 大致分为首页、新闻详情页两大部分。

1.首页需求分析

首页包括新闻列表部分，如图 13-5 所示。首页应展示新闻列表，新闻列表主要由新闻标题组成。

2.新闻详情页需求分析

当在首页单击新闻列表条目时，应该能进入新闻详情界面。新闻详情界面主要用于展示新闻的详细内容，如图 13-6 所示。

图13-5 首页界面

图13-6 新闻详情界面

新闻详情页包含了"返回"按钮、新闻标题、新闻发布时间、新闻正文等内容。其中，单击"返回"按钮，可以返回首页（或前一次访问记录）。

13.6 实现lite-news-ui原型

要开发 Angular 应用，首先需要具备开发环境，同时还需要了解 TypeScript 语言的语法。本节将快速演示如何搭建开发环境，如何初始化 Angular 应用，以及如何实现 Angular 功能。

由于篇幅有限，本节只介绍与本书主题相关的 Angular 核心内容的开发。

13.6.1 开发环境准备

开发 Angular 应用，需要具备 Node.js 和 npm 环境。如果你的电脑里没有 Node.js 和 npm，需要先安装它们。

1. 为什么需要安装Node.js和npm

如果熟悉 Java，就一定知道 Maven。Node.js 与 npm 的关系就如同 Java 与 Maven 的关系。

（1）Node.js 与 Java 都是运行应用的平台，且都运行在虚拟机中。Node.js 基于 Google V8 引擎，而 Java 基于 JVM。

（2）npm 与 Maven 类似，都是用于依赖管理。npm 管理 JavaScript 库，而 Maven 管理 Java 库。

2. 安装步骤

Node.js 的下载地址为 https://nodejs.org/en/download/。为了能够享受 Angular 开发所带来的乐趣，请安装最新版本的 Node.js 和 npm。

Node.js 的安装比较简单，按图 13-7 ~ 图 13-10 所示的步骤进行安装即可。

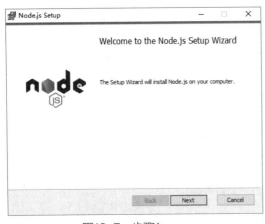

图13-7 步骤1

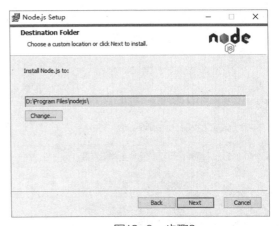

图13-8 步骤2

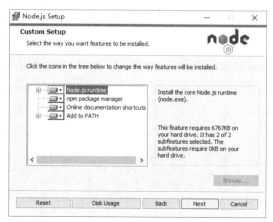

图13-9 步骤3

图13-10 步骤4

安装完成之后，在终端/控制台窗口中运行 node -v 和 npm –v 命令，如图 13-11 所示，以验证安装是否正确。

图13-11 验证安装

3. 设置npm镜像

npm 默认是从国外的 npm 源来获取和下载包信息，由于网络的原因，有时可能无法正常访问源，导致软件安装失败。

可以采用国内的 npm 镜像来解决网速慢的问题。可以在终端通过以下命令来设置 npm 镜像。

```
npm config set registry=http://registry.npm.taobao.org
```

其他更多设置方式，可以参考笔者的博客 https://waylau.com/faster-npm/。

4. 选择合适的IDE

如果你是一名前端工程师，那么不必花太多时间来安装 IDE（Integrated Development Environment，集成开发环境），用平时熟悉的 IDE 来开发 Angular 即可。例如，前端工程师经常会选择 Visual Studio Code、Eclipse、WebStorm、Sublime Text 等来进行开发。理论上，开发 Angular 不会对开发工具有任何限制，甚至可以直接使用文本编辑器来开发。

如果你是一名初级的前端工程师，或者不知道如何来安装 IDE，那么笔者建议你尝试一下 Visual Studio Code，其下载地址为 https://code.visualstudio.com。

提示：Visual Studio Code 与 TypeScript 都是微软出品的，对 TypeScript 和 Angular 编程有着一流的支持，而且 Visual Studio Code 是免费的，可以随时下载使用。选择适合自己的 IDE 有助于提升编程质量和开发效率。

5. 安装Angular CLI

Angular CLI 是一个命令行界面工具，它可以创建项目、添加文件及执行一大堆开发任务，如测试、打包和发布 Angular 应用。可通过 npm 采用全局的方式来安装 Angular CLI，具体命令如下。

```
npm install -g @angular/cli
```

若看到控制台输出如下内容，则说明 Angular CLI 已经安装成功。

```
>npm install -g @angular/cli
C:\Users\Administrator\AppData\Roaming\npm\ng -> C:\Users\Administrator\AppData\Roaming\npm\node_modules\@angular\cli\bin\ng
npm WARN optional SKIPPING OPTIONAL DEPENDENCY: fsevents@1.2.4 (node_modules\@angular\cli\node_modules\fsevents):
npm WARN notsup SKIPPING OPTIONAL DEPENDENCY: Unsupported platform for fsevents@1.2.4: wanted {"os":"darwin","arch":"any"} (current: {"os":"win32","arch":"x64"})

+ @angular/cli@7.0.2
added 248 packages in 81.427s
```

13.6.2 初始化lite-news-ui

通过 Angular CLI 工具可以快速初始化 Angular 应用的骨架，命令如下。

```
ng new lite-news-ui
```

执行 ng serve 命令可以启动该应用，并可以在 http://localhost:4200/ 中访问该应用，其运行界面如图 13-12 所示。

图13-12　运行界面

13.6.3 添加Angular Material

为了提升用户体验，需要在应用中引入一款成熟的 UI 组件。目前，市面上有非常多的 UI 组件可供选择，如 Angular Material、Ant Design，这些 UI 组件各有优势。本实例采用 Angular Material，主要考虑到该 UI 组件是 Angular 官方团队开发的，且其帮助文档、社区资源非常丰富，对于开发者而言非常友好。

为了添加 Angular Material 库到应用中，可以通过 Angular CLI 执行下面的命令。

```
ng add @angular/material
```

安装过程中，命令行会做出如下提示，要求用户选择一套主题。任意选择一套主题后按 "Enter" 键即可。

```
? Choose a prebuilt theme name, or "custom" for a custom theme: (Use arrow keys)
> Indigo/Pink          [ Preview: https://material.angular.io?theme=indigo-pink ]
  Deep Purple/Amber    [ Preview: https://material.angular.io?theme=deep-purple-amber ]
  Pink/Blue Grey       [ Preview: https://material.angular.io?theme=pink-bluegrey ]
  Purple/Green         [ Preview: https://material.angular.io?theme=purple-green ]
  Custom
```

1. 配置动画

安装完 Angular Material 库之后，应用会自动导入 BrowserAnimationsModule 以支持动画。打开 app.module.ts 文件，可以观察到 Angular CLI 生成的如下源码。

```
import { BrowserModule } from '@angular/platform-browser';
import { NgModule } from '@angular/core';

import { AppRoutingModule } from './app-routing.module';
import { AppComponent } from './app.component';
import { HttpClientModule } from '@angular/common/http';
import { BrowserAnimationsModule } from '@angular/platform-browser/animations';

@NgModule({
  declarations: [
    AppComponent
  ],
  imports: [
    BrowserModule,
    AppRoutingModule,
    HttpClientModule,
```

```
    BrowserAnimationsModule // 动画模块
  ],
  providers: [],
  bootstrap: [AppComponent]
})
export class AppModule { }
```

提示：TypeScript 语言也支持模块化的开发。上述代码中的 import 和 export，与 Java 9 中的模块化操作的功能极其类似。

2. 按需导入组件模块

按需将 UI 组件的模块导入应用中。例如，要使用按钮模块，就在 app.module.ts 文件中导入 MatButtonModule。

```
…
import { MatButtonModule } from '@angular/material';
@NgModule({
  declarations: [
    AppComponent
  ],
  imports: [
    BrowserModule,
    AppRoutingModule,
    HttpClientModule,
    BrowserAnimationsModule,
    MatButtonModule // 按钮模块
  ],
  providers: [],
  bootstrap: [AppComponent]
})
export class AppModule { }
```

13.6.4　创建新闻列表组件

如果首页主要展示新闻列表，就要相应地在应用中创建与之对应的新闻列表组件。使用 Angular CLI 执行如下命令以创建组件。

```
ng generate component news
```

用下面的代码替换掉应用初始化时自动生成的 app.component.html 中的内容。

```
<app-news></app-news>
```

上述代码的含义为，应用主模板引用了新闻列表组件的模板。其中，"app-news" 就是新闻列表组件模板的选择器（Selector），可以在 news.component.ts 文件中找到。

```
import { Component, OnInit } from '@angular/core';
@Component({
  selector: 'app-news',
  templateUrl: './news.component.html',
  styleUrls: ['./news.component.css']
})
export class NewsComponent implements OnInit {

  constructor() { }

  ngOnInit() {
  }

}
```

运行以上代码，可以看到图 13-13 所示的效果。

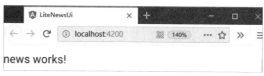

图13-13　运行界面

13.6.5　实现新闻列表原型设计

为了实现新闻列表，需要导入 MatListModule 模块，其代码如下。

```
…
import { MatListModule } from '@angular/material/list';

@NgModule({
  declarations: [
    AppComponent,
    NewsComponent
  ],
  imports: [
    …
    MatListModule// 用于新闻列表
  ],
```

修改 news.component.html 文件，添加如下内容。

```
<mat-nav-list>
  <a mat-list-item href="/" >传三星Galaxy A91即将发布  骁龙855+后置三摄</a>
  <a mat-list-item href="/" >双自拍镜头才好  三星Galaxy S10+竟如此受追捧</a>
  <a mat-list-item href="/" >红魔3S玄铁黑明日首销  2999元</a>
  <a mat-list-item href="/" >阿里巴巴成首个单季营收破千亿中国互联网公司</a>
```

```
  <a mat-list-item href="/" >亿万网民一起参与春晚互动 热度空前</a>
  <a mat-list-item href="/" >这三人摘得诺贝尔经济学奖 他们是谁</a>
  <a mat-list-item href="/" >多姿势 多距离 "神枪手"是这样练成的！</a>
  <a mat-list-item href="/" >红米8|8A发布 699元起标配骁龙439</a>
</mat-nav-list>
```

上述内容是静态数据，用于展示新闻列表的原型。

运行以上代码，可以看到图13-14所示的运行效果。

图13-14 运行界面

13.7 实现路由器

如果需要在首页和新闻详情页之间来回切换，就需要设置路由器。

13.7.1 创建路由

使用 Angular CLI 创建应用的路由，其命令如下。

```
ng generate module app-routing --flat --module=app
```

其中，"* -flat" 把上述命令生成的 app-routing.ts 文件放进了 src/app 中，而不是单独的目录中。"* --module=app" 告诉 CLI 将其注册到 AppModule 的 imports 数组中。

路由器代码修改如下。

```
import { NgModule } from '@angular/core';
import { Routes, RouterModule } from '@angular/router';
```

```
import { NewsComponent } from "./news/news.component";
import { NewsDetailComponent } from './news-detail/news-detail.compo-
nent';

const routes: Routes = [
  { path: '', component: NewsComponent },    // 新闻列表
  { path: 'news', component: NewsDetailComponent} // 新闻详情
];

@NgModule({
  imports: [RouterModule.forRoot(routes)],
  exports: [RouterModule]
})
export class AppRoutingModule { }
```

通过设置该路由器，可方便地实现首页和新闻详情页之间的切换。

13.7.2 添加路由出口

在 app.component.html 页面中添加路由出口，其代码如下。

```
<router-outlet></router-outlet>
```

13.7.3 修改新闻列表组件

修改新闻列表组件模板，当单击新闻列表的条目时，能够从新闻列表组件路由到新闻详情组件，其代码修改如下。

```
<<mat-nav-list>
  <a mat-list-item href="/news" >传三星Galaxy A91即将发布  骁龙855+后置三摄</a>
  <a mat-list-item href="/news" >双自拍镜头才好  三星Galaxy S10+竟如此受追捧</a>
  <a mat-list-item href="/news" >红魔3S玄铁黑明日首销  2999元</a>
  <a mat-list-item href="/news" >阿里巴巴成首个单季营收破千亿中国互联网公司</a>
  <a mat-list-item href="/news" >亿万网民一起参与春晚互动  热度空前</a>
  <a mat-list-item href="/news" >这三人摘得诺贝尔经济学奖  他们是谁</a>
  <a mat-list-item href="/news" >多姿势 多距离  "神枪手"是这样练成的！</a>
  <a mat-list-item href="/news" >红米8|8A发布  699元起标配骁龙439</a>
</mat-nav-list>
```

其中，href 指定要路由的路径，即新闻详情组件。

13.7.4 增加返回事件处理

修改 news-detail.component.html，在返回按钮上增加事件处理，用于返回上一次浏览的界面（一般就是新闻列表界面），其代码如下。

```html
<button mat-raised-button (click)="goback()">返回</button>
<mat-card class="example-card">
  ...
</mat-card>
```

同时，需要在 news-detail.component.ts 中增加 goback() 方法。

```typescript
import { Component, OnInit } from '@angular/core';
import { Location } from '@angular/common'; // 用于回退浏览记录

@Component({
  selector: 'app-news-detail',
  templateUrl: './news-detail.component.html',
  styleUrls: ['./news-detail.component.css']
})
export class NewsDetailComponent implements OnInit {

  constructor(private location: Location) { }

  ngOnInit() {
  }

  // 返回
  goback() {
    // 浏览器回退浏览记录
    this.location.back();
  }
}
```

13.7.5 运行应用

单击新闻列表和返回按钮，就能实现首页和新闻详情页之间的切换。

为了更加真实地反映移动端访问应用的效果，可以通过浏览器模拟移动端界面效果。

Firefox、Chrome 等浏览器均支持模拟移动端界面的效果。以 Firefox 浏览器为例，通过执行"打开菜单"→"Web 开发者"→"响应式设计模式"命令来展示移动端的界面效果，如图 13-15 所示。

在模拟移动端访问应用的效果如图 13-16 所示。

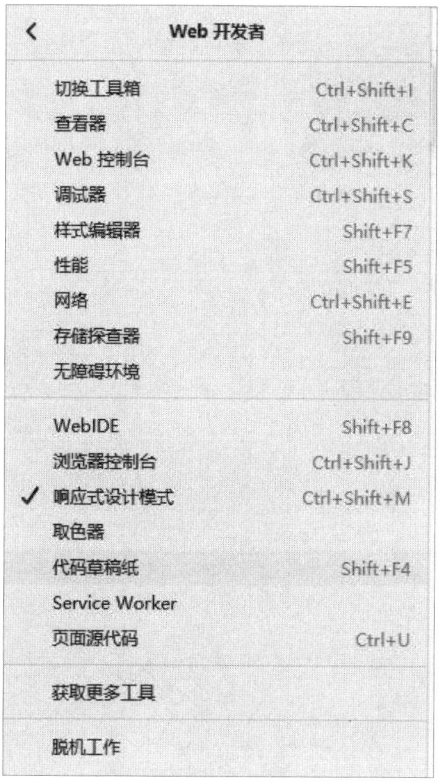

图13-15 设置

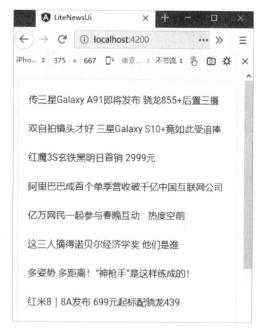

图13-16 移动端运行界面

13.7.6 实现新闻详情页原型设计

接下来是实现新闻详情页原型设计。

新闻详情页用于展示新闻的详细内容。相比首页的新闻条目而言，新闻详情页还多了新闻发布时间、创建人、新闻内容等。

1. 新建新闻详情组件

通过 Angular CLI 编写新闻组件 NewsDetailComponent，其命令如下。

```
ng generate component news-detail
```

2. 导入模块

为了实现新闻列表，需要导入 MatListModule 模块，其代码如下。

```
…
import { MatCardModule } from '@angular/material/card';

@NgModule
  declarations: [
    AppComponent,
```

```
    NewsComponent
  ],
  imports: [
    …
    MatCardModule // 用于新闻详情
  ],
```

3. 修改新闻详情页组件模板

修改新闻详情页组件模板 news-detail.component.html，其代码如下。

```html
<button mat-raised-button>返回</button>
<mat-card class="example-card">
  <mat-card-header>
    <mat-card-title>阿里巴巴成首个单季营收破千亿的中国互联网公司</mat-card-title>
    <mat-card-subtitle>2019-1-31 21:00</mat-card-subtitle>
  </mat-card-header>

  <mat-card-content>
    <p>
        中国青年网北京1月30日电  北京时间1月30日晚  阿里巴巴集团公布2019财年第三季度业绩。
        集团收入同比增长41%，达1172.78亿元  这是中国首个实现单季营收破千亿的互联网公司。
        彰显出中国社会强大的消费信心以及阿里巴巴强劲的"平台效应"。
    </p>
    <img mat-card-image src="https://ss1.baidu.com/6ONXsjip0QIZ8tyhnq/it/u=542868259,2390763485&fm=173&app=49&f=JPEG?w=554&h=369&s=FBA400C-0FAF15E8EA8B54D96030080B0">
    <p>财报显示  淘宝移动端月度活跃用户达到6.99亿  较2018年9月增加了3300万。
        淘宝作为国民级应用价值凸显   "单季营收破千亿"和"人手一辆购物车"的背后是数字经济所激发的旺盛消费需求。超预期增长的数字经济正在激发中国消费者的巨大消费潜力。
    </p>
  </mat-card-content>
</mat-card>
```

4. 修改app.component.html

为了能访问新闻详情组件界面，修改 app.component.html 代码如下。

```html
<!--<app-news></app-news>-->
<app-news-detail></app-news-detail>
```

其中，"app-news-detail"是新闻详情组件模板的选择器（selector），可以在 news-detail.component.ts 文件中找到。

最终，新闻详情界面原型效果如图 13-17 所示。

图13-17　新闻详情界面原型

13.8　实现用户登录

在之前的章节中，已经初步实现了后台权限的管理。本节将进一步实现用户在 lite-news-ui 应用上的登录授权。

13.8.1　创建后台管理组件

后台管理组件主要用于管理新闻的发布。后台管理的使用角色限制为 ADMIN。换言之，要想访问后台管理界面，需要先在前台进行登录授权。使用 Angular CLI 执行如下命令创建组件。

```
ng generate component admin
```

13.8.2　添加组件到路由器

为了使页面能被访问到，需要将后台管理组件添加到路由器中，其代码如下。

```
...
import { AdminComponent } from './admin/admin.component';
const routes: Routes = [
```

```
…
  { path: 'admin', component: AdminComponent} // 后台管理
];
```

启动应用，访问 http://localhost:4200/admin 页面，可以看到后台管理界面，效果如图13-18所示。

图13-18　后台管理界面

后台管理目前还没有任何业务逻辑，只是搭建了一个初级骨架。

13.8.3　注入HTTP客户端

后台有一个只允许 ADMIN 角色访问的接口 http://localhost:8080/admins/hi，只有当用户登录系统获取权限之后，才能够正常访问该接口，否则就会被重定向到登录界面。

当期望用户访问后台管理界面时，该界面组件会发起一个访问 http://localhost:8080/admins/hi 的请求，以获取该接口的数据。为了实现在 Angular 中发起 HTTP 请求的功能，需要使用 Angular HttpClient API。该 API 包含在 HttpClientModule 模块中，因此需要在应用中导入该模块。

```
…
import { HttpClientModule } from '@angular/common/http';
@NgModule
  declarations: [
    AppComponent,
    NewsComponent,
    NewsDetailComponent,
    AdminComponent
  ],
  imports: [
    …
    HttpClientModule // HTTP客户端
  ],
```

同时，需要在 AdminComponent 中注入 HttpClient，其代码如下。

```
import { Component, OnInit } from '@angular/core';
import { HttpClient } from '@angular/common/http';

@Component({
  selector: 'app-admin',
```

```
  templateUrl: './admin.component.html',
  styleUrls: ['./admin.component.css']
})
export class AdminComponent implements OnInit {

  // 注入HttpClient
  constructor(private http: HttpClient) { }

  ngOnInit() {
  }

}
```

13.8.4 客户端访问后台接口

有了 HttpClient，就能远程发起 HTTP 请求到后台的 REST 接口。

1. 设置反向代理

由于本项目是一个前后台分离的应用，若分开部署运行，势必会遇到跨域访问的问题。

解决跨域问题，业界最为常用的方式是设置反向代理。其原理是设置反向代理服务器，让 Angular 应用都访问自己服务器中的 API，而这类 API 都会被反向代理服务器转发到 Java 等后台服务的 API 中，而这个过程 Angular 应用是无法感知的。

业界经常采用 NGINX 服务来承担反向代理的职责。在 Angular 中，使用反向代理将变得更加简单，因为 Angular 自带反向代理服务器。设置方式为：在 Angualr 应用的根目录下，添加配置文件 proxy.config.json，并填写如下格式的内容。

```
{
  "/api/": {
    "target": "http://localhost:8080/",
    "secure": false,
    "pathRewrite": {
      "^/api": ""
    }
  }
}
```

这个配置说明了，任何在 Angular 发起的以 "/api/" 开头的 URL，都会反向代理到以 "http://localhost:8080/" 开头的 URL 中。例如，当在 Angular 应用中发起请求到 "http://localhost:4200/api/admins/hi" URL 时，反向代理服务器会将该 URL 映射到 "http://localhost:8080/admins/hi"。

添加了该配置文件之后，在启动应用时，只要指定该文件即可，其命令如下。

```
ng serve --proxy-config proxy.config.json
```

2. 客户端发起HTTP请求

使用 HttpClient 发起 HTTP 请求。

```
import { Component, OnInit } from '@angular/core';
import { HttpClient } from '@angular/common/http';

@Component({
  selector: 'app-admin',
  templateUrl: './admin.component.html',
  styleUrls: ['./admin.component.css']
})
export class AdminComponent implements OnInit {
  adminUrl = '/api/admins/hi';
  adminData = '';

  // 注入HttpClient
  constructor(private http: HttpClient) { }

  ngOnInit() {
    this.getData();
  }

  // 获取后台接口数据
  getData() {
    return this.http.get(this.adminUrl, { responseType: 'text' })
      .subscribe(data => this.adminData = data);
  }
}
```

在上述代码中，返回的数据会赋值给 adminData 变量。

3. 绑定数据

编辑 admin.component.html，修改代码如下。

```
<p>
  Get data from admin: {{adminData}}.
</p>
```

上述代码意味着将 adminData 变量绑定到了模板中，adminData 的任何赋值都能及时呈现在页面中。

4. 修改后台安全配置

最后，还需要修改后台的安全配置类 WebSecurityConfig。

```
@Override
protected void configure(HttpSecurity http) throws Exception {
    http.authorizeRequests()
      .antMatchers("/css/**", "/js/**", "/fonts/**", "/index")
      .permitAll() // 都可以访问
```

```
.antMatchers("/admins/**").hasRole("ADMIN") // 需要相应的角色才能访问
//.and()
//.formLogin()      //基于 Form 的表单登录验证
.and()
.httpBasic(); // HTTP基本认证
http.csrf().disable(); // 禁用CSRF
}
```

其中，formLogin 方法是 Spring Security 默认生成的登录界面。由于这是一个前后台分离的项目，不会在后台执行登录，因此禁用了该功能。http.csrf().disable() 禁用了 CSRF 功能。CSRF 功能是一种更加复杂的认证方式，在执行 POST、PUT、DELETE 等操作时，会在 HTTP 协议中添加特定的认证信息，实现起来也会相对复杂。此处简单起见，暂时禁用。

5. 测试

将前后台应用都启动后，尝试访问 http://localhost:4200/admin 页面。由于该页面所访问的 http://localhost:8080/admins/hi 接口是需要认证的，因此首次访问时，会出现图 13-19 所示的提示框。

输入正确的用户名和密码，成功登录之后，可以看到图 13-20 所示的界面，说明后台接口已经成功被访问，且返回了"hello"文本。其中，"hello"文本通过绑定机制，被渲染在了界面中。

图13-19　登录界面

图13-20　成为访问接口

13.9　实现新闻编辑器

本节将在 13.8 节的基础上实现新闻编辑器。新闻编辑器用于实现新闻内容在应用中的录入，这样用户才能在应用中看到新闻条目。

由于新闻类的文章内容排版较为简单，因此本书中是以 Markdown 作为新闻内容编辑格式的。

13.9.1 集成ngx-markdown插件

ngx-markdown 是一款 Markdown 插件（项目地址为 https://jfcere.github.io/ngx-markdown/），能够将 Markdown 格式的内容渲染为 HTML 格式的内容。

执行下面的命令，下载安装 ngx-markdown 插件。

```
npm install ngx-markdown --save
```

修改 angular.json 文件，并添加如下内容。

```
"scripts": [
    "node_modules/marked/lib/marked.js"
]
```

13.9.2 导入MarkdownModule模块

在应用中导入 MarkdownModule 模块，以便启用 ngx-markdown 功能。同时，编辑器界面还用到了 Form 表单等模块，因此也需要一并导入，其代码如下。

```
…
import { MatButtonModule } from '@angular/material';
import { MatListModule } from '@angular/material/list';
import { MatCardModule } from '@angular/material/card';
import { MarkdownModule } from 'ngx-markdown';
import { MatFormFieldModule } from '@angular/material/form-field';
import { MatInputModule } from '@angular/material/input';

imports: [
…
MatFormFieldModule, // Form表单
MatInputModule, // Input
MarkdownModule.forRoot(), // Markdown渲染
],
```

13.9.3 编写编辑器界面

1. 编辑模板

编辑 admin.component.html，其代码如下。

```
<form class="lite-form">
  <mat-form-field class="lite-full-width">
    <input matInput placeholder="输入新闻标题"
    maxlength="100" name="title" (keyup)="syncTitle(newTitle.value)"
    value={{markdownTitle}} #newTitle>
```

```
  </mat-form-field>

  <mat-form-field class="lite-full-width lite-field">
    <textarea matInput placeholder="输入新闻内容" rows="16" name="content"
     (keyup)="syncContent(newContent.value)"
     value={{markdownContent}}
       #newContent></textarea>
  </mat-form-field>
</form>
<button mat-raised-button  (click)="submitData()">提交</button>
<markdown [data]="markdownContent"></markdown>
```

其中：

（1）<input> 用于输入新闻标题。

（2）<textarea> 用于输入新闻内容。

（3）<markdown> 用于实时将输入的 Markdown 格式的新闻内容以 HTML 格式显示。

2. 编辑组件

编辑 admin.component.ts，代码如下。

```
import { Component, OnInit } from '@angular/core';
import { HttpClient } from '@angular/common/http';

import { News } from './../news'
@Component({
  selector: 'app-admin',
  templateUrl: './admin.component.html',
  styleUrls: ['./admin.component.css']
})
export class AdminComponent implements OnInit {
  adminUrl = '/api/admins/hi';
  createNewsUrl = '/api/admins/news';

  adminData = '';
  markdownTitle = '';
  markdownContent = '';

  // 注入HttpClient
  constructor(private http: HttpClient) { }

  ngOnInit() {
    this.getData();
  }

  // 获取后台接口数据
  getData() {
    return this.http.get(this.adminUrl, { responseType: 'text' })
      .subscribe(data => this.adminData = data);
```

```typescript
  }

  // 同步编辑器中的内容
  syncContent(content: string) {
    this.markdownContent = content;
  }

  // 同步编辑器中的标题
  syncTitle(title: string) {
    this.markdownTitle = title;
  }

  // 提交新闻内容到后台
  submitData() {
    console.log('ssss');
    this.http.post<News>(this.createNewsUrl,
      new News(null, this.markdownTitle, null, this.markdownContent,
null)).subscribe(
        data => {console.log(data);
         alert("已经成功提交");

         // 清空数据
         this.markdownTitle = '';
         this.markdownContent = '';
        },
        error => {
          console.error(error);
          alert("提交失败");
        }
     );
   ;
  }
}
```

其中，submitData 方法会将新闻内容提交到后台 REST 接口。

News 类是前台新闻结构体，其代码如下。

```typescript
export class News {
    constructor(
        public newsId: number,
        public title: string, // 标题
        public author: string,  // 作者
        public content: string, // 内容
        public creation: Date, // 日期
    ) { }
}
```

13.9.4　调整后台接口路径

为了让后台的接口受到安全管控，将所有"/users/" "/news/"接口转移至"/admins/"路径下。修改方式比较简单，只需将 NewsController 和 UserController 中的 RequestMapping 值进行调整即可，其实现方式如下。

```
@RestController
@RequestMapping("/admins/news")
public class NewsController {
    …
}


@RestController
@RequestMapping("/users")
public class UserController {
    …
}
```

13.9.5　运行

运行应用，进行测试。

访问 http://localhost:4200/admin，可以看到图 13-21 所示的编辑页面。

在编辑页面输入内容，也可以插入图片。在界面下方可以实时看到 Markdown 格式的新闻内容转为了 HTML 格式的预览效果。

图13-21　编辑页面

13.10 实现新闻列表展示

在首页需要展示最新的新闻列表。lite-news-ui 已经提供了原型，本节基于这些原型来对接真实的后台数据。

13.10.1 实现新闻列表查询的接口

1. 定义NewsMapper接口

在 NewsMapper 中添加一个查询的接口。

```
public interface NewsMapper {
    …
    List<News> getNewsList();
}
```

同时，在 NewsMapper.xml 中增加对该接口的 XML 定义。

```
<select id="getNewsList"
    resultType="com.waylau.lite.news.domain.News">
    select news_id as newsId, title, author, content, creation
    from t_news
    order by news_id desc
</select>
```

需要注意的是，查询出来的新闻列表是按照时间进行排序的，最新的排在最前面。

2. 定义服务接口

在 NewsService 中添加一个查询新闻列表的接口。

```
public interface NewsService {
    …
    List<News> getNewsList();
}
```

同时，在 NewsServiceImpl 中实现该接口。

```
@Service
public class NewsServiceImpl implements NewsService {

    @Autowired
    private NewsMapper newsMapper;

    …
```

```
    @Override
    public List<News> getNewsList() {
        return newsMapper.getNewsList();
    }
}
```

3. 定义REST接口

新增 IndexController，用于处理首页接口的请求，其代码如下。

```
@RestController
@RequestMapping("/news")
public class IndexController {

    @Autowired
    private NewsService newsService;

    @GetMapping
    public List<News> getNewsList() {
        return newsService.getNewsList();
    }
}
```

13.10.2 实现客户端访问新闻列表REST接口

在完成后台接口之后，即可在客户端发起对该接口的调用。

1. 修改组件

修改 news.component.ts，其代码如下。

```
import { Component, OnInit } from '@angular/core';
import { HttpClient } from '@angular/common/http';

import { News } from './../news'

@Component({
  selector: 'app-news',
  templateUrl: './news.component.html',
  styleUrls: ['./news.component.css']
})
export class NewsComponent implements OnInit {
  newsListUrl = '/api/news';
  newsList: News[] = [];

  // 注入HttpClient
  constructor(private http: HttpClient) {
```

```
ngOnInit() {
  this.getData();
}

// 获取后台接口数据
getData() {
  return this.http.get<News[]>(this.newsListUrl)
    .subscribe(data => this.newsList = data);
}
```

上述代码实现了对新闻列表 REST 接口的访问。

2. 修改模板

修改 news.component.html，其代码如下。

```
<mat-nav-list>
  <a mat-list-item *ngFor="let news of newsList"
    href="/news/newsId={{news.newsId}}">{{news.title}}</a>
</mat-nav-list>
```

上述代码中，"*ngFor"实现了将模型中的 newsList 迭代生成新闻条目。

13.10.3 运行应用

运行应用，进行测试。

访问首页 http://localhost:4200，可以看到图 13-22 所示的首页内容。

图13-22　预览

13.11 实现新闻详情展示

lite-news-ui 已经提供了新闻详情的原型,本节基于这些原型来对接真实的后台数据。

13.11.1 实现新闻详情查询的接口

在 IndexController 中,增加用于处理查询新闻详情接口的请求,其代码如下。

```
@RestController
@RequestMapping("/news")
public class IndexController {

    @Autowired
    private NewsService newsService;

    @GetMapping("/{newsId}")
    public News getNews(@PathVariable("newsId") Long newsId) {
        return newsService.getNews(newsId);
    }

    …

}
```

13.11.2 实现客户端访问新闻详情REST接口

在完成后台接口之后,就可以在客户端发起对该接口的调用。

1. 修改组件

修改 news-detail.component.ts,其代码如下。

```
import { Component, OnInit } from '@angular/core';
import { Location } from '@angular/common'; // 用于回退浏览记录
import { HttpClient } from '@angular/common/http';
import { ActivatedRoute } from '@angular/router';

import { News } from './../news'

@Component({
  selector: 'app-news-detail',
  templateUrl: './news-detail.component.html',
  styleUrls: ['./news-detail.component.css']
})
export class NewsDetailComponent implements OnInit {
  newsUrl = '/api/news/';
  news: News = new News(null, null, null, null, null);
```

```
// 注入HttpClient
constructor(private location: Location,
  private http: HttpClient,
  private route: ActivatedRoute) { }


ngOnInit() {
  this.getData();
}

// 获取后台接口数据
getData() {
  const newsId = this.route.snapshot.paramMap.get('newsId');
  return this.http.get<News>(this.newsUrl + newsId)
    .subscribe(data => this.news = data);
}

// 返回
goback() {
  // 浏览器回退浏览记录
  this.location.back();
}
}
```

上述代码实现了对新闻详情 REST 接口的访问。

需要注意的是，newsId 是从 ActivatedRoute 对象中获取的。有关路由器的设置，稍后还将介绍。

2. 修改模板

修改 news-detail.component.html，其代码如下。

```
<button mat-raised-button (click)="goback()">返回</button>
<mat-card>
  <mat-card-header>
    <mat-card-title>{{news.title}}</mat-card-title>
    <mat-card-subtitle>{{news.creation | date}}</mat-card-subtitle>
    <mat-card-subtitle>{{news.author}}</mat-card-subtitle>
  </mat-card-header>

  <mat-card-content>
     <markdown [data]=news.content></markdown>
  </mat-card-content>
</mat-card>
```

其中，针对 news.creation 变量，使用了 Angular 的 Date 管道，以方便对时间格式进行转换。

13.11.3 设置路由

由于从新闻列表切换到新闻详情页面时携带了参数，因此针对这种场景，需要设置带参数的路由路径，其代码如下。

```
const routes: Routes = [
  { path: '', component: NewsComponent },   // 新闻列表
  { path: 'news/:newsId', component: NewsDetailComponent}, // 新闻详情;带参数
  { path: 'admin', component: AdminComponent} // 后台管理
];
```

13.11.4 运行应用

运行应用，进行测试。

访问首页 http://localhost:4200，单击任意新闻条目，可以切换至新闻详情页面，如图 13-23 所示。

图13-23　新闻详情

新闻详情显示的还是原型数据，并非数据库的真实数据。接下来将实现新闻详情界面的改造。

在编辑页面输入内容，也可以插入图片。在界面下方可以实时看到 Markdown 格式的新闻内容转换成了 HTML 格式的预览效果，如图 13-24 所示。

图13-24 预览

13.12 总结

有关新闻头条客户端及服务端的代码已经全部开发完成，基本实现了新闻列表查询、新闻详情展示、新闻录入及权限认证。受限于篇幅，书中的代码力求简单易懂，着重将核心的实现方式呈现给读者。如果想将这款应用作为商业软件，还需要进一步完善，主要包括以下几个方面。

（1）用户的管理。

（2）用户信息的修改。

（3）用户角色的分配。

（4）新闻内容的编辑。

（5）新闻分配。

（6）图片服务器的实现。

以上这些内容，还需要读者通过所掌握的基础知识举一反三，将新闻应用做到精益求精。本书最后所罗列的"参考文献"内容，也可以作为读者平时扩展学习之用。

第14章
使用NGINX实现高可用

NGINX是免费的、开源的、高性能的HTTP服务器和反向代理,同时也是IMAP/POP3代理服务器。NGINX以其高性能、高稳定性、丰富的功能集、简单的配置和低资源消耗而闻名。

本章将介绍如何通过NGINX来实现Anuglar应用的部署,同时实现应用的高可用。

14.1 NGINX概述

NGINX 是市面上仅有的几个为解决 C10K 问题[1]而编写的服务器之一。与传统服务器不同，NGINX 不依赖于线程来处理请求；相反，它使用更加可扩展的事件驱动（异步）架构。这种架构在负载下使用小的、可预测的内存量，即使在不需要处理数千个并发请求的场景下，仍然可以从 NGINX 的高性能和占用内存少等方面获益。NGINX 可以说在各个方面都适用，从最小的 VPS 一直到大型服务器集群。

NGINX 的用户包括 Netflix、Hulu、Pinterest、CloudFlare、Airbnb、WordPress.com、GitHub、SoundCloud、Zynga、Eventbrite、Zappos、Media Temple、Heroku、RightScale、Engine、Yard、MaxCDN 等众多高知名度网站。

更多有关 NGINX 的介绍，可以参阅笔者所著的开源电子书《NGINX 教程》[2]。

14.1.1 NGINX特性

NGINX 具有很多非常优越的特性，主要表现在以下几个方面。

（1）作为 Web 服务器：相比 Apache，NGINX 使用更少的资源，支持更多的并发连接，实现更高的效率，这使 NGINX 尤其受到虚拟主机供应商的欢迎。

（2）作为负载均衡服务器：NGINX 既可以在内部直接支持 Rails 和 PHP，也可以支持作为 HTTP 代理服务器对外进行服务。NGINX 用 C 语言编写，系统资源开销小，CPU 使用效率高。

（3）作为邮件代理服务器：NGINX 同时也是一个非常优秀的邮件代理服务器。

14.1.2 下载、安装、运行NGINX

NGINX 下载地址为 http://nginx.org/en/download.html，在这里可以免费下载各个操作系统的 NGINX 安装包。

14.1.3 安装、运行NGINX

以下是 NGINX 在各个操作系统的不同安装方式。

1. Linux和BSD

大多数 Linux 发行版本和 BSD 版本的软件包存储库中都有 NGINX，可以通过常用的安装软件

---

[1] C10K 问题指的是服务器同时支持成千上万个客户端的问题，也就是 "Concurrent 10000 Connection" 的简写。由于硬件成本的大幅度降低和硬件技术的进步，如果一台服务器能够同时服务更多的客户端，那么也就意味着服务每一个客户端的成本将大幅度降低，从这个角度来看，C10K 问题非常有意义。

[2] 开源电子书地址为 https://waylau.com/books/。

的方法进行安装，如在 Debian 平台使用 apt-get、在 Gentoo 平台使用 emerge、在 FreeBSD 平台使用 ports 等。

2. Red Hat和CentOS

首先添加 NGINX 的 yum 库，然后创建名为 /etc/yum.repos.d/nginx.repo 的文件，并粘贴如下配置到文件中。

CentOS 的配置如下。

```
[nginx]
name=nginx repo
baseurl=http://nginx.org/packages/centos/$releasever/$basearch/
gpgcheck=0
enabled=1
```

RHEL 的配置如下。

```
[nginx]
name=nginx repo
baseurl=http://nginx.org/packages/rhel/$releasever/$basearch/
gpgcheck=0
enabled=1
```

由于 CentOS、RHEL 和 Scientific Linux 之间填充的 $releasever 变量的差异，因此有必要根据自身操作系统版本手动将 $releasever 变量替换为 5（5.x）或 6（6.x）。

3. Debian/Ubuntu

此分发页面 http://nginx.org/packages/ubuntu/dists/ 列出了可用的 NGINX Ubuntu 版本支持。

在 /etc/apt/sources.list 中附加适当的脚本。如果担心存储库添加的持久性（即 DigitalOcean Droplets）不足，那么可以将适当的部分添加到 /etc/apt/sources.list.d/ 下的其他列表文件中，如 /etc/apt/sources.list.d/nginx.list。

```
## Replace $release with your corresponding Ubuntu release.
deb http://nginx.org/packages/ubuntu/ $release nginx
deb-src http://nginx.org/packages/ubuntu/ $release nginx
```

比如 Ubuntu 16.04 (Xenial) 版本，设置如下。

```
deb http://nginx.org/packages/ubuntu/ xenial nginx
deb-src http://nginx.org/packages/ubuntu/ xenial nginx
```

要想安装，执行如下脚本。

```
sudo apt-get update
sudo apt-get install nginx
```

安装过程中若有如下错误：

```
W: GPG error: http://nginx.org/packages/ubuntu xenial Release: The
```

```
following signatures couldn't be verified because the public key is not
available: NO_PUBKEY $key
```

则执行下面的命令。

```
## Replace $key with the corresponding $key from your GPG error.
sudo apt-key adv --keyserver keyserver.ubuntu.com --recv-keys $key
sudo apt-get update
sudo apt-get install nginx
```

4. Debian 6

关于在 Debian 6 上安装 NGINX，添加下面的脚本到 /etc/apt/sources.list 中。

```
deb http://nginx.org/packages/debian/ squeeze nginx
deb-src http://nginx.org/packages/debian/ squeeze nginx
```

5. Ubuntu PPA

这个 PPA 由志愿者维护，不由 nginx.org 分发。由于它有一些额外的编译模块，因此可能更适合安装环境。可以从 Launchpad 上的 NGINX PPA 获取最新的稳定版本的 NGINX，但需要具有 Root 权限才能执行以下命令。

安装 Ubuntu 10.04 及更新版本时执行下面的命令。

```
sudo -s
nginx=stable # use nginx=development for latest development version
add-apt-repository ppa:nginx/$nginx
apt-get update
apt-get install nginx
```

如果有关于 add-apt-repository 的错误，那么可能要先安装 python-software-properties。对于其他基于 Debian/Ubuntu 的发行版本，可以尝试使用最可能在旧版套件上工作的 PPA 变体。

```
sudo -s
nginx=stable # use nginx=development for latest development version
echo "deb http://ppa.launchpad.net/nginx/$nginx/ubuntu lucid main" > /etc/apt/sources.list.d/nginx-$nginx-lucid.list
apt-key adv --keyserver keyserver.ubuntu.com --recv-keys C300EE8C
apt-get update
apt-get install nginx
```

6. Win32

在 Windows 环境上安装 NGINX，其命令如下。

```
cd c:\
unzip nginx-1.15.8.zip
ren nginx-1.15.8 nginx
cd nginx
start nginx
```

如果有以上问题，可以查看日志 c:nginxlogserror.log。

此外，NGINX 官网目前只提供了 32 位的安装包，如果想安装 64 位的版本，可以查看由 Kevin Worthington 维护的 Windows 版本 https://kevinworthington.com/nginx-for-windows/。

14.1.4 验证安装

NGINX 正常启动后会占用 80 端口。在打开的任务管理器中，能够看到相关的 NGINX 活动线程，如图 14-1 所示。

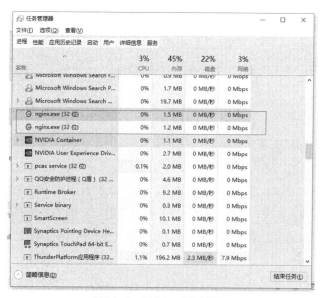

图14-1　NGINX活动线程

打开浏览器，访问 http://localhost:80（其中 80 端口号可以省略），就能看到 NGINX 的欢迎页面，如图 14-2 所示。

图14-2　NGINX的欢迎页面

关闭 NGINX，执行以下命令。

```
nginx -s stop
```

14.1.5 常用命令

NGINX 启动后，有一个主进程（Master Process）和一个或多个工作进程（Worker Process），主进程的作用主要是读入和检查 NGINX 的配置信息，以及维护工作进程；工作进程才是真正处理客户端请求的进程。具体要启动多少个工作进程，可以在 NGINX 的配置文件 nginx.conf 中通过 worker_processes 指令指定。可以通过以下命令来控制 NGINX。

```
nginx -s [ stop | quit | reopen | reload ]
```

其中：

（1）nginx -s stop：强制停止 NGINX，不管工作进程当前是否正在处理用户请求，都会立即退出。

（2）nginx -s quit："优雅地"退出 NGINX，执行这个命令后，工作进程会将当前正在处理的请求处理完毕后退出。

（3）nginx -s reopen：重新打开日志文件。

（4）nginx -s reload：重载配置信息。当 NGINX 的配置文件改变之后，通过执行这个命令，使更改的配置信息生效，而无须重新启动 NGINX。

提示：当重载配置信息时，NGINX 的主进程首先检查配置信息，如果配置信息没有错误，主进程会启动新的工作进程，并发出信息通知旧的工作进程退出，旧的工作进程接收到信号后，会等到处理完当前正在处理的请求后退出。如果 NGINX 检查配置信息发现错误，就会回滚所做的更改，沿用旧的工作进程继续工作。

14.2 部署Angular应用

正如前面所介绍的那样，NGINX 也是高性能的 HTTP 服务器，因此可以部署 Angular 应用。本节将详细介绍部署 Angular 应用的完整流程。

14.2.1 编译Angular应用

执行下面的命令，将 Angular 应用进行编译。

```
ng build
```

编译后的文件默认放在 dist 文件夹下，如图 14-3 所示。

第 14 章 使用 NGINX 实现高可用

图14-3 dist文件夹

14.2.2 部署Angular编译文件

将 Angular 编译文件复制到 NGINX 安装目录的 html 下，如图 14-4 所示。

图14-4 html目录

14.2.3 配置NGINX

打开 NGINX 安装目录下的 conf/nginx.conf 文件，其配置如下。

```
worker_processes  1;

events {
    worker_connections  1024;
}
```

```
http {
    include       mime.types;
    default_type  application/octet-stream;

    sendfile        on;

    keepalive_timeout  65;

    server {
        listen       80;
        server_name  localhost;

        location / {
            root   html;
            index  index.html index.htm;

            #处理Angular路由
            try_files $uri $uri/ /index.html;
        }

        #反向代理
        location /api/ {
            proxy_pass   http://localhost:8080/;
        }

        error_page   500 502 503 504  /50x.html;
        location = /50x.html {
            root   html;
        }
    }
}
```

其修改点主要在于以下两个方面。

（1）新增了"try_files"配置，用于处理 Anuglar 的路由器。

（2）新增了"location"节点，用于执行反向代理，将 Anuglar 应用中的 HTTP 请求转发到后台服务接口上。

14.3 实现负载均衡及高可用

在大型互联网应用中，应用实例通常会部署多个，其好处在于以下两个方面。

（1）实现了负载均衡。让多个实例去分担用户请求的负荷。

（2）实现高可用。当多个实例中的任意一个实例出现故障，剩下的实例仍然能够响应用户的访问请求。因此，从整体上来看，部分实例的故障并不影响整体使用，因此实现了高可用。

本节将演示如何基于 NGINX 来实现负载均衡及高可用。

14.3.1 配置负载均衡

在 NGINX 中，负载均衡配置如下。

```
upstream liteserver {
    server 127.0.0.1:8080;
    server 127.0.0.1:8081;
    server 127.0.0.1:8082;
}

server {
    listen       80;
    server_name  localhost;

    location / {
        root   html;
        index  index.html index.htm;

        #处理Angular路由
        try_files $uri $uri/ /index.html;
    }

    #反向代理
    location /api/ {
        proxy_pass   http://liteserver/;
    }

    error_page   500 502 503 504   /50x.html;
    location = /50x.html {
        root   html;
    }
}
```

其中，proxy_pass 设置了代理服务器，而这个代理服务器是设置在 upstream 中的。upstream 中的每个 server 代表后台服务的一个实例。这里设置了 3 个后台服务实例。

针对 Angular 路由，还需要设置 try_files。

14.3.2　负载均衡常用算法

在 NGINX 中，负载均衡常用算法主要包括以下 5 种。

1. 轮询（默认）

每个请求按时间顺序逐一分配到不同的后端服务器，如果某个后端服务器不可用，就能自动剔除。

以下是轮询的配置。

```
upstream liteserver {
    server 127.0.0.1:8080;
    server 127.0.0.1:8081;
    server 127.0.0.1:8082;
}
```

2. 权重

可以通过 weight 来指定轮询权重，用于后端服务器性能不均的情况。权重值越大，则被分配请求的概率越高。

以下是权重的配置。

```
upstream liteserver {
    server 127.0.0.1:8080 weight=1;
    server 127.0.0.1:8081 weight=2;
    server 127.0.0.1:8082 weight=3;
}
```

3. ip_hash

每个请求按访问 IP 的 hash 值来分配，这样每个访客固定访问一个后端服务器，就可以解决 session 的问题。

以下是 ip_hash 的配置。

```
upstream liteserver {
    ip_hash;
    server 192.168.0.1:8080;
    server 192.168.0.2:8081;
    server 192.168.0.3:8082;
}
```

4. fair

按后端服务器的响应时间来分配请求，响应时间短的优先分配。

以下是 fair 的配置。

```
upstream liteserver {
    fair;
    server 192.168.0.1:8080;
    server 192.168.0.2:8081;
    server 192.168.0.3:8082;
```

}

5. url_hash

按访问 URL 的 hash 结果来分配请求，使每个 URL 定向到同一个后端服务器。因为后端服务器如果被一个客户端访问过，那么就会缓存该客户端的访问记录。当这台服务器再次被同一个客户端访问时，服务器就会从缓存中获取数据返回给客户端，这样整个访问会比较高效。例如，在 upstream 中加入 hash 语句，server 语句中不能写入 weight 等参数，hash_method 使用的是 hash 算法。

以下是 url_hash 的配置。

```
upstream liteserver {
    hash $request_uri;
    hash_method crc32;
    server 192.168.0.1:8080;
    server 192.168.0.2:8081;
    server 192.168.0.3:8082;
}
```

14.3.3 实现后台服务的高可用

执行后台服务器的编译及打包，其命令如下。

```
mvn clean package
```

打包完成之后，就可以在 target 目录下看到 lite-news-server-1.0.0.jar 文件。该文件是一个可执行的 jar 文件。

执行下面的命令，以启动 3 个不同的服务实例。

```
java -jar target/lite-news-server-1.0.0.jar --port=8080

java -jar target/lite-news-server-1.0.0.jar --port=8081

java -jar target/lite-news-server-1.0.0.jar --port=8082
```

这 3 个实例会占用不同的端口，它们独立运行在各自的进程中。

提示：在实际项目中，服务实例往往会部署在不同的主机当中，本示例仅为了简单演示，所以部署在了同一个主机上，但部署方式本质上是类似的。

14.3.4 运行

后台服务启动后，在浏览器的 http://localhost/ 地址访问前台应用，同时观察后台控制台输出的内容，如图 14-5 所示。

图14-5 运行后台服务

可以看到，3个后台服务会轮流接收前台的请求。为了模拟故障，也可以将任意一个后台服务停掉，可以发现前台仍然能够正常响应，这就实现了应用的高可用。

第15章
使用Redis实现高并发

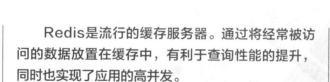

Redis是流行的缓存服务器。通过将经常被访问的数据放置在缓存中,有利于查询性能的提升,同时也实现了应用的高并发。

本章将详细介绍如何基于Redis缓存来实现应用的高并发。

15.1 为什么需要缓存

有时，为了提升整个网站的性能，会将经常访问的数据缓存起来，这样在下次查询时，就能快速找到这些数据。

缓存的使用与系统的时效性有着非常大的关系。当对系统时效性要求不高时，适合使用缓存；当对系统的时效性要求比较高时，则不适合使用缓存。

以新闻应用为例。由于新闻一旦发布，内容不会有大的变化，而用户的访问量又非常高，因此非常适合使用缓存来存储新闻的内容。所以，采用缓存，一方面可以有效降低访问新闻接口服务带来的延时，另一方面也可以减轻新闻接口的负担，提高并发访问量。

接下来演示如何通过集成 Redis 服务器来进行数据的缓存，以提高微服务的并发访问能力。

15.2 了解Redis服务器

Redis 是一个 key-value 模型的内存数据存储系统，与 Memcached 类似，但它支持存储的 value 类型相对更多，包括 string（字符串）、hash（哈希类型）、list（链表）、set（集合）和 sorted set（有序集合）的 range query（范围查询），以及位图、hyperloglogs、空间索引等的 radius query（半径查询）；内置复制、Lua 脚本、LRU 回收、事务及不同级别磁盘持久化功能，同时通过 Redis Sentinel 实现高可用，通过 Redis Cluster 实现自动分区。Redis 是完全开源免费的，并遵守 BSD 的协议。

有关 Redis 的更多介绍，可以参考官方网站 http://redis.io。笔者所著的《分布式系统常用技术及案例分析》对 Redis 也有介绍，读者可以作为参考。

15.2.1 Redis简介

Redis 是一个高性能的 key-value 数据库。Redis 的出现，在很大程度上弥补了 Memcached 这类 key-value 存储的不足，在部分场合可以对关系数据库起到很好的补充作用。它提供了 ActionScript、Bash、C、C#、C++、Clojure、Common Lisp、Crystal、D、Dart、Delphi、Elixir、emacs lisp、Erlang、Fancy、gawk、GNU Prolog、Go、Haskell、Haxe、Io、Java、Julia、Lasso、 Lua、Matlab、mruby、Nim、Node.js、Objective-C、OCaml、Pascal、Perl、PHP、Pure Data、Python、R、Racket、Rebol、Ruby、Rust、Scala、Scheme、Smalltalk、Swift、Tcl、VB、VCL 等众多客户端，使用很方便。有关各种客户端实现库的支持情况可以参考 http://redis.io/clients。

Redis 支持主从同步。可以从主服务器向任意数量的从服务器上同步数据，从服务器可以是关联其他从服务器的主服务器。这使得 Redis 可执行单层树复制。存盘可以有意无意地对数据进行写操作。由于完全实现了发布/订阅机制，使得从数据库在任何地方进行数据同步时，都可订阅一个频道并接收主服务器完整的消息发布记录。而且同步对读取操作的可扩展性和数据冗余很有帮助。

用户可以在 Redis 数据类型上执行原子操作，如追加字符串、增加哈希表中的某个值、在列表中增加一个元素，以及计算集合的交集、并集或差集，获取一个有序集合中最大排名的元素，等等。

为了获取其卓越的性能，Redis 在内存数据集上工作。但是否工作在内存中取决于用户，如果用户想持久化内存数据集的数据，则可以通过偶尔转储内存数据集到磁盘上，或者在一个日志文件中写入每条操作命令来实现。如果用户仅需要一个内存数据库，则持久化操作可以被选择性禁用。

Redis 是用 ANSI C 编写的，在大多数 POSIX 系统中工作，如 Linux、*BSD、OS X 等，而无须添加其他额外的依赖。Linux 和 OS X 系统是 Redis 开发和测试最常用的两个操作系统，所以建议使用 Linux 来部署 Redis。Redis 可以工作在类似于 SmartOS 的 Solaris 派生系统上，但支持是有限的。官方没有对 Windows 的支持，但 Microsoft 开发和维护了 Redis 的 Windows 64 位的接口，可以参见 https://github.com/MSOpenTech/redis。

Redis 具有以下几个特点。

（1）支持事务。
（2）支持发布/订阅模式。
（3）支持使用 Lua 脚本。
（4）key 有生命时间限制。
（5）按照 LRU 机制来清除旧数据。
（6）自动故障转移。

15.2.2　Redis的下载、安装、使用

下面分别介绍如何在 Linux 和 Windows 平台上安装 Redis。

1. Linux平台

在 Linux 平台上安装 Redis 比较简单，下载、解压、编译 Redis 后执行以下命令。

```
$ wget http://download.redis.io/releases/redis-3.2.3.tar.gz
$ tar xzf redis-3.2.3.tar.gz
$ cd redis-3.2.3
$ make
```

之后，就能在 src 目录下看到这个编译文件了。运行 Redis 并执行以下命令。

```
$ src/redis-server
```

使用内置的命令行工具和 Redis 交互。

```
$ src/redis-cli
```

```
redis> set foo bar
OK
redis> get foo
"bar"
```

2. Windows平台

在 Windows 平台，微软特别为 Redis 制作了安装包，下载地址为 https://github.com/Microsoft Archive/redis/releases。本书所使用的案例也是基于该安装包的。双击 redis-server.exe 文件，即可快速启动 Redis 服务器。

安装后，Redis 默认运行在 localhost:6379 地址端口，如图 15-1 所示。

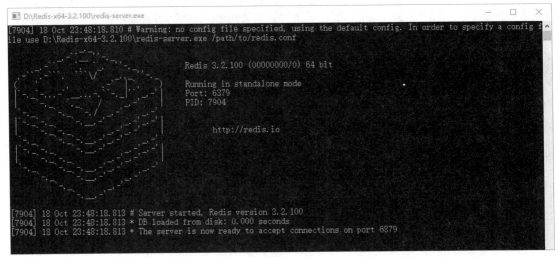

图15-1　运行Redis

15.2.3　Redis的数据类型及抽象

Redis 不仅仅是简单的 key-value 存储，实际上它还是一个 data strutures server（数据结构服务器），用于支持不同的数值类型。在 key-value 中，value 不仅仅局限于 string 类型，它可以是更复杂的数据结构。

（1）二进制安全的 string。

（2）List：一个链表，链表中的元素按照插入顺序排列。

（3）Set：string 集合，集合中的元素是唯一的，没有排序。

（4）Sorted set：与 Set 类似，但是每一个 string 元素关联一个浮点数值，这个数值被称为 Score。元素总是通过 Score 进行排序，与 Set 不同的是，它可以获取一个范围的元素（如获取前 10 个或后 10 个）。

（5）Hash：由关联值字段构成的 Map。字段和值都是 string，这与 Ruby 或 Python 的 Hash 类似。

（6）Bit array（简称为 Bitmap）：像位数值一样通过特别的命令处理字符串，可以设置和清除单独的 bit、统计所有 bit 集合中为 1 的数量、查找第一个设置或没有设置的 bit，等等。

（7）HyperLogLogs：这是一个概率统计用的数据结构，可以被用来估计一个集合的基数。

本章所有的例子都使用 redis-cli 工具来演示。这是一个简单但是非常有用的命令行工具，可以用来给 Redis Server 发送命令。

1. Redis key

Redis key 是二进制安全的。这意味着可以使用任何二进制序列作为 key，如从一个像 "foo" 的字符串到一个 JPEG 文件的内容。空字符串也是一个有效的 key。

关于 key 的其他使用规则如下。

（1）不建议使用非常长的 key。这不仅仅是考虑到内存方面问题，而且在数据集中查找 key 可能需要和多个 key 进行比较。如果当前的任务需要使用一个很大的值，将其进行 Hash 是不错的选择（如使用 SHA1），尤其是从内存和带宽的角度考虑。

（2）非常短的 key 往往也不是一个好主意。如果可以将 key 写成 "user:1000:followers"，就不要使用 "u1000flw"。首先前者更具有可读性，其次增加的空间相比 key 对象本身和值对象占用的空间是很小的。当然，短 key 显然会消耗更少的内存，因此需要找到一个适当的平衡点。

（3）提倡使用模式。类似于 "user:1000" 这样的 "object-type:id" 模式就是一个好主意。点和连接线通常被用在有多个单词的字段中，如 "comment:1234:reply.to" 或 "comment:1234:reply-to"。

（4）允许 key 的最大值为 512 MB。

2. Redis String

Redis String 类型是关联到 Redis key 最简单的值类型，也是 Memcached 中唯一的数据类型，所以对于 Redis 新手来说，使用它是非常自然的。

因为 Redis key 是 String，所以当使用 String 类型作为 value 时，其实就是将一个 String 映射到另一个 String。String 数据类型对于大量的用例是非常有用的，如缓存 HTML 片段或页面。

下面使用 redis-cli 来操作 String 类型，其命令如下。

```
> set mykey somevalue
OK
> get mykey
"somevalue"
```

使用 SET 和 GET 命令可以设置和获取 String 值。注意，SET 会替换已经存入 key 中的任何值，即使这个 key 存入的不是 String 值。所以 SET 执行一次分配，值可以是任何类型的 String（包括二进制数据），如可以存一个 JPEG 图片到一个 key 中，但值不能超过 512 MB。SET 命令有一些有趣的选项，可以通过以下额外的参数来设置。

（1）NX：只在 key 不存在的情况下执行。

（2）XX：只在 key 存在的情况下执行。

下面是操作示例。

```
> set mykey somevalue
OK
> get mykey
"somevalue"
```

String 是 Redis 的基础值，可以对其进行一些有意思的操作，如进行原子递增。

```
> set counter 100
OK
> incr counter
(integer) 101
> incr counter
(integer) 102
> incrby counter 50
(integer) 152
```

INCR 命令将 String 值解析为 Integer，然后将它递增 1，最后将新值作为返回值。这里也有一些类似的命令，如 INCRBY、DECR 和 DCRBY。在内部它们是相同的命令，并且执行方式的差别非常小。

INCR 命令是原子操作，意味着即使多个客户端对同一个 key 发送 INCR 命令，也不会导致 Race Condition（竞争条件）问题。例如，当 client1 和 client2 同时给值加 1 时（旧值为 10），它们不会同时读到 10，最终值一定是 12，因为 read-increment-set 起作用了。

操作 String 有很多命令，如 GETSET 命令将一个 key 设置为新值，并将旧值作为返回值。例如，网站接收到新的访问者，则使用 INCR 命令递增一个 Redis key，这时就可以使用 GETSET 命令来实现。如果想每隔一个小时收集一次信息，并且不丢失每一次的递增，就可以 GETSET 这个 key，将新值 0 赋给它，并将旧值读回。

MSET 和 MGET 命令用于在一条命令中设置或获取多个 key 的值，这对减少网络延时是非常有用的。

下面是操作示例。

```
> mset a 10 b 20 c 30
OK
> mget a b c
1) "10"
2) "20"
3) "30"
```

当使用 MGET 时，Redis 返回的是一个包含值的数组。

3. 修改和查询key空间

还有一些命令没有定义在具体的类型上，但在与 key 空间交互时非常有用。这些命令可以用于任何类型的 key。

例如，EXISTS 命令返回 1 或 0，以标志一个给定的 Key 是否在数据库中存在。又如，DEL 命令用来删除一个 key 和与其关联的值，而不管这个值是什么。下面是操作示例。

```
> set mykey hello
OK
> exists mykey
(integer) 1
> del mykey
(integer) 1
> exists mykey
(integer) 0
```

通过以上示例，也可以看到 DEL 命令根据 key 是否被删除而返回 1 或 0。

这里有很多与 key 空间相关的命令，但是上面两个命令及 TYPE 命令是非常关键的。TYPE 命令返回指定 key 中存放的值的类型。

下面是操作示例。

```
> set mykey x
OK
> type mykey
string
> del mykey
(integer) 1
> type mykey
none
```

4. Redis失效：具有有限生存时间的key

失效是 Redis 的特性之一，且可以用在任何一种值类型中。可以给一个 key 设置一个超时时间，这个超时时间就是有限的生存时间。当这个生存时间过去，这个 key 会自动被销毁。

下面是一些关于 Redis 失效的描述。

（1）在设置失效时间时，可以使用秒（s）或毫秒（ms）精度。

（2）失效时间一般为 1 ms。

（3）失效信息会被复制，并持久化到磁盘中。当 Redis 服务器停止时（Redis 将保存 key 的失效时间），这个时间仍然会在指定的时间内失效。

设置失效时间的命令如下。

```
> set key some-value
OK
> expire key 5
(integer) 1
> get key (immediately)
"some-value"
> get key (after some time)
(nil)
```

这个 key 在两次 GET 调用之间消失了，因为第二次调用延时超过了 5s。在上面的示例中，使用 EXPIRE 命令来设置超时时间（它也可以用来给一个已经设置超时时间的 key 设置一个不同的值。PERSIST 可以用来删除失效时间，并将 key 永远持久化）。当然也可以使用其他 Redis 命令来创建带失效时间的 key。例如，使用 SET 选项。

```
> set key 100 ex 10
OK
> ttl key
(integer) 9
```

上面的示例中设置的 key 值为 String 100，并带有 10s 的超时时间。之后，使用 TTL 命令检测这个 key 的剩余生存时间。

如果想知道如何以毫秒级设置和检测超时时间，查看 PEXPIRE 命令、PTTL 命令及 SET 选项列表，也可以参见 http://redis.io/commands。

15.3 使用Redis

在搭建完成 Redis 服务器之后，如何在应用中使用 Redis 呢？

Spring Data Redis 项目提供了对 Redis 使用的支持。Spring Data Redis 封装了 RedisTemplate 对象来对 Redis 进行各种操作，它支持所有 Redis 原生的 API。使用 Redis 缓存主要分为编程式和声明式这两种方式。

15.3.1 编程式

以下就是编程式使用 RestTemplate 来操作 Redis 的示例。

```
@Service
public class WeatherDataServiceImpl implements WeatherDataService {

    private final static Logger logger = LoggerFactory.getLogger(WeatherDataServiceImpl.class);

    @Autowired
    private RestTemplate restTemplate;

    @Autowired
    private StringRedisTemplate stringRedisTemplate;

    private final String WEATHER_API = "http://wthrcdn.etouch.cn/weather_mini";
```

```java
    private final Long TIME_OUT = 1800L;  // 缓存超时时间

    @Override
    public WeatherResponse getDataByCityId(String cityId) {
        String uri = WEATHER_API + "?citykey=" + cityId;
        return this.doGetWeatherData(uri);
    }

    @Override
    public WeatherResponse getDataByCityName(String cityName) {
        String uri = WEATHER_API + "?city=" + cityName;
        return this.doGetWeatherData(uri);
    }

    private WeatherResponse doGetWeatherData(String uri) {
        ValueOperations<String, String> ops = this.stringRedisTemplate.opsForValue();
        String key = uri; // 将调用的 URI 作为缓存的key
        String strBody = null;

        // 先查缓存  没有再查服务
        if (!this.stringRedisTemplate.hasKey(key)) {
            logger.info("未找到 key " + key);

            ResponseEntity<String> response = restTemplate.getForEntity(uri, String.class);

            if (response.getStatusCodeValue() == 200) {
                strBody = response.getBody();
            }

            ops.set(key, strBody, TIME_OUT, TimeUnit.SECONDS);
        } else {
            logger.info("找到 key " + key + ", value=" + ops.get(key));

            strBody = ops.get(key);
        }

        ObjectMapper mapper = new ObjectMapper();
        WeatherResponse weather = null;

        try {
            weather = mapper.readValue(strBody, WeatherResponse.class);
        } catch (IOException e) {
            logger.error("JSON反序列化异常 ",e);
        }
```

```
        return weather;
    }
}
```

使用编程式来操作 Redis，好处在于代码逻辑比较清晰，并且高度可控；但缺点也比较明显，就是需要非常多的样板代码。

15.3.2 声明式

Spring 提供了如下声明式的缓存注解。

（1）@Cacheable：触发缓存。

（2）@CacheEvict：触发缓存回收。

（3）@CachePut：更新缓存。

（4）@Caching：重新组合要应用于方法的多个缓存操作。

（5）@CacheConfig：在类级别共享一些常见的缓存相关设置。

1. @Cacheable

@Cacheable 声明可缓存的方法。比如：

```
@Cacheable("books")
public Book findBook(ISBN isbn) {…}
```

在上面的代码片段中，findBook 方法与名为 books 的缓存相关联。每次调用该方法时，都会检查缓存以查看调用是否已经执行。虽然在大多数情况下只声明一个缓存，但注解允许指定多个名称，以便使用多个缓存。在这种情况下，将在执行该方法之前检查缓存，如果至少有一个缓存被命中，则返回相关的值。

```
@Cacheable({"books", "isbns"})
public Book findBook(ISBN isbn) {…}
```

2. 默认 key 生成

由于缓存本质上是 key-value 存储，因此每次调用缓存方法时都需要将其转换为适合缓存访问的 key。Spring 缓存抽象使用基于以下算法的简单 KeyGenerator（key 生成器）。

（1）如果没有给出参数，则返回 SimpleKey.EMPTY。

（2）如果只给出一个参数，则返回该实例。

（3）如果给出了一个参数，则返回一个包含所有参数的 SimpleKey。

上述方法适用于大多数情况，只需参数具有 key 并实现有效的 hashCode() 和 equals() 方法。

如果想自定义 key 生成器，可以自行实现 org.springframework.cache.interceptor.KeyGenerator

接口。

3. 自定义key生成声明

由于缓存是通用的，因此目标方法很可能具有不能简单映射到缓存结构顶部的各种签名。特别是当目标方法有多个参数，且其中只有一些适用于缓存（而其余的仅由方法逻辑使用）时，例如：

```
@Cacheable("books")
public Book findBook(ISBN isbn, boolean checkWarehouse, boolean include-Used)
```

对于这种情况，@Cacheable 注解允许用户指定如何通过其 key 属性来生成 key。开发人员可以使用 SpEL 表达式来选择参数、执行操作甚至调用任意方法，而无须编写任何代码或实现任何接口。

下面是各种 SpEL 声明的一些示例。

```
@Cacheable(cacheNames="books", key="#isbn")
public Book findBook(ISBN isbn, boolean checkWarehouse, boolean include-Used)

@Cacheable(cacheNames="books", key="#isbn.rawNumber")
public Book findBook(ISBN isbn, boolean checkWarehouse, boolean include-Used)

@Cacheable(cacheNames="books", key="T(someType).hash(#isbn)")
public Book findBook(ISBN isbn, boolean checkWarehouse, boolean include-Used)
```

可以在操作中定义一个自定义的 KeyGenerator。观察如下示例，"myKeyGenerator" 是自定义 KeyGenerator 的 bean 名称。

```
@Cacheable(cacheNames="books", keyGenerator="myKeyGenerator")
public Book findBook(ISBN isbn, boolean checkWarehouse, boolean include-Used)
```

4. @CachePut

对于需要更新缓存而不干扰方法执行的情况，可以使用 @CachePut 注解。也就是说，该方法将始终执行并将其结果放入缓存中（根据 @CachePut 选项）。它支持与 @Cacheable 相同的选项。

```
@CachePut(cacheNames="book", key="#isbn")
public Book updateBook(ISBN isbn, BookDescriptor descriptor)
```

5. @CacheEvict

从缓存中删除过时或未使用的数据是很有必要的。@CacheEvict 注解定义了删除缓存数据的方法，下面是一个示例。

```
@CacheEvict(cacheNames="books", allEntries=true)
```

```
public void loadBooks(InputStream batch)
```

当需要清除整个缓存区域时，这个 allEntries 选项将会非常方便，相比较逐条清除每个条目而言，这个选项拥有更高的效率。

6. @Caching

在某些情况下，需要指定相同类型的多个注解（如 @CacheEvict 或 @CachePut），此时可以使用 @Caching。@Caching 允许在同一个方法上使用多个嵌套的 @Cacheable、@CachePut 和 @CacheEvict。

```
@Caching(evict = { @CacheEvict("primary"), @CacheEvict(cacheNames="secondary", key="#p0") })
public Book importBooks(String deposit, Date date)
```

7. @CacheConfig

如果某些自定义选项适用于该类的所有操作，就需要使用 @CacheConfig 注解，观察如下示例。

```
@CacheConfig("books")
public class BookRepositoryImpl implements BookRepository {

    @Cacheable
    public Book findBook(ISBN isbn) {…}
}
```

@CacheConfig 是一个类级别注解，允许共享缓存名称，如自定义 KeyGenerator、自定义 CacheManager 及最终的自定义 CacheResolver。将此注解放在类上不会启用任何缓存操作。

方法级别的自定义配置将会覆盖 @CacheConfig 上的配置。因此，可以在以下 3 个层次上自定义缓存配置。

（1）全局配置，可用于 CacheManager、KeyGenerator。

（2）在类上使用 @CacheConfig 注解。

（3）在方法上。

8. 启用缓存

需要注意的是，声明缓存注解并不等同于启用了缓存。就像 Spring 中的许多配置一样，该功能必须声明为启用。

要启用缓存注解，需将 @EnableCaching 注解添加到其中的一个 @Configuration 类中。

```
@Configuration
@EnableCaching
public class AppConfig {
}
```

如果是基于 XML 的配置，则可以使用 cache:annotation-driven 元素。

```
<beans xmlns="http://www.springframework.org/schema/beans"
```

```
    xmlns:xsi="http://www.w3.org/2001/XMLSchema-instance"
    xmlns:cache="http://www.springframework.org/schema/cache"
    xsi:schemaLocation="
        http://www.springframework.org/schema/beans
        http://www.springframework.org/schema/beans/spring-beans.xsd
        http://www.springframework.org/schema/cache
        http://www.springframework.org/schema/cache/spring-cache.xsd">

        <cache:annotation-driven />

</beans>
```

9. 使用自定义缓存

可以使用自定义的缓存注解，观察以下示例。

```
@Retention(RetentionPolicy.RUNTIME)
@Target({ElementType.METHOD})
@Cacheable(cacheNames="books", key="#isbn")
public @interface SlowService {
}
```

在上面的示例中，定义了 SlowService 注解，它本身是基于 @Cacheable 注解的。现在可以将下面的代码进行替换。

```
@Cacheable(cacheNames="books", key="#isbn")
public Book findBook(ISBN isbn, boolean checkWarehouse, boolean include-
Used)
```

替换为自定义的注解。

```
@SlowService
public Book findBook(ISBN isbn, boolean checkWarehouse, boolean include-
Used)
```

15.4　lite-news实现缓存

接下来将演示如何使用 Redis 来缓存 lite-news 服务的数据。

15.4.1　集成Redis客户端

在 lite-news-server 的 pom.xml 文件中，添加 Spring Data Redis 和 Jedis 的依赖。

```
<dependency>
    <groupId>org.springframework.data</groupId>
```

```xml
    <artifactId>spring-data-redis</artifactId>
    <version>${spring-data-redis.version}</version>
</dependency>
<dependency>
    <groupId>redis.clients</groupId>
    <artifactId>jedis</artifactId>
    <version>${jedis.version}</version>
</dependency>
```

15.4.2　增加Spring Data Redis配置

增加配置类 CacheConfig，用于缓存管理的配置，其代码如下。

```java
package com.waylau.lite.news.cache;

import org.springframework.cache.annotation.EnableCaching;
import org.springframework.context.annotation.Bean;
import org.springframework.context.annotation.Configuration;
import org.springframework.data.redis.cache.RedisCacheManager;
import org.springframework.data.redis.connection.jedis.JedisConnectionFactory;

@Configuration
@EnableCaching
public class CacheConfig {

    @Bean
    public RedisCacheManager cacheManager() {
        return RedisCacheManager.create(redisConnectionFactory());
    }

    @Bean
    public JedisConnectionFactory redisConnectionFactory() {
        return new JedisConnectionFactory();
    }

}
```

并将该配置导入 AppConfig 中。

```java
@ComponentScan(basePackages = { "com.waylau.lite.news" })
@Import({WebSecurityConfig.class, CacheConfig.class})
public class AppConfig extends LiteConfig {

}
```

15.4.3　使用Spring缓存注解

在需要进行缓存操作的地方添加相应的注解，其代码如下。

```java
package com.waylau.lite.news.service.impl;

import java.util.List;

import org.springframework.beans.factory.annotation.Autowired;
import org.springframework.cache.annotation.CacheEvict;
import org.springframework.cache.annotation.CachePut;
import org.springframework.cache.annotation.Cacheable;
import org.springframework.stereotype.Service;

import com.waylau.lite.news.domain.News;
import com.waylau.lite.news.mapper.NewsMapper;
import com.waylau.lite.news.service.NewsService;

@Service
public class NewsServiceImpl implements NewsService {

    @Autowired
    private NewsMapper newsMapper;

    @Override
    @Cacheable(cacheNames="news", key="#newsId")
    public News getNews(Long newsId) {
        return newsMapper.getNews(newsId);
    }

    @Override
    public List<News> getNewsList() {
        return newsMapper.getNewsList();
    }

    @Override
    @CachePut(cacheNames="news", key="#news.newsId")
    public void updateNews(News news) {
        newsMapper.updateNews(news);
    }

    @Override
    public void createNews(News news) {
        newsMapper.createNews(news);
    }

    @Override
    @CacheEvict(cacheNames="news", key="#news.newsId")
    public void deleteNews(Long newsId) {
```

```
        newsMapper.deleteNews(newsId);
    }
}
```

15.4.4 序列化

首先，确认 Redis 服务器已经启动。

其次，运行 lite-news-server。

最后，启动 NGINX 服务。这样就能通过客户端应用来访问后台的接口。

初次访问，可能会遇到下面的报错。

```
org.springframework.web.util.NestedServletException: Request processing failed; nested exception is org.springframework.data.redis.serializer.SerializationException: Cannot serialize; nested exception is org.springframework.core.serializer.support.SerializationFailedException: Failed to serialize object using DefaultSerializer; nested exception is java.lang.IllegalArgumentException: DefaultSerializer requires a Serializable payload but received an object of type [com.waylau.lite.news.domain.News]
    at org.springframework.web.servlet.FrameworkServlet.processRequest(FrameworkServlet.java:1013)
    at org.springframework.web.servlet.FrameworkServlet.doGet(FrameworkServlet.java:897)
    at javax.servlet.http.HttpServlet.service(HttpServlet.java:687)
    at org.springframework.web.servlet.FrameworkServlet.service(FrameworkServlet.java:882)
    at javax.servlet.http.HttpServlet.service(HttpServlet.java:790)
    at org.eclipse.jetty.servlet.ServletHolder.handle(ServletHolder.java:867)
    at org.eclipse.jetty.servlet.ServletHandler$CachedChain.doFilter(ServletHandler.java:1623)
    …
```

这是因为 Redis 客户端在进行数据的序列化时失败了。解决方案也比较简单，使用 News 类实现序列化接口即可，其代码如下。

```
package com.waylau.lite.news.domain;

import java.io.Serializable;
import java.util.Date;

public class News implements Serializable {

    private static final long serialVersionUID = 1L;
```

```
private Long newsId;
private String title;
private String author;
private Date creation;
private String content;

// …省略其他getter/setter方法
}
```

15.4.5 运行

启用所有服务和应用后,可以对 NewsServiceImpl.getNews 方法进行断点测试,如图 15-2 所示。

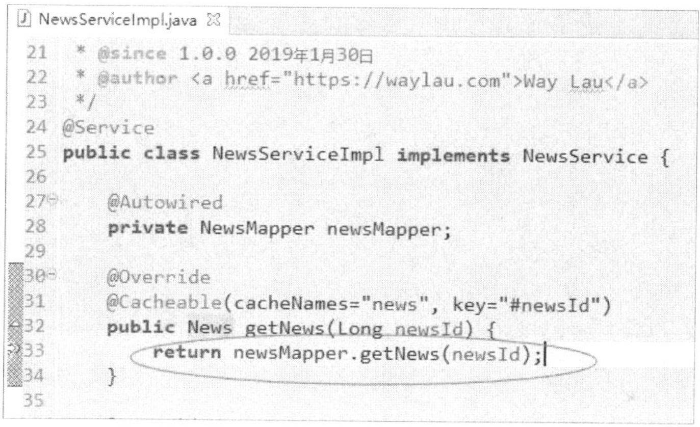

图15-2 断点测试

从断点中可以发现,当首次使用 NewsServiceImpl.getNews 方法时,会触发 newsMapper.getNews 方法。而再次访问相同的 newsId 时(如 newsId 为 17 的数据),newsMapper.getNews 方法将不再执行,这说明数据是从缓存中获取的。

也可以通过 Redis 客户端查到该 key 的数据。

```
127.0.0.1:6379> get news::17
"\xac\xed\x00\x05sr\x00 com.waylau.lite.news.domain.News\x00\x00\x00\
x00\x00\x00\x00\x01\x02\x00\x05L\x00\x06authort\x00\x12Ljava/lang/
String;L\x00\acontentq\x00~\x00\x01L\x00\bcreationt\x00\x10Ljava/util/
Date;L\x00\x06newsIdt\x00\x10Ljava/lang/Long;L\x00\x05titleq\x00~\x00\
x01xpt\x00\x06waylaut\x01\x11\xe6\x98\xa5\xe8\x8a\x82\xe5\x90\x8e\xe6\
x9c\x80\xe4\xbb\xa4\xe4\xba\xba\xe6\x9c\x9f\xe5\xbe\x85\xe7\x9a\x84\
xe6\x97\x97\xe8\x88\xb0\xe6\x9c\xba\xe8\x8e\xab\xe8\xbf\x87\xe4\xba\
x8e\xe4\xb8\x89\xe6\x98\x9fS10\xe7\xb3\xbb\xe5\x88\x97\xef\xbc\x8c\xe8\
x80\x8c\xe5\xae\x98\xe6\x96\xb9\xe6\xb6\x88\xe6\x81\xaf\xe6\x8c\x87\
xe6\x98\x8e\xef\xbc\x8c**S10\xe7\xb3\xbb\xe5\x88\x97**\xe5\xb0\x86\xe5\
x9c\xa82\xe6\x9c\x8821\xe5\x8f\xb7\xe6\xad\xa3\xe5\xbc\x8f\xe5\x8f\x91\
xe5\xb8\x83\xe3\x80\x82\n\n\n![](https://ss1.baidu.com/6ONXsjip0QIZ-
```

```
8tyhnq/it/u=4111162492,3551590482&fm=173&app=49&f=-
JPEG?w=513&h=266&s=3C2086187426C70B0720B7D90300C0A0)sr\x00\x0ejava.
util.Datehj\x81\x01KYt\x19\x03\x00\x00xpw\b\x00\x00\x01h\xac}\x01 xsr\
x00\x0ejava.lang.Long;\x8b\xe4\x90\xcc\x8f#\xdf\x02\x00\x01J\x00\
x05valuexr\x00\x10java.lang.Number\x86\xac\x95\x1d\x0b\x94\xe0\x8b\x02\
x00\x00xp\x00\x00\x00\x00\x00\x00\x00\x11t\x00C\xe4\xb8\x89\xe6\x98\
x9fS10\xe7\xb3\xbb\xe5\x88\x97\xe9\xa6\x96\xe6\x89\xb9\xe5\xae\x98\xe6\
x96\xb9\xe6\xb8\xb2\xe6\x9f\x93\xe5\x9b\xbe\xe4\xba\xae\xe7\x9b\xb8 \
xe4\xb8\x8d\xe4\xb8\x80\xe6\xa0\xb7\xe7\x9a\x84\xe6\xb8\x90\xe5\x8f\
x98\xe8\xae\xbe\xe8\xae\xa1"
```

第16章 Spring Boot概述

在Java开发领域，Spring Boot算是一颗耀眼的明星了。自Spring Boot诞生以来，秉着简化Java企业级应用的宗旨，受到广大Java开发者的好评。特别是微服务架构的兴起，使Spring Boot被称为构建Spring应用中的微服务的最有力工具之一。Spring Boot中众多开箱即用的Starter，为广大开发者尝试开启一个新服务提供了最快捷的方式。

在之前的章节中，展现了Lite框架中很多技术的底层细节。如果熟悉Spring Boot，就会发现Spring Boot与Lite有着众多相似之处。它们的目的都是简化配置，加快开发。本章将介绍Spring Boot的核心功能，由于篇幅有限，不会对Spring Boot进行深入讲解。相关内容，读者可以查阅笔者所著的《Spring Boot 企业级应用开发实战》。

16.1 构建RESTful服务

在前面的章节中,已经对 RESTful 的架构风格及其实现做了介绍。本节将演示如何使用 Spring Boot 快速构建 RESTful 服务。

本节示例源码可以在 spring-boot-rest 目录下找到。

16.1.1 配置环境

演示本例需要依赖 Spring Boot Web Starter。

Spring Boot Web Starter 集成了 Spring MVC,可以方便地构建 RESTful Web 应用,并使用 Tomcat 作为默认的内嵌 Servlet 容器。

spring-boot-rest 的 pom.xml 内容如下。

```xml
<?xml version="1.0" encoding="UTF-8"?>
<project xmlns="http://maven.apache.org/POM/4.0.0" xmlns:xsi="http://www.w3.org/2001/XMLSchema-instance"
    xsi:schemaLocation="http://maven.apache.org/POM/4.0.0 http://maven.apache.org/xsd/maven-4.0.0.xsd">
    <modelVersion>4.0.0</modelVersion>
    <parent>
        <groupId>org.springframework.boot</groupId>
        <artifactId>spring-boot-starter-parent</artifactId>
        <version>2.1.2.RELEASE</version>
        <relativePath/> <!-- lookup parent from repository -->
    </parent>
    <groupId>com.waylau</groupId>
    <artifactId>spring-boot-rest</artifactId>
    <version>1.0.0</version>
    <packaging>jar</packaging>
    <name>spring-boot-rest</name>

    <properties>
        <java.version>1.8</java.version>
    </properties>

    <dependencies>
        <dependency>
            <groupId>org.springframework.boot</groupId>
            <artifactId>spring-boot-starter-web</artifactId>
        </dependency>

        <dependency>
            <groupId>org.springframework.boot</groupId>
            <artifactId>spring-boot-starter-test</artifactId>
            <scope>test</scope>
```

```xml
            </dependency>
        </dependencies>

        <build>
            <plugins>
                <plugin>
                    <groupId>org.springframework.boot</groupId>
                    <artifactId>spring-boot-maven-plugin</artifactId>
                </plugin>
            </plugins>
        </build>
</project>
```

16.1.2　RESTful API设计

本节将实现一个简单版本的"用户管理"RESTful 服务。通过"用户管理"的 API，能够方便地进行用户的增、删、改、查等操作。

用户管理的整体 API 设计如下。

（1）GET /users : 获取用户列表。

（2）POST /users : 保存用户。

（3）GET /users/{id} : 获取用户信息。

（4）PUT /users/{id} : 修改用户。

（5）DELETE /users/{id} : 删除用户。

相应的控制器可以进行如下定义。

```java
@RestController
@RequestMapping("/users")
public class UserController {

    /**
     * 获取用户列表
     *
     * @return
     */
    @GetMapping
    public List<User> getUsers() {
        return null;
    }

    /**
     * 获取用户信息
     *
     * @param id
```

```java
 * @return
 */
@GetMapping("/{id}")
public User getUser(@PathVariable("id") Long id) {
    return null;
}

/**
 * 保存用户
 *
 * @param user
 */
@PostMapping
public User createUser(@RequestBody User user) {
    return null;
}

/**
 * 修改用户
 *
 * @param id
 * @param user
 */
@PutMapping("/{id}")
public void updateUser(@PathVariable("id") Long id, @RequestBody User user) {
}

/**
 * 删除用户
 *
 * @param id
 * @return
 */
@DeleteMapping("/{id}")
public void deleteUser(@PathVariable("id") Long id) {
}
}
```

16.1.3 编写程序代码

下面进行后台编码实现，编码涉及实体类、仓库接口、仓库实现类及控制器类。

1. 实体类

com.waylau.spring.boot.domain 包用于放置实体类。下面定义一个保存用户信息的实体 User。

```
public class User {

    private Long id;

    private String name;

    private String email;

    public User() {
    }

    public User(String name, String email) {
        this.name = name;
        this.email = email;
    }

    // 省略 getter/setter 方法

    @Override
    public String toString() {
        return String.format("User[id=%d, name='%s', email='%s']", id, name, email);
    }
}
```

2. 仓库接口及实现

com.waylau.spring.boot.repository 包用于放置仓库接口及其仓库实现类，也就是数据存储。用户仓库接口 UserRepository 如下。

```
public interface UserRepository {
    /**
     * 新增或修改用户
     *
     * @param user
     * @return
     */
    User saveOrUpateUser(User user);

    /**
     * 删除用户
     *
     * @param id
     */
    void deleteUser(Long id);

    /**
     * 根据用户id获取用户
     *
```

```
     * @param id
     * @return
     */
    User getUserById(Long id);

    /**
     * 获取所有用户的列表
     *
     * @return
     */
    List<User> listUser();
}
```

UserRepository 的实现类如下。

```
@Repository
public class UserRepositoryImpl implements UserRepository {

    private static AtomicLong counter = new AtomicLong();

    private final ConcurrentMap<Long, User> userMap = new ConcurrentHash-
Map<Long, User>();

    @Override
    public User saveOrUpateUser(User user) {
        Long id = user.getId();
        if (id == null || id <= 0) {
            id = counter.incrementAndGet();
            user.setId(id);
        }
        this.userMap.put(id, user);
        return user;
    }

    @Override
    public void deleteUser(Long id) {
        this.userMap.remove(id);
    }

    @Override
    public User getUserById(Long id) {
        return this.userMap.get(id);
    }

    @Override
    public List<User> listUser() {
        return new ArrayList<User>(this.userMap.values());
    }

}
```

其中，"ConcurrentMap<Long, User> userMap" 用来模拟数据的存储，AtomicLong counter 用来生成一个递增的 id，作为用户的唯一编号，@Repository 注解用于标识 UserRepositoryImpl 类是一个可注入的 Bean。

3. 控制器类

com.waylau.spring.boot.controller 包用于放置控制器类，也就是需要实现的 API。
UserController 实现如下。

```java
@RestController
@RequestMapping("/users")
public class UserController {

    @Autowired
    private UserRepository userRepository;

    /**
     * 获取用户列表
     *
     * @return
     */
    @GetMapping
    public List<User> getUsers() {
        return userRepository.listUser();
    }

    /**
     * 获取用户信息
     *
     * @param id
     * @return
     */
    @GetMapping("/{id}")
    public User getUser(@PathVariable("id") Long id) {
        return userRepository.getUserById(id);
    }

    /**
     * 保存用户
     *
     * @param user
     */
    @PostMapping
    public User createUser(@RequestBody User user) {
        return userRepository.saveOrUpateUser(user);
    }

    /**
     * 修改用户
```

```java
     *
     * @param id
     * @param user
     */
    @PutMapping("/{id}")
    public void updateUser(@PathVariable("id") Long id, @RequestBody User user) {
        User oldUser = this.getUser(id);

        if (oldUser != null) {
            user.setId(id);
            userRepository.saveOrUpateUser(user);
        }

    }

    /**
     * 删除用户
     *
     * @param id
     * @return
     */
    @DeleteMapping("/{id}")
    public void deleteUser(@PathVariable("id") Long id) {
        userRepository.deleteUser(id);
    }
}
```

16.1.4　安装REST客户端

为了测试 REST 接口，需要安装一个 REST 客户端。

有非常多的 REST 客户端可供选择，如 Chrome 浏览器的 Postman 插件，或者 Firefox 浏览器的 RESTClient 及 HttpRequester 插件，都能方便地进行 RESTful API 的调试。

这里以 RESTClient 及 HttpRequester 插件的安装为例，进行简单的介绍。

1. Firefox安装REST客户端插件

为了方便测试 REST API，需要一个 REST 客户端来协助。由于笔者使用 Firefox 浏览器居多，所以推荐安装 RESTClient 或 HttpRequester 插件。当然，也可以根据个人喜好来安装其他软件。

在 Firefox 安装插件的界面中，输入关键字 "restclinet"，就能看到这两个插件的信息，单击 "安装" 按钮，如图 16-1 所示。

第 16 章　Spring Boot 概述

图16-1　Firefox 安装 REST 客户端插件

2. 使用 HttpRequester 进行测试

运行程序后，可以对 http://localhost:8080/users/1 接口进行测试。

在 HttpRequester 的请求 URL 中填写接口地址，然后单击"Submit"按钮提交测试请求。在右侧响应界面中，能看到返回的 JSON 数据。图 16-2 所示为 HttpRequester 的使用过程。

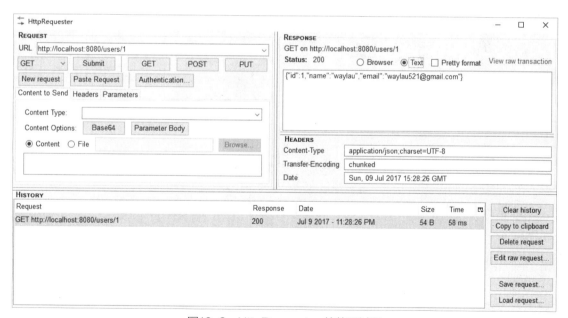

图16-2　HttpRequester 的使用过程

16.1.5　运行、测试程序

运行程序，项目启动在 8080 端口。

403

首先，发送 GET 请求到 http://localhost:8080/users，可以看到，响应返回的是一个空列表 []。

下面发送 POST 请求到 http://localhost:8080/users，用来创建一个用户。请求内容如下。

```
{"name":"waylau","email":"waylau521@gmail.com"}
```

发送成功后即可看到响应的状态是 200，响应的数据如下。

```
{
    "id": 1,
    "name": "waylau",
    "email": "waylau521@gmail.com"
}
```

下面通过该接口再创建几条测试数据，并发送 GET 请求到 http://localhost:8080/users，可以看到，响应返回的是一个有数据的列表。

```
[
    {
        "id": 1,
        "name": "waylau",
        "email": "waylau521@gmail.com"
    },
    {
        "id": 2,
        "name": "老卫",
        "email": "waylau521@163.com"
    }
]
```

可以发送 PUT 方法到 http://localhost:8080/users/2 来修改 id 为 2 的用户信息，修改内容如下。

```
{"name":"柳伟卫","email":"778907484@qq.com"}
```

发送成功后，即可看到响应的状态是 200。然后通过 GET 方法到 http://localhost:8080/users/2 来查看 id 为 2 的用户信息。

```
{
    "id": 2,
    "name": "柳伟卫",
    "email": "778907484@qq.com"
}
```

可以看到，用户数据已经变更了。

至此，这个简单的"用户管理"RESTful 服务已经全部调试完毕。

16.2 Spring Boot的配置详解

本节将重点讲解 Spring Boot 的配置。

16.2.1 理解Spring Boot的自动配置

按照"约定大于配置"的原则，Spring Boot 通过扫描依赖关系来使用类路径中可用的库。对于每个 pom 文件中的"spring-boot-starter-*"依赖，Spring Boot 会执行默认的 AutoConfiguration 类。

AutoConfiguration 类使用"*AutoConfiguration"语法模式，其中"*"代表类库。例如，JPA 存储库的自动配置是通过 JpaRepositoriesAutoConfiguration 来实现的。

使用"--debug"运行应用程序可以查看自动配置的相关报告。下面的命令用于显示"spring-boot-rest"应用的自动配置报告。

```
$ java -jar build/libs/spring-boot-rest-1.0.0.jar --debug
```

以下是自动配置类的一些示例。

（1）ServerPropertiesAutoConfiguration。

（2）RepositoryRestMvcAutoConfiguration。

（3）JpaRepositoriesAutoConfiguration。

（4）JmsAutoConfiguration。

如果应用程序有特殊的要求，如需要排除某些库的自动配置，也是能完全实现的。以下是排除 DataSourceAutoConfiguration 的示例。

```
@EnableAutoConfiguration(exclude={DataSourceAutoConfiguration.class})
```

16.2.2 重写默认的配置值

也可以使用应用程序覆盖默认配置值。重写的配置值放在 application.properties 文件中即可。例如，要想更改应用启动的端口号，可以在 application.properties 文件中添加如下内容。

```
server.port=8081
```

这样，当这个应用程序再次启动时，就会使用端口 8081。

16.2.3 更换配置文件的位置

默认情况下，Spring Boot 将所有配置外部化到 application.properties 文件中。但是，它们仍然是应用程序的一部分。

此外，可以通过以下设置来实现从外部读取属性。

（1）spring.config.name：配置文件名。

（2）spring.config.location：配置文件的位置。

其中，spring.config.location 可以是本地文件位置。

以下命令可从外部启动 Spring Boot 应用程序来配置文件。

```
$ java -jar build/libs/spring-boot-rest-1.0.0.jar --spring.config.name=-bootrest.properties
```

16.2.4 自定义配置

开发者可以将自定义属性添加到 application.properties 文件中。

例如，在 application.properties 文件中自定义一个名为 "file.server.url" 的属性。在 Spring Boot 启动后，就能将该属性自动注入应用中。

下面是完整的示例。

```
@Controller
@RequestMapping("/u")
public class UserspaceController {

    @Value("${file.server.url}")
    private String fileServerUrl;

    @GetMapping("/{username}/blogs/edit")
    public ModelAndView createBlog(@PathVariable("username") String username, Model model) {
        model.addAttribute("blog", new Blog(null, null, null));
        model.addAttribute("fileServerUrl", fileServerUrl);       // 将文件服务器的地址返回客户端
        return new ModelAndView("/userspace/blogedit", "blogModel", model);
    }
}
```

16.2.5 使用.yaml作为配置文件

.yaml 文件是 application.properties 文件的一种替代。YAML 提供了类似于 JSON 的结构化配置。YAML 数据结构可以用类似于大纲的缩排方式呈现，结构通过缩进来表示，连续的项目通过减号 "–" 来表示，map 结构中的 key/value 对用冒号 ":" 来分隔。示例如下。

```
spring:
  application:
```

```
        name: waylau
    datasource:
        driverClassName: com.mysql.jdbc.Driver
        url: jdbc:mysql://localhost/test
server:
    port: 8081
```

16.2.6　profiles的支持

Spring Boot 支持 profiles，即不同的环境使用不同的配置。通常需要设置一个系统属性（spring.profiles.active）或 OS 环境变量（SPRING_PROFILES_ACTIVE）。例如，使用"-D"参数启动应用程序（记住将其放在主类或 jar 之前）。

```
$ java -jar -Dspring.profiles.active=production build/libs/spring-boot-rest-1.0.0.jar
```

在 Spring Boot 中，还可以在 application.properties 中设置激活的配置文件，例如：

```
spring.profiles.active=production
```

YAML 文件实际上是由"---"行分隔的文档序列，每个文档分别解析为平坦化的映射。

如果一个 YAML 文档包含一个 spring.profiles 键，那么配置文件的值（逗号分隔的配置文件列表）将被反馈到 Spring Environment.acceptsProfiles() 中。并且如果这些配置文件中的任何一个被激活，那么文档将被包含在最终的合并中。例如：

```
server:
    port: 9000
---
spring:
    profiles: development
server:
    port: 9001

---
spring:
    profiles: production
server:
    port: 0
```

在以上示例中，默认端口为 9000，但是若 Spring profile 的"development"处于激活状态，则端口为 9001；若"production"处于激活状态，则端口为 0。

要使用".properties"文件做同样的事情，可以使用"application-${profile}.properties"的方式来指定特定于配置文件的值。

16.3 内嵌 Servlet 容器

Spring Boot Web Starter 内嵌了 Tomcat 服务器。在应用中使用嵌入式的 Servlet 容器，可以方便地进行项目的启动和调试。

Spring Boot 支持嵌入式 Tomcat、Jetty 和 Undertow 服务器。默认情况下，嵌入式服务器将侦听端口 8080 上的 HTTP 请求。

16.3.1 注册Servlet、过滤器和监听器

当使用嵌入式 Servlet 容器时，可以通过使用 Spring bean 或扫描 Servlet 组件，从 Servlet 规范（如 HttpSessionListener）中注册 Servlet、过滤器和所有监听器。

默认情况下，如果上下文只包含一个 Servlet，它将映射到 "/"。在多个 Servlet bean 的情况下，bean 名称将被用作路径前缀，过滤器将映射到 "/*"。

如果觉得基于惯例的映射不够灵活，可以使用 ServletRegistrationBean、FilterRegistrationBean 和 ServletListenerRegistrationBean 类进行完全控制。

16.3.2 Servlet上下文初始化

嵌入式 Servlet 容器不会直接执行 Servlet 3.0+ 的 javax.servlet.ServletContainerInitializer 接口或 Spring 的 org.springframework.web.WebApplicationInitializer 接口。这是因为在 war 中运行的第三方库会带来破坏 Spring Boot 应用程序的风险。

如果需要在 Spring Boot 应用程序中执行 Servlet 上下文初始化，就应注册一个实现 org.springframework.boot.web.servlet.ServletContextInitializer 接口的 bean。onStartup 方法提供对 ServletContext 的访问，并且如果需要，可以轻松地将其用作现有 WebApplicationInitializer 的适配器。

当使用嵌入式容器时，可以使用 @ServletComponentScan 自动注入启用了 @WebServlet、@WebFilter 和 @WebListener 注解的类。

需要注意的是，@ServletComponentScan 在独立部署的容器中不起作用，这是因为在独立部署的容器中，将使用容器内置的发现机制。

16.3.3 ServletWebServerApplicationContext

Spring Boot 使用一种新型的 ApplicationContext 来支持嵌入式的 Servlet 容器。ServletWebServerApplicationContext 就是这样一种特殊类型的 WebApplicationContext，它通过搜索单个 ServletWebServerFactory bean 来引导自身。通常，TomcatServletWebServerFactory、JettyServletWebServerFactory 或 UndertowServletWebServerFactory 将被自动配置。

通常，开发者并不需要关心这些实现类。因为在大多数应用程序中，它们将被自动配置，并会创建适当的 ApplicationContext 和 ServletWebServerFactory。

16.3.4　更改内嵌Servlet容器

Spring Boot Web Starter 默认使用 Tomcat 来作为内嵌的容器，在依赖中加入相应 Servlet 容器的 Starter 就能实现默认容器的替换，比如以下 2 个示例。

（1）spring-boot-starter-jetty：使用 Jetty 作为内嵌容器，可以替换 spring-boot-starter-tomcat。

（2）spring-boot-starter-undertow：使用 Undertow 作为内嵌容器，可以替换 spring-boot-starter-tomcat。

可以使用 Spring Environment 属性配置常见 Servlet 容器的相关设置，通常在 application.properties 文件中定义属性。

常见的 Servlet 容器设置如下。

（1）网络设置：监听 HTTP 请求的端口（server.port）、绑定到 server.address 的接口地址等。

（2）会话设置：会话是否持久（server.session.persistence）、会话超时（server.session.timeout）、会话数据的位置（server.session.store-dir）和会话 cookie 配置（server.session.cookie.*）。

（3）错误管理：错误页面的位置（server.error.path）等。

（4）SSL。

（5）HTTP 压缩。

Spring Boot 尽可能地尝试公开这些常见设置，但也会有一些特殊的配置。对于这些例外的情况，Spring Boot 提供了专用命名空间来对应特定于服务器的配置（如 server.tomcat 和 server.undertow）。

16.4　实现安全机制

本节将介绍基于 Spring Security 实现的基本认证及 OAuth2。

16.4.1　实现基本认证

如果 Spring Security 位于类路径上，那么所有 HTTP 端点都默认使用基本认证，这样就能使 Web 应用程序得到一定的安全保障。最为快捷的方式是在依赖中添加 Spring Boot Security Starter。

```
<dependencies>
  <dependency>
      <groupId>org.springframework.boot</groupId>
      <artifactId>spring-boot-starter-security</artifactId>
```

```xml
        </dependency>

        <dependency>
            <groupId>org.springframework.security</groupId>
            <artifactId>spring-security-test</artifactId>
            <scope>test</scope>
        </dependency>
</dependencies>
```

如果要向 Web 应用程序中添加方法级别的安全保障,可以在 Spring Boot 应用中添加 @Enable-GlobalMethodSecurity 注解来实现,如下面的示例。

```java
@EnableGlobalMethodSecurity
@SpringBootApplication
public class Application {
    public static void main(String[] args) {
        SpringApplication.run(Application.class, args);
    }
}
```

注意,@EnableGlobalMethodSecurity 可以配置多个参数。

(1) prePostEnabled:决定 Spring Security 的前注解 @PreAuthorize、@PostAuthorize 等是否可用。

(2) secureEnabled:决定 Spring Security 的保障注解 @Secured 是否可用。

(3) jsr250Enabled:决定 JSR-250 的注解 @RolesAllowed 等是否可用。

配置方式分别如下。

```java
@EnableGlobalMethodSecurity(securedEnabled = true)
public class MethodSecurityConfig {
// …
}

@EnableGlobalMethodSecurity(jsr250Enabled = true)
public class MethodSecurityConfig {
// …
}

@EnableGlobalMethodSecurity(prePostEnabled = true)
public class MethodSecurityConfig {
// …
}
```

在同一个应用程序中可以启用多个类型的注解,但是应该只设置一个注解对应行为类的接口或类。如果将两个注解同时应用于某一特定方法,那么只有其中一个将被应用。

1. @Secured

此注释是用来定义业务方法的安全配置属性的列表。可以在需要安全角色/权限的方法上指定 @Secured，只有有那些角色/权限的用户才可以调用该方法。如果有人不具备要求的角色/权限却试图调用此方法，将会抛出 AccessDenied 异常。

@Secured 源于 Spring 之前的版本，它有一个局限是不支持 Spring EL 表达式。可以看看下面的示例。

如果想指定 AND（和）这个条件，也就是让 deleteUser 方法同时拥有 ADMIN & DBA，仅仅通过使用 @Secured 注解是无法实现的。这时可以使用 Spring 的新注解 @PreAuthorize/@PostAuthorize（支持 Spring EL），使实现这一功能成为可能，而且无限制。

2. @PreAuthorize/@PostAuthorize

Spring 的 @PreAuthorize/@PostAuthorize 注解更适合方法级的安全配置，也支持 Spring EL 表达式语言，提供了基于表达式的访问控制。

（1）@PreAuthorize 注解：适合进入方法前的权限验证，@PreAuthorize 可以将登录用户的角色/权限参数传入方法中。

（2）@PostAuthorize 注解：使用并不多，在方法执行后再进行权限验证。

以下是一个使用了 @PreAuthorize 注解的示例。

```
@PreAuthorize("hasAuthority('ROLE_ADMIN')")  // 指定角色权限才能操作方法
@GetMapping(value = "delete/{id}")
public ModelAndView delete(@PathVariable("id") Long id, Model model) {
    userService.removeUser(id);
    model.addAttribute("userList", userService.listUsers());
    model.addAttribute("title", "删除用户");
    return new ModelAndView("users/list", "userModel", model);
}
```

3. 登录账号和密码

默认的 AuthenticationManager 具有单个用户账号，其中用户名称为 "user"，密码是一个随机码，在应用程序启动时以 INFO 级别打印，具体内容如下。

```
Using default security password: 78fa195d-3f4c-48b1-ad50-e24c31d5cf36
```

当然，也可以在配置文件中自定义用户名和密码。

```
security.user.name=guest
security.user.password=guest123
```

16.4.2　实现OAuth 2.0认证

如果类路径中有 spring-security-oauth2，就可以利用某些自动配置来轻松设置授权或资源服务器，即可实现 OAuth 2.0 认证。

1. 什么是OAuth 2.0

OAuth 2.0 的规范可以参考 RFC 6749（http://tools.ietf.org/html/rfc6749）。

OAuth 是一个开放标准，允许用户让第三方应用访问该用户在某一网站上存储的私密资源（如照片、视频、联系人列表等），而无须将用户名和密码提供给第三方应用。目前，OAuth 的最新版本为 2.0。

OAuth 允许用户提供一个令牌（而不是用户名和密码）来访问他们存放在特定服务提供者处的数据。每一个令牌授权一个特定的网站（如视频编辑网站）在特定的时段（如接下来的 2 小时内）访问特定的资源（如仅仅是某一相册中的视频）。这样，OAuth 允许用户授权第三方网站访问他们存储在另外的服务提供者处的信息，而无须分享他们的访问许可或所有的数据。

2. OAuth 2.0的核心概念

OAuth 2.0 中主要有以下 4 类角色。

（1）Resource Owner：资源所有者，指终端的"用户"（user）。

Resource Server：资源服务器，服务提供者用其存放受保护资源。访问这些资源，需要获得访问令牌（Access Token）。它与认证服务器可以是同一台服务器，也可以是不同的服务器。如果使用新浪微博的账号来登录新浪博客网站，新浪博客的资源和新浪微博的认证是同一家，就可以认为是同一个服务器。如果使用新浪博客的账号登录知乎，那么显然知乎的资源和新浪的认证不是一个服务器。

（2）Client：客户端，代表向受保护资源进行资源请求的第三方应用程序。

（3）Authorization Server：授权服务器，在验证资源所有者并获得授权后，将发放访问令牌给客户端。

3. OAuth 2.0的认证流程

OAuth 2.0 的认证流程如下。

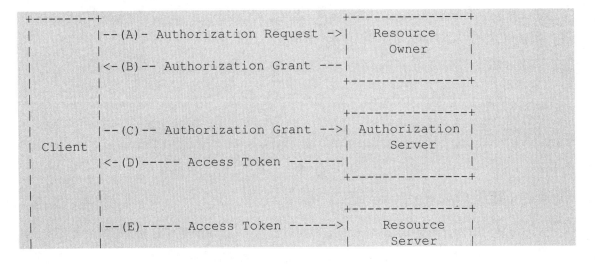

```
|          |<-(F)--- Protected Resource ---|                 |
+--------+                                  +---------------+
```

- （A）用户打开客户端后，客户端请求资源所有者（用户）的授权
- （B）用户同意给予客户端授权
- （C）客户端使用上一步获得的授权，向认证服务器申请访问令牌
- （D）认证服务器对客户端进行认证并确认无误后，同意发放访问令牌
- （E）客户端使用访问令牌向资源服务器申请获取资源
- （F）资源服务器确认令牌无误，同意向客户端开放资源

其中，用户授权有以下 4 种模式。

- 授权码模式（Authorization Code）
- 简化模式（Implicit）
- 密码模式（Resource Owner Password Credentials）
- 客户端模式（Client Credentials）

4. 配置

项目的核心配置如下。

```
github.client.clientId=ad2abbc19b6c5f0ed117
github.client.clientSecret=26db88a4dfc34cebaf196e68761c1294ac4ce265
github.client.accessTokenUri=https://github.com/login/oauth/access_to-
ken
github.client.userAuthorizationUri=https://github.com/login/oauth/
authorize
github.client.clientAuthenticationScheme=form
github.client.tokenName=oauth_token
github.client.authenticationScheme=query
github.resource.userInfoUri=https://api.github.com/user
```

以上代码包括了作为一个 client 所需要的大部分参数。其中，clientId、clientSecret 是在 GitHub 注册一个应用时生成的。如果读者不想注册应用，则可以直接用上面的配置。如果要注册，可参看本章最后的注册流程。

5. 项目的安全配置

安全配置需要加上 @EnableWebSecurity、@EnableOAuth2Client 注解，用来启用 Web 安全认证机制，并表明这是一个 OAuth 2.0 客户端。其中，@EnableGlobalMethodSecurity 注解说明项目采用了基于方法的安全设置。

```
@EnableWebSecurity
@EnableOAuth2Client   // 启用 OAuth 2.0 客户端
@EnableGlobalMethodSecurity(prePostEnabled = true) // 启用方法安全设置
```

```java
public class SecurityConfig extends WebSecurityConfigurerAdapter {
```

使用 Spring Security,需要继承 org.springframework.security.config.annotation.web.configuration. WebSecurityConfigurerAdapter,并重写以下 configure 方法。

```java
@Override
protected void configure(HttpSecurity http) throws Exception {
    http.addFilterBefore(ssoFilter(), BasicAuthenticationFilter.class)
        .antMatcher("/**")
        .authorizeRequests()
        .antMatchers("/", "/index", "/403","/css/**", "/js/**", "/fonts/**").permitAll() // 不设限制 都允许访问
        .anyRequest()
        .authenticated()
        .and().logout().logoutSuccessUrl("/").permitAll()
        .and().csrf().csrfTokenRepository(CookieCsrfTokenRepository.withHttpOnlyFalse())
        ;
}
```

上面的配置设置了一些过滤策略,除了静态资源及不需要授权的页面允许访问外,其他资源都是需要授权访问的。

其中设置了一个过滤器 ssoFilter,用于在 BasicAuthenticationFilter 之前进行拦截。如果拦截的是 /login,就是访问认证服务器。

```java
private Filter ssoFilter() {
    OAuth2ClientAuthenticationProcessingFilter githubFilter = new OAuth2ClientAuthenticationProcessingFilter("/login");
    OAuth2RestTemplate githubTemplate = new OAuth2RestTemplate(github(), oauth2ClientContext);
    githubFilter.setRestTemplate(githubTemplate);
    UserInfoTokenServices tokenServices = new UserInfoTokenServices(githubResource().getUserInfoUri(), github().getClientId());
    tokenServices.setRestTemplate(githubTemplate);
    githubFilter.setTokenServices(tokenServices);
    return githubFilter;
}

@Bean
public FilterRegistrationBean oauth2ClientFilterRegistration(
        OAuth2ClientContextFilter filter) {
    FilterRegistrationBean registration = new FilterRegistrationBean();
    registration.setFilter(filter);
    registration.setOrder(-100);
    return registration;
}
```

```
@Bean
@ConfigurationProperties("github.client")
public AuthorizationCodeResourceDetails github() {
    return new AuthorizationCodeResourceDetails();
}

@Bean
@ConfigurationProperties("github.resource")
public ResourceServerProperties githubResource() {
    return new ResourceServerProperties();
}
```

6. 资源服务器

下面写了两个控制器来提供相应的资源。

（1）MainController.java。

```
@Controller
public class MainController {

    @GetMapping("/")
    public String root() {
        return "redirect:/index";
    }

    @GetMapping("/index")
    public String index(Principal principal, Model model) {
        if(principal == null ){
            return "index";
        }
        System.out.println(principal.toString());
        model.addAttribute("principal", principal);
        return "index";
    }

    @GetMapping("/403")
    public String accesssDenied() {
        return "403";
    }
}
```

在 index 页面，如果认证成功，就会显示一些认证信息。

（2）UserController.java。

```
@RestController
@RequestMapping("/")
public class UserController {
    /**
     * 查询所用用户
```

```
 * @return
 */
@GetMapping("/users")
@PreAuthorize("hasAuthority('ROLE_USER')")    // 指定角色权限才能操作方法
public ModelAndView list(Model model) {

    List<User> list = new ArrayList<>();          // 当前所在页面数据列表
    list.add(new User("waylau",29));
    list.add(new User("老卫",30));
    model.addAttribute("title", "用户管理");
    model.addAttribute("userList", list);
    return new ModelAndView("users/list", "userModel", model);
}
}
```

7. 前端页面

前端页面主要是采用 Thymeleaf 及 Bootstrap 编写的。

首页用于显示用户的基本信息。

```
<body>
    <div class="container">
        <div class="mt-3">
            <h2>Hello Spring Security</h2>
        </div>
        <div sec:authorize="isAuthenticated()" th:if="${principal}" th:-object="${principal}">
            <p>已有用户登录</p>
            <p>登录的用户为: <span sec:authentication="name"></span></p>
            <p>用户权限为: <span th:text="*{userAuthentication.authorities}"></span></p>
            <p>用户头像为: <img alt="" class="avatar width-full rounded-2" height="230"
                th:src="*{userAuthentication.details.avatar_url}" width="230"></p>

        </div>
        <div sec:authorize="isAnonymous()">
            <p>未有用户登录</p>
        </div>
    </div>
</body>
```

用户管理界面显示用户的列表:

```
<body>
<div class="container">
```

```html
<div class="mt-3">
    <h2 th:text="${userModel.title}">Welcome to waylau.com</h2>
</div>

<table class="table table-hover">
    <thead>
    <tr>
        <td>Age</td>
        <td>Name</td>
        <td sec:authorize="hasRole('ADMIN')">Operation</td>
    </tr>
    </thead>
    <tbody>
    <tr th:if="${userModel.userList.size()} eq 0">
        <td colspan="3">没有用户信息    </td>
    </tr>
    <tr th:each="user : ${userModel.userList}">

        <td th:text="${user.age}">11</td>
        <td th:text="${user.name}">waylau</a></td>
        <td sec:authorize="hasRole('ADMIN')">
            <div >
                我是管理员
            </div>
        </td>
    </tr>
    </tbody>
</table>

</body>
```

8. 运行效果

图 16-3 所示为没有授权的首页。

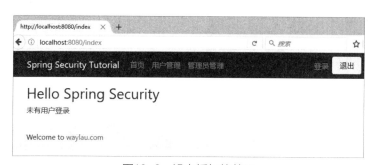

图16-3 没有授权的首页

单击"登录"按钮后，会重定向到 GitHub，在此界面登录并进行授权，如图 16-4 所示。

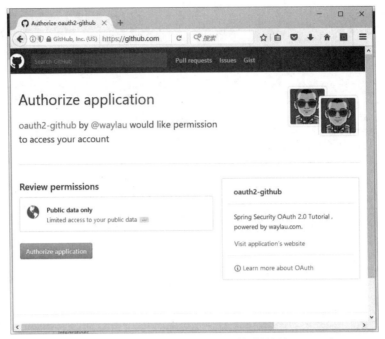

图16-4　登录 GitHub 界面并进行授权

图 16-5 所示为授权后的首页。

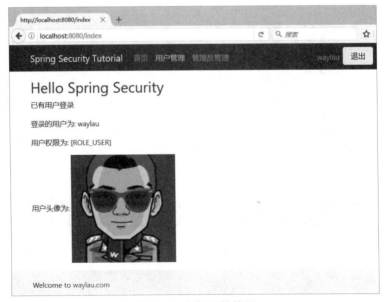

图16-5　授权后的首页

授权后即可进入用户管理界面，如图 16-6 所示。

第 16 章 Spring Boot 概述

图16-6 用户管理界面

9. 注册GitHub 应用

如果需要注册，可按下面的流程生成 Client ID 和 Client Secret，访问 https://github.com/settings/applications/new 页面注册应用，并生成客户端 ID 和密码。

```
Client ID      ad2abbc19b6c5f0ed117
Client Secret  26db88a4dfc34cebaf196e68761c1294ac4ce265
```

将客户端 ID 和密码写入程序配置即可。图 16-7 所示为注册应用信息的界面。

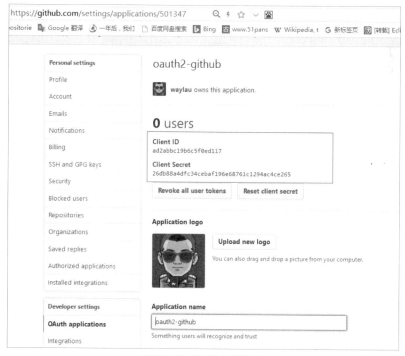

图16-7 注册应用信息

16.4.3 示例源码

本章内容大多选自笔者所著的开源电子书《Spring Security 教程》。如果想了解更多关于 Spring Security 安全方面的内容及示例源码，可参阅 https://github.com/waylau/spring-security-tutorial（该电子书地址）。

第17章 基于Spring Boot的Lite框架

本章将对Lite框架进行改造,基于Spring Boot技术来实现一个Lite Spring Boot Starter。

17.1　Lite Spring Boot Starter项目搭建

搭建 Spring Boot 应用，最快的方式莫过于使用 Spring Initializr 自动生成应用的骨架。

Spring Initializr 是用于初始化 Spring Boot 项目的可视化平台。虽然通过 Maven 或 Gradle 来添加 Spring Boot 提供的 Starter 非常简单，但是由于组件和关联部分太多，能有这样一个可视化的配置构建管理平台对于用户来说非常友好。下面将演示如何通过 Spring Initializr 初始化一个 Spring Boot 项目原型。

17.1.1　访问Spring Initializr

访问网站 https://start.spring.io/，这是 Spring 提供的官方 Spring Initializr 网站。当然，也可以搭建自己的 Spring Initializr 平台。按照页面提示，输入相应的项目元数据（Project Metadata）资料，并选择依赖。由于要初始化一个 Web 项目，因此在依赖搜索框中输入关键字"web"，并且选择"Web:Full-stack web development with Tomcat and Spring MVC"选项。该项目将会采用 Spring MVC 作为 MVC 的框架，并且集成 Tomcat 作为内嵌的 Web 容器。图 17-1 所示为 Spring Initializr 的管理界面。

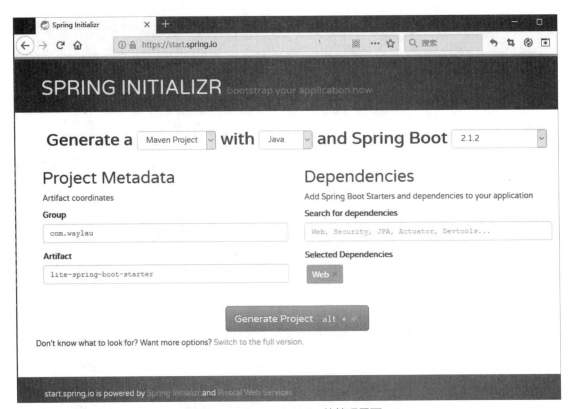

图17-1　Spring Initializr的管理界面

这里采用 Maven 作为项目管理工具，Spring Boot 选择最新的 2.1.2 版本，将 Group 的信息填写为 "com.waylau"，Artifact 填写为 "lite-spring-boot-starter"。最后，单击 "Generate Project" 按钮，即可下载以项目 "lite-spring-boot-starter" 命名的 zip 包。该压缩包包含了这个原型项目的所有源码及配置。

解压 lite-spring-boot-starter.zip 文件，可以看到该目录下的文件结构，是一个典型的 Maven 项目结构，如图 17-2 所示。

图17-2　项目结构

17.1.2　编译项目

在项目 lite-spring-boot-starter 的根目录下，执行 mvn clean package 来对项目进行编译、构建，构建过程如下。

```
>mvn clean package
[INFO] Scanning for projects…
[INFO]
[INFO] ----------------< com.waylau:lite-spring-boot-starter >----------------
[INFO] Building lite-spring-boot-starter 0.0.1-SNAPSHOT
[INFO] --------------------------------[ jar ]---------------------------------
Downloading from nexus-aliyun: http://maven.aliyun.com/nexus/content/groups/public/org/apache/maven/plugins/maven-clean-plugin/3.1.0/maven-clean-plugin-3.1.0.pom
Downloaded from nexus-aliyun: http://maven.aliyun.com/nexus/content/groups/public/org/apache/maven/plugins/maven-clean-plugin/3.1.0/maven-clean-plugin-3.1.0.pom (5.2 kB at 2.3 kB/s)
[INFO]
[INFO] --- maven-clean-plugin:3.1.0:clean (default-clean) @ lite-spring-boot-starter ---
[INFO]
[INFO] --- maven-resources-plugin:3.1.0:resources (default-resources) @ lite-spring-boot-starter ---
[INFO] Using 'UTF-8' encoding to copy filtered resources.
[INFO] Copying 1 resource
[INFO] Copying 0 resource
```

```
[INFO]
[INFO] --- maven-compiler-plugin:3.8.0:compile (default-compile) @ lite-
spring-boot-starter ---
[INFO] Changes detected - recompiling the module!
[INFO] Compiling 1 source file to D:\workspaceGithub\lite-spring-boot-
starter\target\classes
[INFO]
[INFO] --- maven-resources-plugin:3.1.0:testResources (default-testRe-
sources) @ lite-spring-boot-starter ---
[INFO] Using 'UTF-8' encoding to copy filtered resources.
[INFO] skip non existing resourceDirectory D:\workspaceGithub\lite-
spring-boot-starter\src\test\resources
[INFO]
[INFO] --- maven-compiler-plugin:3.8.0:testCompile (default-testCom-
pile) @ lite-spring-boot-starter ---
[INFO] Changes detected - recompiling the module!
[INFO] Compiling 1 source file to D:\workspaceGithub\lite-spring-boot-
starter\target\test-classes
[INFO]
[INFO] --- maven-surefire-plugin:2.22.1:test (default-test) @ lite-
spring-boot-starter ---
…
[INFO] Replacing main artifact with repackaged archive
[INFO] ------------------------------------------------------------------------
[INFO] BUILD SUCCESS
[INFO] ------------------------------------------------------------------------
[INFO] Total time: 12:04 min
[INFO] Finished at: 2019-02-14T00:09:08+08:00
[INFO] ------------------------------------------------------------------------
```

为了节省篇幅，这里省去了大部分的下载过程。最后，如果看到"BUILD SUCCESS"字样，就说明已经编译成功。

回到项目的根目录下，可以发现多出了一个 target 目录，在该目录下可以看到一个 lite-spring-boot-starter-0.0.1-SNAPSHOT.jar 文件，该文件就是项目编译后的可执行文件。通过下面的命令运行该文件。

```
java -jar target/lite-spring-boot-starter-0.0.1-SNAPSHOT.jar
```

成功运行后，可以在控制台看到如下输出。

```
>java -jar target/lite-spring-boot-starter-0.0.1-SNAPSHOT.jar
```

```
  .   ____          _            __ _ _
 /\\ / ___'_ __ _ _(_)_ __  __ _ \ \ \ \
( ( )\___ | '_ | '_| | '_ \/ _` | \ \ \ \
 \\/  ___)| |_)| | | | | || (_| |  ) ) ) )
  '  |____| .__|_| |_|_| |_\__, | / / / /
 =========|_|==============|___/=/_/_/_/
 :: Spring Boot ::        (v2.1.2.RELEASE)

...

2019-02-14 00:12:05.644  INFO 3336 --- [           main] o.
a.c.c.C.[Tomcat].[localhost].[/]         : Initializing Spring embedded
WebApplicationContext
2019-02-14 00:12:05.645  INFO 3336 --- [           main] o.s.web.con-
text.ContextLoader                       : Root WebApplicationContext: initializa-
tion completed in 2055 ms
2019-02-14 00:12:05.969  INFO 3336 --- [           main] o.
s.s.concurrent.ThreadPoolTaskExecutor    : Initializing ExecutorService
'applicationTaskExecutor'
2019-02-14 00:12:06.271  INFO 3336 --- [           main] o.
s.b.w.embedded.tomcat.TomcatWebServer    : Tomcat started on port(s): 8080
(http) with context path ''
2019-02-14 00:12:06.277  INFO 3336 --- [           main] c.
w.l.LiteSpringBootStarterApplication     : Started LiteSpringBootStarter-
Application in 3.373 seconds (JVM running for 3.936)
```

从输出内容可以看出，该项目使用的是 Tomcat 容器，使用的端口号是 8080。

在控制台按"Ctrl+C"组合键，可以关闭该程序。

17.1.3 探索项目

启动项目后，在浏览器的地址栏输入 http://localhost:8080/，可以得到如下信息。

```
Whitelabel Error Page

This application has no explicit mapping for /error, so you are seeing
this as a fallback.
Fri Jun 30 00:56:44 CST 2017
There was an unexpected error (type=Not Found, status=404).
No message available
```

由于在项目中还没有任何对请求的处理程序，因此 Spring Boot 会返回上述默认错误提示信息。稍后将实现对请求的处理。

17.1.4 编写RESTful服务

编写一个控制器 LiteController，代码如下。

```
package com.waylau.lite.spring.boot.starter.mvc.controller;

import org.springframework.web.bind.annotation.GetMapping;
import org.springframework.web.bind.annotation.RequestMapping;
import org.springframework.web.bind.annotation.RestController;

import com.waylau.lite.spring.boot.starter.Lite;

@RestController
@RequestMapping("/lite")
public class LiteController {

    @GetMapping
    public Lite sayHi() {
        return new Lite("waylau.com", "1.0.0");
    }

}
```

其中,Lite 类代码如下。

```
package com.waylau.lite.spring.boot.starter;

public class Lite {

    private String author;
    private String version;

    public Lite(String author, String version) {
        this.author = author;
        this.version = version;
    }

    public String getAuthor() {
        return author;
    }

    public void setAuthor(String author) {
        this.author = author;
    }

    public String getVersion() {
        return version;
    }

    public void setVersion(String version) {
        this.version = version;
    }

}
```

启动应用后，访问 http://localhost:8080/lite，可以看到页面内容如图 17-3 所示。

图17-3　页面内容

17.1.5　示例源码

Lite Spring Boot Starter 源码已经开源，可访问 https://github.com/waylau/lite-spring-boot-starter 进行获取。

17.2　集成Jetty

正如前面章节中介绍的那样，Spring Boot Web Starter 默认使用 Tomcat 来作为内嵌的容器。只要在依赖中加入相应 Servlet 容器的 Starter，就能实现默认容器的替换。例如，想要使用 Jetty 作为内嵌容器，就添加 spring-boot-starter-jetty 来替换 spring-boot-starter-tomcat。

pom.xml 文件修改如下。

```xml
<dependencies>
   <dependency>
      <groupId>org.springframework.boot</groupId>
      <artifactId>spring-boot-starter-web</artifactId>
      <exclusions>
         <exclusion>
            <groupId>org.springframework.boot</groupId>
            <artifactId>spring-boot-starter-tomcat</artifactId>
         </exclusion>
      </exclusions>
   </dependency>
   <dependency>
      <groupId>org.springframework.boot</groupId>
      <artifactId>spring-boot-starter-jetty</artifactId>
   </dependency>
```

```xml
<dependency>
    <groupId>org.springframework.boot</groupId>
    <artifactId>spring-boot-starter-test</artifactId>
    <scope>test</scope>
</dependency>
</dependencies>
```

运行应用，观察控制台输出内容，可以看到已经使用了 Jetty 作为容器。

```
  .   ____          _            __ _ _
 /\\ / ___'_ __ _ _(_)_ __  __ _ \ \ \ \
( ( )\___ | '_ | '_| | '_ \/ _` | \ \ \ \
 \\/  ___)| |_)| | | | | || (_| |  ) ) ) )
  '  |____| .__|_| |_|_| |_\__, | / / / /
 =========|_|==============|___/=/_/_/_/
 :: Spring Boot ::        (v2.1.2.RELEASE)

2019-02-14 00:36:23.638  INFO 16348 --- [           main] l.s.b.s.LiteSpringBootStarterApplication : Starting LiteSpringBootStarterApplication on AGOC3-705091335 with PID 16348 (D:\workspaceGithub\lite-spring-boot-starter\target\classes started by Administrator in D:\workspaceGithub\lite-spring-boot-starter)
…

2019-02-14 00:36:26.685  INFO 16348 --- [           main] o.e.jetty.server.AbstractConnector       : Started ServerConnector@4aa21f-9d{HTTP/1.1,[http/1.1]}{0.0.0.0:8080}
2019-02-14 00:36:26.688  INFO 16348 --- [           main] o.s.b.web.embedded.jetty.JettyWebServer  : Jetty started on port(s) 8080 (http/1.1) with context path '/'
2019-02-14 00:36:26.692  INFO 16348 --- [           main] l.s.b.s.LiteSpringBootStarterApplication : Started LiteSpringBootStarterApplication in 3.511 seconds (JVM running for 3.977)
```

17.3 集成Spring Security

16.4 节中已经介绍过如何集成 Spring Security，这里只需添加如下依赖即可。

```xml
<dependencies>
    …
    <!-- Security相关 -->
    <dependency>
        <groupId>org.springframework.boot</groupId>
```

```xml
        <artifactId>spring-boot-starter-security</artifactId>
    </dependency>
    <dependency>
        <groupId>org.springframework.security</groupId>
        <artifactId>spring-security-test</artifactId>
        <scope>test</scope>
    </dependency>
</dependencies>
```

接下来演示如何启用 Spring Security。

17.3.1 添加安全配置类

添加安全配置类 WebSecurityConfig，代码如下。

```java
package com.waylau.lite.spring.boot.starter.security;

import org.springframework.security.config.annotation.web.builders.HttpSecurity;
import org.springframework.security.config.annotation.web.configuration.EnableWebSecurity;
import org.springframework.security.config.annotation.web.configuration.WebSecurityConfigurerAdapter;

@EnableWebSecurity // 启用Spring Security功能
public class LiteSecurityConfig
    extends WebSecurityConfigurerAdapter {

    /**
     * 自定义配置
     */
    @Override
    protected void configure(HttpSecurity http) throws Exception {

        // 允许所有人访问
        http.authorizeRequests().anyRequest().anonymous();
    }

}
```

该配置类与 Lite 框架中的 LiteSecurityConfig 完全一致。

为了演示安全认证的效果，现将该配置更改如下。

```java
package com.waylau.spring.config;

import org.springframework.context.annotation.Bean;
```

```java
import org.springframework.security.config.annotation.web.builders.HttpSecurity;
import org.springframework.security.config.annotation.web.configuration.EnableWebSecurity;
import org.springframework.security.config.annotation.web.configuration.WebSecurityConfigurerAdapter;
import org.springframework.security.core.userdetails.User;
import org.springframework.security.core.userdetails.UserDetailsService;
import org.springframework.security.provisioning.InMemoryUserDetailsManager;

@EnableWebSecurity // 启用Spring Security功能
public class WebSecurityConfig
    extends WebSecurityConfigurerAdapter {

    /**
     * 自定义配置
     */
    @Override
    protected void configure(HttpSecurity http) throws Exception {
        http.authorizeRequests().anyRequest().authenticated() //所有请求都需认证
            .and()
            .formLogin() // 使用form表单登录
            .and()
            .httpBasic(); // HTTP基本认证
    }

    @SuppressWarnings("deprecation")
    @Bean
    public UserDetailsService userDetailsService() {
        InMemoryUserDetailsManager manager =
                new InMemoryUserDetailsManager();

        manager.createUser(
                User.withDefaultPasswordEncoder()   // 密码编码器
                .username("waylau")    // 用户名
                .password("123")       // 密码
                .roles("USER")         // 角色
                .build()
                );
        return manager;
    }
}
```

WebSecurityConfig 配置的含义已经在第 8 章进行了详细介绍，此处不再赘述。

17.3.2 运行应用

运行应用类后，便可以在浏览器中访问应用的接口进行测试。

使用 GET 请求访问 http://localhost:8080/lite 时，可以看到请求被安全机制拦截了，并重定向到了图 17-4 所示的登录界面。

图17-4　登录界面

当输入账号密码成功登录之后，即可正常访问该接口，如图 17-5 所示。

图17-5　访问接口

17.4 集成MyBatis

MyBatis 社区也提供了基于 Spring Boot 的 Starter——MyBatis Spring Boot Starter。通过 MyBatis Spring Boot Starter 可以大大减少 MyBatis 的配置工作。

17.4.1 添加MyBatis Spring Boot Starter依赖

在 pom.xml 中添加 MyBatis Spring Boot Starter 及 MySQL 驱动的依赖，代码如下。

```xml
<!-- MyBatis相关 -->
<dependency>
    <groupId>org.mybatis.spring.boot</groupId>
    <artifactId>mybatis-spring-boot-starter</artifactId>
    <version>${mybatis-spring-boot-starter.version}</version>
</dependency>
<dependency>
    <groupId>mysql</groupId>
    <artifactId>mysql-connector-java</artifactId>
    <version>${mysql-connector-java.version}</version>
</dependency>
```

17.4.2 配置数据库连接

在 application.properties 中添加相关配置，代码如下。

```
spring.datasource.driverClassName = com.mysql.cj.jdbc.Driver
spring.datasource.url = jdbc:mysql://localhost:3306/lite_news?useSSL=false&serverTimezone=UTC
spring.datasource.username = root
spring.datasource.password = 123456
```

从以上代码中可以发现，application.properties 与 lite.properties 中的内容高度吻合。

17.4.3 配置Mapper

1. 定义UserMapper

定义 UserMapper，代码如下。

```java
package com.waylau.lite.spring.boot.starter.mapper;

import com.waylau.lite.spring.boot.starter.domain.User;

public interface UserMapper {

    void createUser(User user);

    void deleteUser(Long userId);

    void updateUser(User user);

    User getUser(Long userId);
```

```java
    User getUserByUsername(String username);
}
```

2. 定义UserMapper XML文件

定义 UserMapper XML 文件，代码如下。

```xml
<?xml version="1.0" encoding="UTF-8"?>
<!DOCTYPE mapper
    PUBLIC "-//mybatis.org//DTD Mapper 3.0//EN"
    "http://mybatis.org/dtd/mybatis-3-mapper.dtd">

<mapper namespace="com.waylau.lite.spring.boot.starter.mapper.UserMapper">

    <insert id="createUser"
        parameterType="com.waylau.lite.spring.boot.starter.domain.User">
        insert into t_user (username, password, email)
        values( #{username}, #{password}, #{email})
    </insert>

    <delete id="deleteUser" parameterType="long">
        delete from t_user
        where user_id = #{userId}
    </delete>

    <update id="updateUser"
        parameterType="com.waylau.lite.spring.boot.starter.domain.User">
        update t_user set
        password = #{password},
        email = #{email}
        where user_id = #{userId}
    </update>

    <select id="getUser" parameterType="long"
        resultType="com.waylau.lite.spring.boot.starter.domain.User">
        select user_id as userId, username, password, email
        from t_user where user_id = #{userId}
    </select>

    <select id="getUserByUsername" parameterType="string"
        resultType="com.waylau.lite.spring.boot.starter.domain.User">
        select user_id as userId, username, password, email
        from t_user where username = #{username}
    </select>
</mapper>
```

3. 配置扫描位置

涉及 MyBatis 描述文件位置的配置主要有以下两个。

（1）在 application.properties 中添加如下配置。

```
...
mybatis.type-aliases-package = com.waylau.lite.spring.boot.starter.domain
```

上述配置定义了领域对象的位置。

（2）在应用的入口处进行如下配置。

```
package com.waylau.lite.spring.boot.starter;

import org.mybatis.spring.annotation.MapperScan;
import org.springframework.boot.SpringApplication;
import org.springframework.boot.autoconfigure.SpringBootApplication;

@SpringBootApplication
@MapperScan("com.waylau.lite.spring.boot.starter.mapper") // MyBatis扫描Mapper的位置
public class LiteSpringBootStarterApplication {

    public static void main(String[] args) {
        SpringApplication.run(LiteSpringBootStarterApplication.class, args);
    }

}
```

MapperScan 定义了 Mapper 的位置。

17.4.4 编写测试用例

编写测试用例，代码如下。

```
package com.waylau.lite.spring.boot.starter.mapper;

import static org.junit.Assert.assertEquals;
import static org.junit.Assert.assertNull;

import org.junit.Test;
import org.junit.runner.RunWith;
import org.slf4j.Logger;
import org.slf4j.LoggerFactory;
import org.springframework.beans.factory.annotation.Autowired;
import org.springframework.boot.test.context.SpringBootTest;
import org.springframework.test.context.junit4.SpringRunner;
import org.springframework.transaction.annotation.Transactional;

import com.waylau.lite.spring.boot.starter.domain.User;
import com.waylau.lite.spring.boot.starter.mapper.UserMapper;
```

```java
@RunWith(SpringRunner.class)
@SpringBootTest
@Transactional
public class UserMapperTests {

    static final Logger logger = LoggerFactory.getLogger(UserMapperTests.class);

    @Autowired
    private UserMapper userMapper;

    @Test
    public void testCreatetUser() {
        User user = new User();
        user.setUsername("waylau");
        user.setPassword("123456");
        user.setEmail("waylau521@gmail.com");
        userMapper.createUser(user);
        logger.info(user.toString());
    }

    @Test
    public void testDeleteUser() {
        userMapper.deleteUser(1L);

        User userNew = userMapper.getUser(1L);
        assertNull(userNew);
    }

    @Test
    public void testUpdateUser() {
        User user = userMapper.getUser(1L);
        user.setPassword("12345678");
        user.setEmail("waylau521@gmail.com");
        userMapper.updateUser(user);

        User userNew = userMapper.getUser(1L);
        assertEquals(user.getPassword(), userNew.getPassword());
    }

    @Test
    public void testGetUser() {
        User user = userMapper.getUser(1L);
        assertEquals("waylau", user.getUsername());
    }

    @Test
```

```
public void testGetUserByUsername() {
    User user = userMapper.getUserByUsername("waylau");
    assertEquals("waylau", user.getUsername());
}
}
```

其中，@SpringBootTest 是 Spring Boot 专有的注解，用于加载 Spring TestContext 框架。使用 JUnit 命令运行该测试用例，绿色表示该代码测试通过。

17.5 总结

在使用 Spring Boot 的过程中，会发现 Spring Boot 与 Lite 框架有许多相似的地方，包括配置内容、资源加载方式等。其实，Spring Boot 本身并没有提供太多的"新鲜"功能，其底层实现还是 Spring 框架。Spring Boot 并不是要成为 Spring 平台中众多"基础层"（Foundation）项目的替代者，它的目标不是为已解决的问题提供新的解决方案，而是为平台带来另一种开发体验，从而简化对这些已有技术的使用方法。对于已经熟悉 Spring 生态系统的开发人员来说，Spring Boot 是一个很理想的选择，而对于采用 Spring 技术的新人来说，Spring Boot 提供了一种更简洁的方式来使用这些技术。

所以，本书一开始并没有从 Spring Boot 讲起，而是从 Servlet、Spring 讲起，因为这些都是基础。只有掌握了这些基础，再回到"上层"类似于 Spring Boot 的框架时，才不至于"迷失"。

无论是 Lite 还是 Spring Boot，都能够兑现简化企业级应用开发的承诺。但对于新手而言，学习搭建 Lite 框架的过程却是另外一份收获，毕竟 Spring Boot 虽然好用，但也隐藏了太多的技术细节，不利于开发者学习和掌握底层技术。

读者如果想了解 Spring Boot、Spring Cloud 等高级框架方面的应用，也可以参阅笔者所著的"Spring 三剑客"作为提升和扩展，它们是《Spring 5 开发大全》《Spring Boot 企业级应用开发实战》和《Spring Cloud 微服务架构开发实战》。

附录

本书所采用的技术及相关版本

本书所采用的技术及相关版本较新，读者可将相关开发环境设置为与本书所采用的一致，或者不低于本书所列的配置。

- JDK 9
- Eclipse Java EE IDE for Web Developers 4.8.0
- Tomcat 9.0.13
- Servlet 4.0.1
- Jetty 9.4.14.v20181114
- Spring 5.1.5.RELEASE
- Spring Web MVC 5.1.5.RELEASE
- Jackson JSON 2.9.7
- Spring Security 5.2.0.BUILD-SNAPSHOT
- MyBatis 3.4.6
- Mybatis Spring 1.3.2
- MySQL Community Server 8.0.12
- Apache Commons DBCP 2.5.0
- JUnit Jupiter Engine 5.3.2
- logback 1.2.3
- SLF4J 1.7.25
- NGINX 1.15.8
- Angular 7.0.6
- Redis 3.2.100
- Spring Data Redis 2.1.4.RELEASE
- Jedis 2.10.2
- Spring Boot 2.1.2.RELEASE
- MyBatis Spring Boot Starter 2.0.0

参考文献

[1] JCP. JSR 153: Enterprise JavaBeans 2[EB/OL].https://jcp.org/en/jsr/detail?id=153，2002-07-19.

[2] JCP. JSR 345: Enterprise JavaBeans 3.2[EB/OL].https://jcp.org/en/jsr/detail?id=345，2013-04-04.

[3] JOHNSON R. Expert One-on-One J2EE Design and Development[M]. UK：Wrox，2002.

[4] JOHNSON R，JUERGEN HOELLER. Expert One-on-One J2EE Development without EJB[M]. Indiana：Wiley Publishing，2004.

[5] JCP. JSR 369: JavaTM Servlet 4.0 Specification2[EB/OL].https://jcp.org/en/jsr/detail?id=369，2017-09-05.

[6] 柳伟卫.Java Servlet 3.1 规范 [EB/OL].https://github.com/waylau/servlet-3.1-specification，2018-11-29.

[7] The Jetty Definitive Reference[EB/OL].https://www.eclipse.org/jetty/documentation/current/，2018-11-29.

[8] 柳伟卫. Spring 5 开发大全 [M]. 北京：北京大学出版社，2018.

[9] 柳伟卫. Cloud Native 分布式架构原理与实践 [M]. 北京：北京大学出版社，2019.

[10] 柳伟卫. Spring Cloud 微服务架构开发实战 [M]. 北京：北京大学出版社，2018.

[11] FOWLER M. Inversion of Control Containers and the Dependency Injection pattern[EB/OL].https://martinfowler.com/articles/injection.html，2004-01-23.

[12] IETF. Forwarded HTTP Extension[EB/OL].https://tools.ietf.org/html/rfc7239，2014-04-01.

[13] 柳伟卫. Spring Security 教程 [EB/OL].https://github.com/waylau/spring-security-tutorial，2018-12-12.

[14] Spring Security Reference[EB/OL].https://docs.spring.io/spring-security/site/docs/5.2.0.BUILD-SNAPSHOT/reference/htmlsingle/，2018-12-12.

[15] 柳伟卫. Spring Boot 企业级应用开发实战 [M]. 北京：北京大学出版社，2018.

[16] MyBatis Reference Documentation[EB/OL].http://www.mybatis.org/mybatis-3/index.html，2018-12-02.

[17] 柳伟卫. 分布式系统常用技术及案例分析 [M]. 北京：电子工业出版社，2017.

[18] 柳伟卫. Angular 入门、进阶、商业实战 [M]. 北京：电子工业出版社，2019.

[19] 柳伟卫.NGINX 教程 [EB/OL].https://github.com/waylau/nginx-tutorial，2019-02-11.

[20] 柳伟卫. 分布式系统常用技术及案例分析 [M]. 2 版 . 北京：电子工业出版社，2019.

[21] Redis Commands[EB/OL].https://redis.io/commands，2019-02-13.

[22] Spring Data Redis Reference[EB/OL].https://docs.spring.io/spring-data/data-redis/docs/current/reference/html/，2019-02-13.